AF290123

FUNDAMENTALS OF THE THEORY OF OPERATOR ALGEBRAS

SPECIAL TOPICS

VOLUME III

Elementary Theory—An Exercise Approach

FUNDAMENTALS OF THE THEORY OF OPERATOR ALGEBRAS

SPECIAL TOPICS

VOLUME III

Elementary Theory—An Exercise Approach

Richard V. Kadison

Department of Mathematics
University of Pennsylvania
Philadelphia, Pennsylvania

John R. Ringrose

School of Mathematics
University of Newcastle
Newcastle upon Tyne, England

BIRKHÄUSER BOSTON
675 Massachusetts Avenue, Cambridge, MA 02139-3309

Library of Congress Cataloging in Publication Data

Kadison, Richard V., 7-25-25
    Fundamentals of the theory of operator algebras
    Special topics
    Elementary theory—an exercise approach
    Includes preface, exercise groupings
    bibliography, index.
    1. Operator algebras. I. Ringrose, John R.

II. Title

Printed on acid free paper.
ISBN-13:978-1-4612-7834-4   e-ISBN-13:978-1-4612-3212-4
DOI: 10.1007/978-1-4612-3212-4

Printed by Quinn-Woodbine, Inc., Woodbine, New Jersey
9 8 7 6 5 4 3 2 1

CONTENTS

Preface   vii
Contents of Volume IV   ix
Exercise Groupings   xi

Chapter 1.   **Linear Spaces — Exercises and**   1
             Solutions

Chapter 2.   **Basics of Hilbert Space and**   40
             Linear Operators — Exercises
             and Solutions

Chapter 3.   **Banach Algebras — Exercises**   84
             and Solutions

Chapter 4.   **Elementary C^*-Algebra Theory**   139
             — Exercises and Solutions

Chapter 5.   **Elementary von Neumann**   207
             Algebra Theory — Exercises
             and Solutions

Bibliography   263

Index   268

PREFACE

These volumes are companions to the treatise; "Fundamentals of the Theory of Operator Algebras," which appeared as Volume 100 — I and II in the series, Pure and Applied Mathematics, published by Academic Press in 1983 and 1986, respectively. As stated in the preface to those volumes, "Their primary goal is to teach the subject and lead the reader to the point where the vast recent research literature, both in the subject proper and in its many applications, becomes accessible." No attempt was made to be encyclopædic; the choice of material was made from among the fundamentals of what may be called the "classical" theory of operator algebras.

By way of supplementing the topics selected for presentation in "Fundamentals," a substantial list of exercises comprises the last section of each chapter. An equally important purpose of those exercises is to develop "hand-on" skills in use of the techniques appearing in the text. As a consequence, each exercise was carefully designed to depend only on the material that precedes it, and separated into segments each of which is realistically capable of solution by an attentive, diligent, well-motivated reader.

The process by which the exercises were designed involved solving each of them completely and then subjecting the solutions to detailed scrutiny. It became apparent, in the course of this operation, that the written solutions could be of considerable value, if they were made generally available, as models with which a reader's solutions could be compared, as indicators of methods and styles for producing further solutions on an individual basis, and as a speedy route through one or another of the many special topics that supplement those in the text proper of "Fundamentals" (for the reader without the time or inclination to develop it as an exercise set). The present texts contain those written solutions; the first of these texts has the solutions to the exercises appearing in Volume I of "Fundamentals" and the second has the solutions to those appearing in Volume II. The statements of the exercises precede their solutions, for the obvious convenience of the reader.

In most instances, where an exercise or group of exercises develops a topic, the solutions have been given in what the authors feel is optimal form. Solutions are, of course, geared to the state of knowledge developed at the point in the book where the exercise occurs. Very occasionally, this necessitates an approach that is slightly less than optimal (for example, Exercise 2.8.10 occurs before square roots of positive operators, which could be used to advantage in its solution, are introduced). From time to time, an exercise reappears, with the task of finding a solution involving newly acquired information.

Of course, knowledge of and long experience with the literature of the subject has had a major influence on both the content of the exercises and the form of their solutions. In many cases, there is no specific source for the exercise or its solution. In virtually no instance was the solution of an exercise copied directly from the literature; the solutions were constructed with the path and stage of development of "Fundamentals" where the exercise occurs, very much in mind. Often, where a set of exercises develops a special topic, the solutions present a new and simpler route to the results appearing in the set.

References are placed after the solutions. As in "Fundamentals," no attempt is made to be thorough in referencing. The references appearing are chosen with a few goals in mind: to supply the reader with additional material, closely related to the exercise and its solution, that may be of interest for further study, to provide a very sketchy historical context that may be deepened by consulting the papers cited and their bibliographies. Where a set of consecutive exercises is largely inspired by a single article, for the most part, the first of the set and/or the highpoints of the topic note the article.

A guide to the topics treated by sets of related exercises follows this preface.

CONTENTS OF VOLUME IV

Chapter 6. **Comparison Theory of Projections**
    — Exercises and Solutions

Chapter 7. **Normal States and Unitary Equivalence of von Neumann Algebras**
    — Exercises and Solutions

Chapter 8. **The Trace**
    — Exercises and Solutions

Chapter 9. **Algebra and Commutant**
    — Exercises and Solutions

Chapter 10. **Special Representations of C^*-Algebras**
    — Exercises and Solutions

Chapter 11. **Tensor Products**
    — Exercises and Solutions

Chapter 12. **Approximation by Matrix Algebras**
    — Exercises and Solutions

Chapter 13. **Crossed Products**
    — Exercises and Solutions

Chapter 14. **Direct Integrals and Decompositions**
    — Exercises and Solutions

Bibliography
Index

EXERCISE GROUPINGS

Algebras of affiliated operators
6.9.53–6.9.55, 8.7.60

Approximate identities in C^*-algebras
4.6.35–4.6.37, 4.6.60, 10.5.6

β-compactification of $\mathbb{N}$
3.5.5, 3.5.6, 4.6.56, 5.7.16–5.7.19

Canonical anticommutation relations
10.5.88–10.5.90, 12.4.39, 12.4.40

Characterizations of von Neumann algebras among C^*-algebras
   (i)   in terms of order structure
   7.6.35–7.6.40
   (ii)   as dual spaces
   7.6.41–7.6.45, 10.5.87

Compact linear operators
2.8.20–2.8.29, 2.8.37–2.8.39, 3.5.12, 3.5.14, 3.5.15, 3.5.17–3.5.20

Completely positive mappings
11.5.15–11.5.24

Conditional expectations
8.7.23–8.7.30, 8.7.50, 10.5.85–10.5.87, 13.4.1, 13.4.24, 13.4.25

The Connes T-invariant
   (i)   calculation for certain matricial factors
   13.4.9–13.4.15
   (ii)   general properties
   14.4.14–14.4.16, 14.4.19, 14.4.20

Coupling constant and operator
8.7.57, 8.7.58, 9.6.3–9.6.7, 9.6.30

Derivations and automorphisms
  (i) continuity of derivations
    4.6.65, 4.6.66, 7.6.15, 10.5.12, 10.5.13
  (ii) inner and universally weakly inner derivations
    8.7.51–8.7.55, 10.5.61–10.5.64, 10.5.71, 10.5.72, 10.5.76–
    10.5.79, 12.4.38
  (iii) inner and universally weakly inner automorphisms
    10.5.1, 10.5.14, 10.5.60–10.5.75

Diagonalization of abelian self-adjoint subsets of $n \otimes \mathcal{R}$
    6.9.18–6.9.35

Diximier approximation theorem
    8.7.4–8.7.13, 10.5.2, 10.5.80

Extreme point examples
  (i) simple examples
    1.9.19, 2.8.13, 2.8.14, 5.7.8, 5.7.12, 5.7.13
  (ii) projections (see also Proposition 7.4.6)
    12.4.13–12.4.15
  (iii) unitary elements (see also Theorem 7.3.1)
    See the listing under "Unitary elements of C^*-algebras," sec-
    tions (ii) and (iii).

Extremely disconnected spaces
    5.7.14–5.7.21

Flip automorphisms
  (i) algebras generated by two projections
    12.4.11, 12.4.12
  (ii) free action
    12.4.17, 12.4.18
  (iii) a Liapunov theorem for operator algebras
    12.4.13–12.4.15
  (iv) the flip in various algebras
    11.5.25, 12.4.16, 12.4.19–12.4.27, 12.4.34, 12.4.35, 14.4.11–
    14.4.13

Friedrichs extension
    7.6.52–7.6.55

Fundamental group (of a factor of type II_1)
13.4.4–13.4.8

Generalized Schwarz inequality and applications
10.5.7–10.5.10, 11.5.23

Harmonic analysis on certain groups
3.5.33–3.5.42

Ideals
(i)   ideals in C^*-algebras
4.6.41, 4.6.42, 4.6.60–4.6.64, 10.5.11
(ii)   ideals in von Neumann algebras
6.9.46–6.9.51, 8.7.14–8.7.22
(iii)   primitive ideals and the Dauns–Hofmann theorem
10.5.81–10.5.84

Isometries and Jordan homomorphisms
7.6.16–7.6.18, 10.5.21–10.5.36

Modular theory
(i)   states that satisfy the modular condition
9.6.14–9.6.17
(ii)   dual cones
9.6.51–9.6.65, 13.4.19

Non-normal tracial weight
8.7.42–8.7.45

Relative commutants
(i)   commutation formulae for tensor products
12.4.34–12.4.37
(ii)   maximal matricial subfactors
12.4.29, 12.4.30
(iii)   non-normal factors
12.4.22, 12.4.23, 12.4.28–12.4.31, 14.4.17, 14.4.18

Representations
(i)   function representations of partially ordered vector spaces
and C^*-algebras
4.6.26, 4.6.27, 4.6.48–4.6.54
(ii)   states, representations; quasi-equivalence, type
7.6.33, 7.6.34, 10.5.38–10.5.50

Stone–Weierstrass theorems for C^*-algebras
4.6.70, 10.5.52–10.5.59

Strong continuity of operator functions
5.7.35–5.7.37, 12.4.32

Tensor products
  (i)   center of a tensor product of C^*-algebras
  11.5.1–11.5.4
  (ii)   inductive limits and infinite tensor products
  11.5.26–11.5.30
  (iii)   simple C^*-algebras
  11.5.5, 11.5.6
  (iv)   slice maps and relative commutant formulae for von Neumann algebras
  12.4.34–12.4.37
  (v)   type I C^*-algebras
  11.5.8, 11.5.9

Unitary implementation
  (i)   * automorphisms and * isomorphisms
  9.6.18–9.6.33, 13.4.3
  (ii)   continuous groups of * automorphisms
  9.6.65, 13.4.19–13.4.23, 14.4.8–14.4.10
  (iii)   norm-continuous groups of * automorphisms
  See the listing under "Derivatives and automorphisms," section (iii)
  (iv)   trace-preserving isomorphisms
  8.7.2, 9.6.35–9.6.39

Unitary elements of C^*-algebras
  (i)   exponential unitaries and connectivity
  4.6.2, 4.6.3, 4.6.5–4.6.9, 4.6.59
  (ii)   Russo–Dye theorem
  10.5.3–10.5.5
  (iii)   convex combinations of unitary elements
  10.5.91–10.5.100

Vectors and vector states
7.6.19, 7.6.21–7.6.28, 7.6.33, 9.6.54, 9.6.60, 9.6.63

CHAPTER 1

LINEAR SPACES

1.9 Exercises

1.9.1.   Show that, if q is a sublinear functional on a (real or complex) vector space $\mathcal{V}$, and V is the convex set $\{x \in \mathcal{V} : q(x) < 1\}$, then the support functional p of V is given by

$$p(x) = \max\{q(x), 0\} \qquad (x \in \mathcal{V}).$$

Solution.   For each x in $\mathcal{V}$
$$
\begin{aligned}
p(x) &= \inf\{c : \ c \in \mathbb{R}, \ c > 0, \ x \in cV\} \\
&= \inf\{c : \ c \in \mathbb{R}, \ c > 0, \ c^{-1}x \in V\} \\
&= \inf\{c : \ c \in \mathbb{R}, \ c > 0, \ q(c^{-1}x) < 1\} \\
&= \inf\{c : \ c \in \mathbb{R}, \ c > 0, \ c^{-1}q(x) < 1\} \\
&= \inf\{c : \ c \in \mathbb{R}, \ c > 0, \ q(x) < c\} \\
&= \max\{q(x), 0\}. \qquad \blacksquare
\end{aligned}
$$

1.9.2.   Suppose that q_1 and q_2 are semi-norms on a (real or complex) vector space $\mathcal{V}$, ρ is a linear functional on $\mathcal{V}$, and

$$|\rho(x)| \le q_1(x) + q_2(x) \qquad (x \in \mathcal{V}).$$

Show that ρ can be expressed in the form $\rho_1 + \rho_2$, where ρ_1, ρ_2 are linear functionals on $\mathcal{V}$, and

$$\rho_j(x) \le q_j(x) \qquad (x \in \mathcal{V}; \ j = 1, 2).$$

[*Hint.* Define a semi-norm p on the vector space $\mathcal{V} \times \mathcal{V}$, and a linear functional σ_0 on the "diagonal" subspace $\{(x, x) : x \in \mathcal{V}\}$ of $\mathcal{V} \times \mathcal{V}$, by

$$p((x, y)) = q_1(x) + q_2(y), \qquad \sigma_0((x, x)) = \rho(x).]$$

Solution. With σ_0 and p defined as in the hint, we have

$$|\sigma_0((x,x))| = |\rho(x)| \le q_1(x) + q_2(x) = p((x,x)),$$

for all (x,x) in the diagonal subspace of $\mathcal{V} \times \mathcal{V}$. By Theorem 1.1.7, σ_0 extends to a linear functional σ on $\mathcal{V} \times \mathcal{V}$, for which

$$|\sigma((x,y))| \le p((x,y)) \qquad ((x,y) \in \mathcal{V} \times \mathcal{V}).$$

We can now define linear functionals ρ_1, ρ_2 on $\mathcal{V}$ by

$$\rho_1(x) = \sigma((x,0)), \qquad \rho_2(x) = \sigma((0,x)).$$

Then
$$\rho_1(x) + \rho_2(x) = \sigma((x,x)) = \sigma_0((x,x)) = \rho(x),$$
$$|\rho_1(x)| = |\sigma((x,0))| \le p((x,0)) = q_1(x),$$
$$|\rho_2(x)| = |\sigma((0,x))| \le p((0,x)) = q_2(x),$$

for each x in $\mathcal{V}$.   ■

    1.9.3.   Suppose that $\mathcal{V}$ is a (real or complex) vector space, on which two locally convex topologies, τ_1 and τ_2, are specified. Let Γ_j be the set of all τ_j-continuous semi-norms on $\mathcal{V}$, for $j = 1, 2$; and let

$$\Gamma = \{q_1 + q_2 : \ q_1 \in \Gamma_1, \ q_2 \in \Gamma_2\}.$$

Prove that Γ is a separating family of semi-norms on $\mathcal{V}$, and that the corresponding locally convex topology τ is the coarsest locally convex topology on $\mathcal{V}$ that is finer than both τ_1 and τ_2. Show also that a linear functional ρ on $\mathcal{V}$ is τ-continuous if and only if it has the form $\rho_1 + \rho_2$, where ρ_j is a τ_j-continuous linear functional on $\mathcal{V}$, for $j = 1, 2$.

    Solution. Since Γ_j is a separating family of semi norms on $\mathcal{V}$ and gives rise to the locally convex topology τ_j (see Theorem 1.2.6 and its proof), it follows that $\Gamma\,(\supseteq \Gamma_1 \cup \Gamma_2)$ is separating and gives rise to a locally convex topology τ on $\mathcal{V}$ which is finer than both τ_1 and τ_2.

    If a locally convex topology τ_0 on $\mathcal{V}$ is finer than both τ_1 and τ_2, then each semi-norm in Γ is τ_0-continuous. This implies that the basic τ-neighborhoods, defined from elements of Γ as in Theorem 1.2.6, are τ_0 open. Thus τ is coarser than τ_0.

Let ρ be a τ-continuous linear functional on $\mathcal{V}$. Since a finite linear combination, with positive coefficients, of elements of Γ is again an element of Γ, it follows from Proposition 1.2.8(iii) that there is an element p of Γ such that $|p(x)| \le p(x)$ $(x \in \mathcal{V})$. We may set $p = q_1 + q_2$, where $q_1 \in \Gamma_1$ and $q_2 \in \Gamma_2$. From the result of Exercise 1.9.2, together with Proposition 1.2.8(iii), ρ has the form $\rho_1 + \rho_2$, where ρ_j is τ_j-continuous $(j = 1, 2)$. Conversely, each such ρ is τ-continuous. $\blacksquare$

1.9.4.   Show that, if $\{x_1, \ldots, x_n\}$ are elements of a normed space $\mathcal{X}$ such that 0 is in the closure of

$$\{a_1 x_1 + \cdots + a_n x_n : a_j \text{ scalars}, \prod_{j=1}^{n} (a_j - 1) = 0\},$$

then $\{x_1, \ldots, x_n\}$ are linearly dependent.

Solution. We argue by induction on the number of elements in the set $\{x_1, \ldots, x_n\}$. If there is just one element x_1, then $a_1 = 1$ and $x_1 = 0$. Suppose that we have established the result for sets with $n - 1$ or fewer elements. The one-dimensional space $\mathcal{V}_0$ generated by x_n is closed (Theorem 1.2.17). The quotient space $\mathcal{X}/\mathcal{V}_0$ is again a normed space, and the quotient mapping, $x \to \bar{x}$, is continuous (Theorem 1.5.3). If 0 is not in the closure of

$$\{a_1 x_1 + \cdots + a_n x_n : \prod_{j=1}^{n-1} (a_j - 1) = 0\},$$

then 0 is in the closure of $\{a_1 x_1 + \cdots + a_{n-1} x_{n-1} + x_n : a_j \text{ scalars}\}$. It follows that $x_n \in [x_1, \ldots, x_{n-1}]$ and that $\{x_1, \ldots, x_n\}$ are linearly dependent in this case. We may assume that 0 is in the closure of

$$\{a_1 x_1 + \cdots + a_n x_n : \prod_{j=1}^{n-1} (a_j - 1) = 0\}.$$

Thus $\bar{0}$ is in the closure of

$$\{a_1 \bar{x}_1 + \cdots + a_{n-1} \bar{x}_{n-1} : a_j \text{ scalars}, \prod_{j=1}^{n-1} (a_j - 1) = 0\}.$$

Our inductive hypothesis applies to complete the proof. $\blacksquare$

1.9.5.   Suppose that Y and Z are closed subspaces of a Banach space $\mathfrak{X}$, such that $Y \cap Z = \{0\}$, and let

$$k = \inf\{\|y - z\| : y \in Y,\ z \in Z,\ \|y\| = \|z\| = 1\}.$$

Show that the subspace $Y + Z$ of $\mathfrak{X}$ is closed if and only if $k > 0$. [See Corollary 1.8.8 and the discussion following it.]

Solution.   If $Y + Z$ is closed, there is a positive real number C such that $\|y\| \leq C\|y+z\|$, for all y in Y and z in Z, by Corollary 1.8.8. Thus $\|y - z\| \geq C^{-1}$ whenever $y \in Y$, $z \in Z$ and $\|y\| = \|z\| = 1$; and $k \geq C^{-1} > 0$.

If $Y + Z$ is not closed, there is no positive real number C with the above property. Thus, for each positive integer n, we can choose y_n in Y and z_n in Z so that $\|y_n\| > n\|y_n + z_n\|$. By homogeneity, we may suppose that

$$\|y_n\| = 1, \qquad \|y_n + z_n\| < \frac{1}{n}.$$

Since

$$\left|1 - \|z_n\|\right| = \left|\|y_n\| - \|-z_n\|\right| \leq \|y_n + z_n\| < \frac{1}{n},$$

we have

$$\left\|\|z_n\|^{-1}z_n - z_n\right\| = \left|\|z_n\|^{-1} - 1\right|\|z_n\| = \left|1 - \|z_n\|\right| < \frac{1}{n},$$

and thus

$$\left\|y_n + \|z_n\|^{-1}z_n\right\| \leq \left\|y_n + z_n\right\| + \left\|\|z_n\|^{-1}z_n - z_n\right\| < \frac{2}{n}.$$

This shows that $k < \frac{2}{n}$, for each $n = 1, 2, \ldots$; so $k = 0$.   ∎

1.9.6.   Show that, if $\mathfrak{X}$ is a real Banach space, then $\mathfrak{X} \times \mathfrak{X}$ becomes a complex Banach space $\mathfrak{X}_{\mathbb{C}}$ when its linear structure and norm are defined by

$$(x, y) + (u, v) = (x + u, y + v),$$
$$(a + ib)(x, y) = (ax - by, bx + ay),$$
$$\|(x, y)\| = \sup\{\|(\cos\theta)x + (\sin\theta)y\| : 0 \leq \theta \leq 2\pi\},$$

for all x, y, u, v in $\mathfrak{X}$ and a, b in $\mathbb{R}$.

Prove also that the set $\{(x, 0) : x \in \mathfrak{X}\}$ is a closed real-linear subspace $\mathfrak{X}_{\mathbb{R}}$ of $\mathfrak{X}_{\mathbb{C}}$, that $\mathfrak{X}_{\mathbb{C}} = \{h + ik : h, k \in \mathfrak{X}_{\mathbb{R}}\}$, and that the mapping $x \to (x, 0)$ is an isometric isomorphism from $\mathfrak{X}$ onto (the real Banach space) $\mathfrak{X}_{\mathbb{R}}$.

Solution. It is apparent that, with the given algebraic structure, $\mathfrak{X}_{\mathbb{C}}$ is an abelian group with respect to addition, that the mapping $(c,z) \to cz : \mathbb{C} \times \mathfrak{X}_{\mathbb{C}} \to \mathfrak{X}_{\mathbb{C}}$ is distributive in both variables, that $c_1(c_2 z) = (c_1 c_2)z$, and that $1 \cdot z = z$. Hence $\mathfrak{X}_{\mathbb{C}}$ is a complex vector space.

With the given definition of $\|\ \|_{\mathbb{C}}$, when $z, z_1, z_2 \in \mathfrak{X}_{\mathbb{C}}$ and $a \in \mathbb{R}$, we have $\|z\|_{\mathbb{C}} \geq 0$, $\|z\|_{\mathbb{C}} = 0$ if and only if $z = 0$, $\|z_1 + z_2\|_{\mathbb{C}} \leq \|z_1\|_{\mathbb{C}} + \|z_2\|_{\mathbb{C}}$ and $\|az\|_{\mathbb{C}} = |a|\,\|z\|_{\mathbb{C}}$. When α is real and $z = (x,y) \in \mathfrak{X}_{\mathbb{C}}$,

$$
\begin{aligned}
\left\| e^{-i\alpha} z \right\|_{\mathbb{C}} &= \left\| \big((\cos\alpha)x + (\sin\alpha)y, -(\sin\alpha)x + (\cos\alpha)y \big) \right\|_{\mathbb{C}} \\
&= \sup\{ \| \cos\theta[(\cos\alpha)x + (\sin\alpha)y] \\
&\qquad + \sin\theta[-(\sin\alpha)x + (\cos\alpha)y]\| : 0 \leq \theta \leq 2\pi \} \\
&= \sup\{ \|[\cos(\theta+\alpha)]x + [\sin(\theta+\alpha)]y\| : 0 \leq \theta \leq 2\pi \} \\
&= \sup\{ \|(\cos\theta)x + (\sin\theta)y\| : 0 \leq \theta \leq 2\pi \} = \|z\|_{\mathbb{C}}.
\end{aligned}
$$

Thus $\|\ \|_{\mathbb{C}}$ is a norm on $\mathfrak{X}_{\mathbb{C}}$.

Since

$$
\max\{\|x\|, \|y\|\} \leq \|(x,y)\|_{\mathbb{C}} \leq \|x\| + \|y\|,
$$

the topology and uniform structure induced by the norm $\|\ \|_{\mathbb{C}}$ on $\mathfrak{X}_{\mathbb{C}}$ $(= \mathfrak{X} \times \mathfrak{X})$ are simply the products of those induced on $\mathfrak{X}$ by $\|\ \|$. Hence $\mathfrak{X}_{\mathbb{C}}$ is complete, and is thus a complex Banach space.

The remaining assertions are immediate.   ∎

1.9.7.   Suppose that $\mathfrak{X}$ is an infinite-dimensional normed space, and V is a neighborhood of 0 in the weak topology on $\mathfrak{X}$. Show that V contains a closed subspace of finite codimension in $\mathfrak{X}$. Deduce that the weak topology on $\mathfrak{X}$ is strictly coarser than the norm topology.

Solution.   There is a finite set $\{\rho_1, \ldots, \rho_n\}$ of bounded linear functionals on $\mathfrak{X}$, and a positive real number ε, such that V contains the set

$$
\{x \in \mathfrak{X} : |\rho_j(x)| < \varepsilon \ (j = 1, \ldots, n)\}.
$$

In particular, therefore, V contains the null space $\mathcal{N}$ of the continuous linear operator $T : \mathfrak{X} \to \mathbb{K}^n$ (where $\mathbb{K}$ is the scalar field) defined by

$$
Tx = \big(\rho_1(x), \rho_2(x), \ldots, \rho_n(x) \big).
$$

Since $\mathfrak{X}/\mathcal{N}$ is isomorphic to a subspace of $\mathbb{K}^n$, and T is continuous, $\mathcal{N}$ has finite codimension and is closed. Since $\mathfrak{X}$ is infinite-dimensional, $\mathcal{N} \neq \{0\}$.

The norm (initial) topology on $\mathfrak{X}$ is finer than the weak topology, and we have to show that it is strictly finer. The open unit ball is a neighborhood of 0 for the norm topology on $\mathfrak{X}$. It contains no nonzero subspace of $\mathfrak{X}$, and thus (by the previous paragraph) contains no neighborhood of 0 in the weak topology on $\mathfrak{X}$. $\blacksquare$

1.9.8. Show that, if $\mathfrak{X}$ is a separable Banach space, each bounded sequence $\{\rho_n\}$ in $\mathfrak{X}^\#$ has a subsequence that is weak* convergent to an element of $\mathfrak{X}^\#$.

Solution. Let $\{x_1, x_2, x_3, \ldots\}$ be an everywhere dense subset of $\mathfrak{X}$. The bounded sequence $\{\rho_n(x_1)\}$ of scalars has a convergent subsequence $\{\rho_{n,1}(x_1)\}$. The bounded sequence $\{\rho_{n,1}(x_2)\}$ of scalars has a convergent subsequence $\{\rho_{n,2}(x_2)\}$. The bounded sequence $\{\rho_{n,2}(x_3)\}$ of scalars has a convergent subsequence $\{\rho_{n,3}(x_3)\}$. Continuing in the same way, for each positive integer k we obtain a subsequence $\{\rho_{1,k}, \rho_{2,k}, \rho_{3,k}, \ldots\}$ of $\{\rho_n\}$, such that $\{\rho_{n,k}\}$ is a subsequence of $\{\rho_{n,k-1}\}$, and $\{\rho_{n,k}(x_k)\}$ converges. The diagonal sequence $\{\rho_{n,n}\}$ is a subsequence of $\{\rho_n\}$, and coincides from its kth term onward with a subsequence of $\{\rho_{n,k}\}$; so $\{\rho_{n,n}(x_k)\}$ converges (as $n \to \infty$), for each $k = 1, 2, \ldots$.

Suppose that $x \in \mathfrak{X}$. Let K be $\sup\{\|\rho_n\| : n \in \mathbb{N}\}$, ε be a positive real number, and choose k so that $\|x - x_k\| < \varepsilon/(3K)$. We can then choose n_0 so that $|\rho_{m,m}(x_k) - \rho_{n,n}(x_k)| < \frac{1}{3}\varepsilon$ whenever $m > n \geq n_0$; and we have

$$|\rho_{m,m}(x) - \rho_{n,n}(x)| \leq |\rho_{m,m}(x - x_k)| + |\rho_{m,m}(x_k) - \rho_{n,n}(x_k)|$$
$$+ |\rho_{n,n}(x_k - x)| < \varepsilon \qquad (m > n \geq n_0).$$

Thus $\{\rho_{n,n}(x)\}$ converges, for each x in $\mathfrak{X}$.

Let $\rho(x)$ be $\lim_{n\to\infty} \rho_{n,n}(x)$ $(x \in \mathfrak{X})$. Since $|\rho_{n,n}(x)| \leq K\|x\|$, we have $|\rho(x)| \leq K\|x\|$. From this, it follows easily that ρ is a bounded linear functional on $\mathfrak{X}$, and $\{\rho_{n,n}\}$ is weak* convergent to ρ. $\blacksquare$

1.9.9. Prove that, if $\mathfrak{X}$ and $\mathfrak{Y}$ are normed spaces, $\mathfrak{X} \neq \{0\}$, and $\mathcal{B}(\mathfrak{X}, \mathfrak{Y})$ is complete, then $\mathfrak{Y}$ is complete.

Solution. Suppose that $\{y_n\}$ is a Cauchy sequence in $\mathfrak{Y}$. Let x_0 be a unit vector in $\mathfrak{X}$, and choose ρ in $\mathfrak{X}^{\#}$ so that $\rho(x_0) = \|\rho\| = 1$. The equation $T_n x = \rho(x) y_n$ defines an element T_n of $\mathcal{B}(\mathfrak{X}, \mathfrak{Y})$, and $T_n x_0 = y_n$. Moreover

$$
\begin{aligned}
\|(T_m - T_n)x\| &= \|\rho(x)(y_m - y_n)\| \\
&= |\rho(x)|\,\|y_m - y_n\| \le \|y_m - y_n\|\,\|x\|,
\end{aligned}
$$

for each x in $\mathfrak{X}$; so $\|T_m - T_n\| \le \|y_m - y_n\|$, and $\{T_n\}$ is a Cauchy sequence in $\mathcal{B}(\mathfrak{X}, \mathfrak{Y})$. Since $\mathcal{B}(\mathfrak{X}, \mathfrak{Y})$ is complete, it has an element T such that $\|T_n - T\| \to 0$. Thus

$$
\|y_n - Tx_0\| = \|(T_n - T)x_0\| \le \|T_n - T\| \to 0,
$$

and $\{y_n\}$ converges to the element Tx_0 of $\mathfrak{Y}$.   ∎

1.9.10.   Suppose that $\mathfrak{X}_0$ is a closed subspace of a Banach space $\mathfrak{X}$, $Q : \mathfrak{X} \to \mathfrak{X}/\mathfrak{X}_0$ is the quotient mapping, and $\mathfrak{X}_0^{\perp}$ ($\subseteq \mathfrak{X}^{\#}$) is the closed subspace consisting of all bounded linear functionals on $\mathfrak{X}$ that vanish on $\mathfrak{X}_0$.

(i)   Show that the (Banach) adjoint operator $Q^{\#}$ is an isometric linear mapping from $(\mathfrak{X}/\mathfrak{X}_0)^{\#}$ onto $\mathfrak{X}_0^{\perp}$.

(ii)   Show that the mapping $T : \rho + \mathfrak{X}_0^{\perp} \to \rho|\mathfrak{X}_0$ is an isometric isomorphism from $\mathfrak{X}^{\#}/\mathfrak{X}_0^{\perp}$ onto $\mathfrak{X}_0^{\#}$.

Solution. (i)   Since the mapping $Q : \mathfrak{X} \to \mathfrak{X}/\mathfrak{X}_0$ has norm 1, $Q^{\#} : (\mathfrak{X}/\mathfrak{X}_0)^{\#} \to \mathfrak{X}^{\#}$ also has norm 1. For each ρ in $(\mathfrak{X}/\mathfrak{X}_0)^{\#}$,

$$
(1) \qquad\qquad (Q^{\#}\rho)(x) = \rho(Qx) \qquad (x \in \mathfrak{X}).
$$

Since Q maps the open unit ball of $\mathfrak{X}$ onto that of $\mathfrak{X}/\mathfrak{X}_0$, it follows from (1) that $\|Q^{\#}\rho\| = \|\rho\|$. Again from (1), $Q^{\#}\rho$ vanishes on $\mathfrak{X}_0$; so $Q^{\#}((\mathfrak{X}/\mathfrak{X}_0)^{\#}) \subseteq \mathfrak{X}_0^{\perp}$. Each ω in $\mathfrak{X}_0^{\perp}$ has a factorization $\omega_0 Q$, where ω_0 is a bounded linear functional on $\mathfrak{X}/\mathfrak{X}_0$, by Theorem 1.5.8 (with $\mathfrak{Y}$ the scalar field); and $\omega = Q^{\#}\omega_0$. This proves (i).

(ii)   The mapping $S : \rho \to \rho|\mathfrak{X}_0$ is a bounded linear operator from $\mathfrak{X}^{\#}$ into $\mathfrak{X}_0^{\#}$, with $\|S\| \le 1$, and has null space $\mathfrak{X}_0^{\perp}$. By Theorem 1.5.8, it has a factorization

$$
\mathfrak{X}^{\#} \to \mathfrak{X}^{\#}/\mathfrak{X}_0^{\perp} \xrightarrow{T} \mathfrak{X}_0^{\#},
$$

where the mapping $T : \rho + \mathfrak{X}_0^{\perp} \to S\rho = \rho|\mathfrak{X}_0$ is one-to-one, with $\|T\| \leq 1$. Given any ω in $\mathfrak{X}_0^{\#}$, ω extends (without increase of norm) to an element ρ of $\mathfrak{X}^{\#}$. Since $T(\rho + \mathfrak{X}_0^{\perp}) = S\rho = \rho|\mathfrak{X}_0 = \omega$, and

$$\|\rho\| = \|\omega\| = \|T(\rho + \mathfrak{X}_0^{\perp})\| \leq \|\rho + \mathfrak{X}_0^{\perp}\| \leq \|\rho\|,$$

it now follows that T is isometric, and maps $\mathfrak{X}^{\#}/\mathfrak{X}_0$ onto $\mathfrak{X}_0^{\#}$.   ■

1.9.11.   Show that, if $\mathfrak{X}_0$ is a closed subspace of a Banach space $\mathfrak{X}$, than $\mathfrak{X}$ is reflexive if and only if both $\mathfrak{X}_0$ and $\mathfrak{X}/\mathfrak{X}_0$ are reflexive.

Solution.   First, suppose that $\mathfrak{X}$ is reflexive. Suppose that $\Omega \in \mathfrak{X}_0^{\#\#}$. Since the mapping $\rho \to \rho|\mathfrak{X}_0$ is a bounded linear operator from $\mathfrak{X}^{\#}$ onto $\mathfrak{X}_0^{\#}$, the equation $\Omega_1(\rho) = \Omega(\rho|\mathfrak{X}_0)$ defines a bounded linear functional Ω_1 on $\mathfrak{X}^{\#}$. Since $\mathfrak{X}$ is reflexive, there exists x_1 in $\mathfrak{X}$, such that

$$\rho(x_1) = \Omega_1(\rho) = \Omega(\rho|\mathfrak{X}_0) \qquad (\rho \in \mathfrak{X}^{\#}).$$

Since $\rho(x_1) = 0$ whenever $\rho|\mathfrak{X}_0 = 0$, it follows from Corollary 1.2.13 that $x_1 \in \mathfrak{X}_0$. It now follows, from the displayed equation above, that

$$\Omega(\sigma) = \sigma(x_1)$$

for each $\sigma \; (= \rho|\mathfrak{X}_0)$ in $\mathfrak{X}_0^{\#}$. Thus $\mathfrak{X}_0$ is reflexive.

Let $\Lambda \in (\mathfrak{X}/\mathfrak{X}_0)^{\#\#}$. With $Q : \mathfrak{X} \to \mathfrak{X}/\mathfrak{X}_0$ the quotient mapping, it follows from the result of Exercise 1.9.10(i) that the equation

$$\Lambda_1(Q^{\#}\rho) = \Lambda(\rho) \qquad (\rho \in (\mathfrak{X}/\mathfrak{X}_0)^{\#})$$

defines a bounded linear functional Λ_1 on $\mathfrak{X}_0^{\perp}$ $(\subseteq \mathfrak{X}^{\#})$ and this extends to a bounded linear functional Λ_2 on $\mathfrak{X}^{\#}$. Since $\mathfrak{X}$ is reflexive, there exists x_2 in $\mathfrak{X}$, such that $\Lambda_2(\sigma) = \sigma(x_2)$ for each σ in $\mathfrak{X}^{\#}$. With x_0 the element Qx_2 of $\mathfrak{X}/\mathfrak{X}_0$, and ρ in $(\mathfrak{X}/\mathfrak{X}_0)^{\#}$, we have

$$\Lambda(\rho) = \Lambda_1(Q^{\#}\rho) = \Lambda_2(Q^{\#}\rho)$$
$$= (Q^{\#}\rho)(x_2) = \rho(Qx_2) = \rho(x_0).$$

Thus $\mathfrak{X}/\mathfrak{X}_0$ is reflexive.

Conversely, suppose that both $\mathfrak{X}_0$ and $\mathfrak{X}/\mathfrak{X}_0$ are reflexive, and let $\Omega \in \mathfrak{X}^{\#\#}$. Since $Q^{\#\#}\Omega \in (\mathfrak{X}/\mathfrak{X}_0)^{\#\#}$, there is an element x_1 of $\mathfrak{X}/\mathfrak{X}_0$ such that

$$(Q^{\#\#}\Omega)(\rho) = \rho(x_1) \qquad (\rho \in (\mathfrak{X}/\mathfrak{X}_0)^{\#}).$$

Moreover, $x_1 = Qx_2$, for some x_2 in $\mathfrak{X}$. With $\hat{x}_2$ the canonical image of x_2 in $\mathfrak{X}^{\#\#}$, we have

$$\hat{x}_2(Q^{\#}\rho) = (Q^{\#}\rho)(x_2) = \rho(Qx_2) = \rho(x_1)$$
$$= (Q^{\#\#}\Omega)(\rho) = \Omega(Q^{\#}\rho),$$

for each ρ in $(\mathfrak{X}/\mathfrak{X}_0)^{\#}$. It now follows, from the result of Exercise 1.9.10(i), that $(\Omega - \hat{x}_2)|\mathfrak{X}_0^{\perp} = 0$. From this, and since the mapping $\rho \to \rho|\mathfrak{X}_0 : \mathfrak{X}^{\#} \to \mathfrak{X}_0^{\#}$ carries the unit ball of $\mathfrak{X}^{\#}$ onto that of $\mathfrak{X}_0^{\#}$, the equation

$$\Lambda(\rho|\mathfrak{X}_0) = (\Omega - \hat{x}_2)(\rho) \qquad (\rho \in \mathfrak{X}^{\#})$$

defines (unambiguously) a bounded linear functional Λ on $\mathfrak{X}_0^{\#}$. Since the space $\mathfrak{X}_0$ is reflexive, there is an element x_3 of $\mathfrak{X}_0$ ($\subseteq \mathfrak{X}$) such that $\Lambda(\rho|\mathfrak{X}_0) = \rho(x_3)$, for each ρ in $\mathfrak{X}^{\#}$; and we have

$$\hat{x}_3(\rho) = \rho(x_3) = \Lambda(\rho|\mathfrak{X}_0) = (\Omega - \hat{x}_2)(\rho).$$

Thus $\Omega = \hat{x}_2 + \hat{x}_3$, and $\mathfrak{X}$ is reflexive. $\qquad\blacksquare$

1.9.12. Show that a Banach space $\mathfrak{X}$ is reflexive if and only if its dual space $\mathfrak{X}^{\#}$ is reflexive.

Solution. Elements of $\mathfrak{X}$, $\mathfrak{X}^{\#}$, $\mathfrak{X}^{\#\#}$, $\mathfrak{X}^{\#\#\#}$ will be denoted by x, ω, X, Ω, respectively (if necessary, with subscripts), and

$$\begin{aligned}
x \to \hat{x} : &\qquad \mathfrak{X} \to \widehat{\mathfrak{X}} \subseteq \mathfrak{X}^{\#\#}, \\
\omega \to \hat{\omega} : &\qquad \mathfrak{X}^{\#} \to \widehat{\mathfrak{X}^{\#}} \subseteq \mathfrak{X}^{\#\#\#}
\end{aligned}$$

are the canonical (isometric linear) imbeddings.

Suppose that $\mathfrak{X}$ is reflexive. Given Ω in $\mathfrak{X}^{\#\#\#}$, the equation $\omega(x) = \Omega(\hat{x})$ ($x \in \mathfrak{X}$) defines an element ω of $\mathfrak{X}^{\#}$, and

$$\Omega(\hat{x}) = \omega(x) = \hat{x}(\omega) = \hat{\omega}(\hat{x}) \qquad (x \in \mathfrak{X}).$$

Since $\widehat{\mathfrak{X}} = \mathfrak{X}^{\#\#}$, it follows that $\Omega = \hat{\omega}$; so $\mathfrak{X}^{\#}$ is reflexive.

Now suppose that $\mathfrak{X}^{\#}$ is reflexive. If $\mathfrak{X}$ is not reflexive, there is a bounded linear functional Ω on $\mathfrak{X}^{\#\#}$, non-zero, but vanishing on the proper closed subspace $\widehat{\mathfrak{X}}$ of $\mathfrak{X}^{\#\#}$. Thus $\Omega \in \mathfrak{X}^{\#\#\#}$, and $\Omega = \hat{\omega}$ for some ω in $\mathfrak{X}^{\#}$, since $\mathfrak{X}^{\#}$ is reflexive. For each x in $\mathfrak{X}$,

$$\omega(x) = \hat{x}(\omega) = \hat{\omega}(\hat{x}) = \Omega(\hat{x}) = 0.$$

Thus $\omega = 0$, and $\Omega = \hat{\omega} = 0$, a contradiction. Hence $\mathfrak{X}$ is reflexive, when $\mathfrak{X}^{\#}$ is reflexive. $\blacksquare$

1.9.13.   Suppose that $\mathfrak{X}$ is a Banach space with the following property: given any positive real number ε, there is a positive real number $\delta(\varepsilon)$ such that $\|x - y\| < \varepsilon$ whenever x and y lie in the unit ball $(\mathfrak{X})_1$ and $\|\frac{1}{2}(x + y)\| > 1 - \delta(\varepsilon)$ (such a Banach space is said to be *uniformly convex*). Prove that $\mathfrak{X}$ is reflexive [*Hint.* Suppose that $\Omega_0 \in \mathfrak{X}^{\#\#}$ and $\|\Omega_0\| = 1$. Choose ρ_0 in $\mathfrak{X}^{\#}$ so that $\|\rho_0\| = 1$ and $|\Omega_0(\rho_0) - 1| < \delta(\varepsilon)$; and let

$$\mathcal{K} = \{\hat{x} : x \in (\mathfrak{X})_1, \ |\rho_0(x) - 1| < \delta(\varepsilon)\},$$

where $x \to \hat{x}$ is the natural isometric isomorphism from $\mathfrak{X}$ into $\mathfrak{X}^{\#\#}$. Prove that Ω_0 lies in the weak* closure of $\mathcal{K}$, and that $\|\hat{x} - \hat{y}\| < \varepsilon$ whenever $\hat{x}, \hat{y} \in \mathcal{K}$. Deduce that $\|\Omega_0 - \hat{x}\| \le \varepsilon$ for each $\hat{x}$ in $\mathcal{K}$.]

Solution.   If $\mathcal{V}_0$ is a weak* neighborhood of Ω_0 in $\mathfrak{X}^{\#\#}$, then so is the set
$$\mathcal{V}_1 = \{\Omega \in \mathcal{V}_0 : |\Omega(\rho_0) - 1| < \delta(\varepsilon)\}.$$

By Theorem 1.6.5(ii), $\mathcal{V}_1$ meets $(\widehat{\mathfrak{X}})_1$; so we can choose x in $(\mathfrak{X})_1$ such that $\hat{x} \in \mathcal{V}_1$, and then $\hat{x} \in \mathcal{V}_0 \cap \mathcal{K}$. Since $\mathcal{K}$ meets each weak* neighborhood of Ω_0, Ω_0 lies in the weak* closure of $\mathcal{K}$.

If $\hat{x}, \hat{y} \in \mathcal{K}$, we have $x, y \in (\mathfrak{X})_1$ and

$$\left\|\tfrac{1}{2}(x + y)\right\| \ge \left|\rho_0\left(\tfrac{1}{2}(x + y)\right)\right| \ge 1 - \left|1 - \rho_0\left(\tfrac{1}{2}(x + y)\right)\right|$$
$$\ge 1 - \tfrac{1}{2}|1 - \rho_0(x)| - \tfrac{1}{2}|1 - \rho_0(y)| > 1 - \delta(\varepsilon).$$

Thus $\|\hat{x} - \hat{y}\| = \|x - y\| < \varepsilon$.

From the preceding paragraph, if $\hat{x} \in \mathcal{K}$, then $\mathcal{K} \subseteq \hat{x} + \varepsilon(\mathfrak{X}^{\#\#})_1$, a weak* compact set. Since Ω_0 lies in the weak* closure of $\mathcal{K}$, $\Omega_0 \in \hat{x} + \varepsilon(\mathfrak{X}^{\#\#})_1$; that is $\|\Omega_0 - \hat{x}\| \le \varepsilon$. Since $\operatorname{dist}(\Omega_0, \widehat{\mathfrak{X}}) \le \varepsilon$ (for each positive ε), and $\widehat{\mathfrak{X}}$ is norm closed in $\mathfrak{X}^{\#\#}$, $\Omega_0 \in \widehat{\mathfrak{X}}$. $\blacksquare$

1.9.14.   Suppose that $\mathfrak{X}$ is a normed space, $\mathcal{M}$ is a linear subspace of the dual space $\mathfrak{X}^{\#}$, and $\mathcal{M}$ separates the points of $\mathfrak{X}$. Let ρ be a linear functional on $\mathfrak{X}$.

(i)   Suppose that the restriction $\rho|(\mathfrak{X})_1$ of ρ to the unit ball $(\mathfrak{X})_1$ is continuous in the weak topology $\sigma(\mathfrak{X}, \mathcal{M})$. Show that $\rho \in \mathfrak{X}^{\#}$. Prove also that, if $\varepsilon > 0$, there is a finite subset $\{\omega_1, \ldots, \omega_n\}$ of $\mathcal{M}$ such that

$$|\rho(x)| < \varepsilon \quad \text{whenever} \quad x \in (\mathfrak{X})_1 \quad \text{and} \quad \sum_{j=1}^{n} |\omega_j(x)| < 1.$$

Deduce that

$$|\rho(x)| \le \varepsilon \|x\| + \|\rho\| \sum_{j=1}^{n} |\omega_j(x)| \qquad (x \in \mathfrak{X}).$$

From this inequality, together with the result of Exercise 1.9.2, deduce that ρ has the form $\rho_1 + \rho_2$, where $\rho_1 \in \mathfrak{X}^{\#}$, $\|\rho_1\| \le \varepsilon$, and $\rho_2 \in \mathcal{M}$.

(ii)   Prove that $\rho|(\mathfrak{X})_1$ is $\sigma(\mathfrak{X}, \mathcal{M})$-continuous if and only if ρ lies in the norm closure $\mathcal{M}^{=}$ of $\mathcal{M}$ in $\mathfrak{X}^{\#}$.

Solution. (i)   Since $\rho|(\mathfrak{X})_1$ is $\sigma(\mathfrak{X}, \mathcal{M})$-continuous, it is norm continuous, and this implies that ρ is bounded; so $\rho \in \mathfrak{X}^{\#}$. Given ε (> 0), the $\sigma(\mathfrak{X}, \mathcal{M})$-continuity of $\rho|(\mathfrak{X})_1$ at 0 entails the existence of a finite set $\{\omega_1, \ldots, \omega_n\}$ ($\subseteq \mathcal{M}$) satisfying the first displayed condition in (i). To prove the second condition it suffices, by homogeneity, to consider only the case in which $\|x\| = 1$. In this case, the second condition follows from the first when $\sum_1^n |\omega_j(x)| < 1$, and is self-evident when $\sum_1^n |\omega_j(x)| \ge 1$.

It now follows from Exercise 1.9.2 that $\rho = \rho_1 + \rho_2$, where the linear functionals ρ_1 and ρ_2 satisfy

$$|\rho_1(x)| \le \varepsilon \|x\|, \qquad |\rho_2(x)| \le \|\rho\| \sum_{j=1}^{n} |\omega_j(x)| \qquad (x \in \mathfrak{X}).$$

Thus $\rho_1 \in \mathfrak{X}^{\#}$, $\|\rho_1\| \le \varepsilon$; and $\rho_2 \in \mathcal{M}$, since ρ_2 vanishes on the intersection of the null-spaces of $\omega_1, \ldots, \omega_n$, and is therefore a linear combination of $\omega_1, \ldots, \omega_n$ by Proposition 1.1.1.

(ii)   If $\rho|(\mathfrak{X})_1$ is $\sigma(\mathfrak{X}, \mathcal{M})$-continuous, it follows from (i) that $\rho \in \mathfrak{X}^{\#}$, and $\mathrm{dist}(\rho, \mathcal{M}) \leq \varepsilon$ for every positive ε. Hence $\rho \in \mathcal{M}^{=}$.

If $\rho \in \mathcal{M}^{=}$, then $\rho|(\mathfrak{X})_1$ is the uniform limit of a sequence $\rho_n|(\mathfrak{X})_1$, where each ρ_n lies in $\mathcal{M}$. Since ρ_n is $\sigma(\mathfrak{X}, \mathcal{M})$-continuous, so is $\rho_n|(\mathfrak{X})_1$, and hence so is $\rho|(\mathfrak{X})_1$.   ∎

1.9.15.   Suppose that $\mathfrak{X}$ is a Banach space, $x \to \hat{x}$ is the natural isometric isomorphism from $\mathfrak{X}$ into $\mathfrak{X}^{\#\#}$, and $\omega \in \mathfrak{X}^{\#\#}$. Show that, if the restriction $\omega|(\mathfrak{X}^{\#})_1$ of ω to the unit ball of $\mathfrak{X}^{\#}$ is continuous in the weak* topology $\sigma(\mathfrak{X}^{\#}, \mathfrak{X})$, then $\omega = \hat{x}$ for some x in $\mathfrak{X}$. [*Hint.* Use the result of Exercise 1.9.14(ii).]

Solution.   The weak* topology $\sigma(\mathfrak{X}^{\#}, \mathfrak{X})$ is in fact $\sigma(\mathfrak{X}^{\#}, \widehat{\mathfrak{X}})$, where $\widehat{\mathfrak{X}}$ is the closed subspace $\{\hat{x} : x \in \mathfrak{X}\}$ of $\mathfrak{X}^{\#\#}$. The result now follows at once from Exercise 1.9.14(ii), applied with $\mathfrak{X}^{\#}$ and $\widehat{\mathfrak{X}}$ ($\subseteq \mathfrak{X}^{\#\#}$) in place of $\mathfrak{X}$ and $\mathcal{M}$ ($\subseteq \mathfrak{X}^{\#}$).   ∎

1.9.16.   Suppose that $\mathfrak{X}$ and $\mathfrak{Y}$ are Banach spaces, and $S \in \mathcal{B}(\mathfrak{Y}^{\#}, \mathfrak{X}^{\#})$.

(i)   Prove that, if S is continuous relative to the weak* topologies on $\mathfrak{Y}^{\#}$ and $\mathfrak{X}^{\#}$, then $S = T^{\#}$ for some T in $\mathcal{B}(\mathfrak{X}, \mathfrak{Y})$.

(ii)   By using the result of Exercise 1.9.15, show that (i) remains valid when the weak* continuity of S is replaced by weak* continuity of $S|(\mathfrak{Y}^{\#})_1$.

Solution.   Given x in $\mathfrak{X}$, the equation

$$\omega_x(\rho) = (S\rho)(x) \qquad (\rho \in \mathfrak{Y}^{\#})$$

defines a bounded linear functional ω_x on $\mathfrak{Y}^{\#}$, and $\|\omega_x\| \leq \|S\|\,\|x\|$. Moreover, ω_x (under the conditions of (i)), or $\omega_x|(\mathfrak{Y}^{\#})_1$ (under the conditions of (ii)), is weak* continuous. By Proposition 1.3.5 or Exercise 1.9.15, there is an element y_x of $\mathfrak{Y}$, such that $\omega_x = \hat{y}_x$; and $\|y_x\| = \|\hat{y}_x\| = \|\omega_x\| \leq \|S\|\,\|x\|$. It is clear that the mapping $T : x \to y_x : \mathfrak{X} \to \mathfrak{Y}$ is a bounded linear operator. When $x \in \mathfrak{X}$ and $\rho \in \mathfrak{Y}^{\#}$ we have

$$(S\rho)(x) = \omega_x(\rho) = \rho(y_x) = \rho(Tx) = (T^{\#}\rho)(x);$$

so $S = T^{\#}$.   ∎

1.9.17. Suppose that $\mathfrak{X}, \mathfrak{Y}$ are Banach spaces and $T \in \mathcal{B}(\mathfrak{X}, \mathfrak{Y})$. Prove that

(i) the set $\{T^{\#}\rho : \rho \in (\mathfrak{Y}^{\#})_1\}$ is weak* compact, and hence norm closed in $\mathfrak{X}^{\#}$;

(ii) if $\mathfrak{X}$ is reflexive, the set $\{Tx : x \in (\mathfrak{X})_1\}$ is weakly compact, and hence norm closed in $\mathfrak{Y}$.

Solution. By Propositions 1.3.3 and 1.6.8, $T : \mathfrak{X} \to \mathfrak{Y}$ is weakly continuous, while $T^{\#} : \mathfrak{Y}^{\#} \to \mathfrak{X}^{\#}$ is continuous relative to the weak* topologies. From Theorem 1.6.5(i), $(\mathfrak{Y}^{\#})_1$ is weak* compact; and $(\mathfrak{X})_1$ is weakly compact when $\mathfrak{X}$ is reflexive, by Theorem 1.6.7. This establishes the compactness assertions, in both (i) and (ii). In each case, the compact set is closed (in the weak*, or weak, topology), and is therefore closed in the (finer) norm topology. ∎

1.9.18. Suppose that $\mathfrak{X}$ and $\mathfrak{Y}$ are Banach spaces, $T \in \mathcal{B}(\mathfrak{X}, \mathfrak{Y})$, and the image $T(\mathfrak{X})$ $(= \{Tx : x \in \mathfrak{X}\})$ is a closed subspace of $\mathfrak{Y}$. Prove that $T^{\#}(\mathfrak{Y}^{\#})$ is the closed subspace of $\mathfrak{X}^{\#}$ consisting of all bounded linear functionals on $\mathfrak{X}$ that vanish on the null space of T. [*Hint.* If $\rho \in \mathfrak{X}^{\#}$ and ρ vanishes on the null space of T, the equation $\omega_0(Tx) = \rho(x)$ defines (unambiguously) a linear functional ω_0 on $T(\mathfrak{X})$. By applying the open mapping theorem to T, as an operator from $\mathfrak{X}$ onto $T(\mathfrak{X})$, prove that ω_0 is bounded. Deduce that $\rho = T^{\#}\omega$, for some ω in $\mathfrak{Y}^{\#}$.]

Solution. Given ρ as in the "hint," if $x_1, x_2 \in \mathfrak{X}$ and $Tx_1 = Tx_2$, then $T(x_1 - x_2) = 0$; so $\rho(x_1 - x_2) = 0$, and $\rho(x_1) = \rho(x_2)$. This shows that ω_0 (as in the "hint") is defined unambiguously, and it is clear that ω_0 is a linear functional on $T(\mathfrak{X})$. Since ρ is bounded on $(\mathfrak{X})_1$, ω_0 is bounded on $T((\mathfrak{X})_1)$. By the open mapping theorem, $T((\mathfrak{X})_1^0)$ is an open subset of $T(\mathfrak{X})$, where $(\mathfrak{X})_1^0$ is the open unit ball with center 0 in $\mathfrak{X}$, and is thus a neighborhood of 0 (in the norm topology of the Banach space $T(\mathfrak{X})$). Thus ω_0 is a bounded linear functional on $T(\mathfrak{X})$. By the Hahn–Banach theorem, it extends to a bounded linear functional ω on $\mathfrak{Y}$. When $x \in \mathfrak{X}$, $(T^{\#}\omega)(x) = \omega(Tx) = \omega_0(Tx) = \rho(x)$; and $\rho = T^{\#}\omega$.

Conversely, if $\rho = T^{\#}\omega$ for some ω in $\mathfrak{Y}^{\#}$, then $\rho(x) = \omega(Tx) = 0$ whenever x lies in the null space of T. Hence $T^{\#}(\mathfrak{Y}^{\#})$ consists exactly of those ρ in $\mathfrak{X}^{\#}$ that vanish on the null space of T, and is thus a closed subspace of $\mathfrak{X}^{\#}$. ∎

1.9.19. Let l_∞ denote the Banach space $l_\infty(\mathbb{N}, \mathbb{C})$ of Example 1.7.1, where $\mathbb{N}$ is the set of positive integers; an element of l_∞ is a bounded complex sequence $\{x_1, x_2, \ldots\}$, and $\|\{x_n\}\| = \sup\{|x_n| : n \in \mathbb{N}\}$. Let c and c_0 be the linear subspaces of l_∞ defined by

$$c = \left\{ \{x_n\} \in l_\infty : \lim_{n \to \infty} x_n \text{ exists} \right\},$$

$$c_0 = \left\{ \{x_n\} \in l_\infty : \lim_{n \to \infty} x_n = 0 \right\}.$$

Prove that

(i) c and c_0 are closed subspaces of l_∞;

(ii) the sequence $\{1, 1, 1, \ldots\}$ is an extreme point in the closed unit ball of l_∞, and also in the closed unit ball of c;

(iii) the closed unit ball of c_0 has no extreme point.

Deduce that c_0 is isometrically isomorphic neither to c, nor to any Banach dual space.

Solution. (i) Suppose that $X \in l_\infty$ and $X = \lim_{r \to \infty} X_r$, where $X_1, X_2, \ldots \in c$. We may suppose that

$$X = \{x_1, x_2, \ldots\}, \quad X_r = \{x_1^{(r)}, x_2^{(r)}, \ldots\},$$

and write y_r for $\lim_{n \to \infty} x_n^{(r)}$. Given $\varepsilon\ (> 0)$, we can choose a positive integer $N(\varepsilon)$ such that $\|X - X_r\| \le \varepsilon$ whenever $r \ge N(\varepsilon)$; that is,

$$|x_n - x_n^{(r)}| \le \varepsilon \qquad (n = 1, 2, \ldots;\ r \ge N(\varepsilon)).$$

Thus

$$|x_n^{(r)} - x_n^{(s)}| \le 2\varepsilon \qquad (n = 1, 2, \ldots;\ r, s \ge N(\varepsilon));$$

and, as $n \to \infty$, we obtain

$$|y_r - y_s| \le 2\varepsilon \qquad (r, s \ge N(\varepsilon)).$$

Hence $\{y_r\}$ is a Cauchy sequence in $\mathbb{C}$ and, if y denotes its limit, we have

$$|y_r - y| \le 2\varepsilon \qquad (r \ge N(\varepsilon)).$$

Choose and fix an integer $r\ (\ge N(\varepsilon))$. For $n = 1, 2, \ldots$,

$$|x_n - y| \le |x_n - x_n^{(r)}| + |x_n^{(r)} - y_r| + |y_r - y|$$
$$\le 3\varepsilon + |x_n^{(r)} - y_r|;$$

and the right hand side is less than 4ε, when n is sufficiently large, since $x_n^{(r)} \to y_r$ as $n \to \infty$. Hence $\{x_n\}$ converges to y. This shows that $X \in c$. If each X_r lies in c_0, then each y_r is 0; so $y = 0$, and $X \in c_0$, in this case. Thus c and c_0 are closed in l_∞.

(ii)   The sequence $U = \{1, 1, 1, \ldots\}$ lies in the closed unit ball of l_∞ (and that of c). If $U = (1 - \alpha)X + \alpha Y$, where $0 < \alpha < 1$ and $X\ (= \{x_n\})$, $Y\ (= \{y_n\})$ lie in that ball, then x_n, y_n lie in the unit disk in $\mathbb{C}$, and $(1 - \alpha)x_n + \alpha y_n = 1$, for each n. Since 1 is an extreme point of the disk, $x_n = y_n = 1$; so $X = Y = U$. This shows that U is an extreme point of the closed unit ball of l_∞ (or c).

(iii)   If $Z = \{z_n\}$ lies in the closed unit ball of c_0, then $|z_n| \le 1$ for all n, and $|z_n| \le \frac{1}{2}$ for sufficiently large n (say, for $n \ge k$). Upon replacing z_k by $z_k \pm \frac{1}{2}$, while leaving the other z_n's alone, we obtain distinct elements Z', Z'' in the unit ball of c_0, such that $Z = \frac{1}{2}(Z' + Z'')$. Hence Z is not an extreme point of that ball.

From (iii), c_0 is not isometrically isomorphic to a Banach space whose unit ball has an extreme point. By (ii), and by Corollary 1.6.6 (and Theorem 1.6.5(i)) it is isometrically isomorphic neither to c nor to a Banach dual space.   ∎

1.9.20.   With the notation of Exercise 1.9.19, let U be the element $\{1, 1, 1, \ldots\}$ of c; and, for $k = 1, 2, \ldots$, let E_k (in c_0) be the sequence that has 1 in the kth position and zeros elsewhere.

(i)   Prove that, if $X = \{x_n\} \in c_0$, then $X = \sum_{k=1}^{\infty} x_k E_k$, the series converging in the norm topology of c_0 (that is, $\|X - \sum_{k=1}^{m} x_k E_k\| \to 0$ as $m \to \infty$).

(ii)   Prove that, if $X = \{x_n\} \in c$ and $x = \lim_{n \to \infty} x_n$, then $X - xU \in c_0$ and $X = xU + \sum_{k=1}^{\infty} (x_k - x)E_k$, the series converging in the norm topology of c.

Solution. (i)   For $m = 1, 2, \ldots$, $x - \sum_{k=1}^{m} x_k E_k$ is the sequence $\{0, 0, \ldots, 0, x_{m+1}, x_{m+2}, \ldots\}$, and

$$\left\| X - \sum_{k=1}^{m} x_k E_k \right\| = \sup\{|x_n| : n > m\}.$$

Since $X \in c_0$, $x_n \to 0$ as $n \to \infty$; and thus

$$\left\| X - \sum_{k=1}^{m} x_k E_k \right\| \to 0 \quad \text{as } m \to \infty.$$

(ii) $X - xU$ is the sequence $\{x_n - x\}$, which converges to 0. Thus $X - xU \in c_0$ and, by (i),

$$\left\| X - xU - \sum_{k=1}^{m} (x_k - x)E_k \right\| \to 0 \quad \text{as } m \to \infty. \qquad \blacksquare$$

1.9.21.   Adopt the notation of Exercises 1.9.19 and 1.9.20; in addition, let l_1 denote the Banach space $l_1(\mathbb{N}, \mathbb{C})$ of Example 1.7.3, so that an element of l_1 is a complex sequence $Y = \{y_1, y_2, \ldots\}$ such that $(\|Y\| =) \sum_{n=1}^{\infty} |y_n| < \infty$.

(i)   Show that, if $Y = \{y_1, y_2, \ldots\} \in l_1$, the equation

$$\rho(X) = \sum_{m=1}^{\infty} y_n x_n \qquad (X = \{x_n\} \in c_0)$$

defines a bounded linear functional ρ on c_0, and $\|\rho\| = \|Y\|$. Prove also that $y_n = \rho(E_n)$.

(ii)   Show that each bounded linear functional on c_0 arises, as in (i), from an element $Y = \{y_1, y_2, \ldots\}$ of l_1.

(iii)   Deduce that the Banach dual space $c_0^{\#}$ is isometrically isomorphic to l_1.

Solution. (i)   Suppose that $Y = \{y_1, y_2, \ldots\} \in l_1$. When $X = \{x_1, x_2, \ldots\} \in c_0$, we have $|x_n| \le \|X\|$ for each n, and thus

$$\sum_{n=1}^{\infty} |y_n x_n| \le \|X\| \sum_{n=1}^{\infty} |y_n| = \|X\| \, \|Y\| < \infty.$$

Hence the equation $\rho(X) = \sum_{n=1}^{\infty} y_n x_n$ defines a mapping $\rho : c_0 \to \mathbb{C}$, and $|\rho(X)| \le \|X\| \, \|Y\|$. It is apparent that ρ is linear, so it is a bounded linear functional on c_0, with $\|\rho\| \le \|Y\|$.

For $n = 1, 2, \ldots$, let z_n be a complex number such that $|z_n| = 1$ and $y_n z_n = |y_n|$. For $k = 1, 2, \ldots$, let Z_k be $\{z_1, \ldots, z_k, 0, 0, \ldots\}$. Then Z_k is in the unit ball of c_0; so

$$\|\rho\| \ge |\rho(Z_k)| = \left| \sum_{n=1}^{k} y_n z_n \right| = \sum_{n=1}^{k} |y_n|.$$

Hence $\|\rho\| \geq \sum_{n=1}^{\infty} |y_n| = \|Y\|$; and $\|\rho\| = \|Y\|$.

From the definition of ρ, and of the sequence E_n, we have $\rho(E_n) = y_n$.

(ii)   Suppose that ρ is a bounded linear functional on c_0, and let $Y = \{y_1, y_2, \ldots\}$ be the complex sequence defined by $\rho(E_n) = y_n$. With $Z_1, Z_2, \ldots$ in c_0 defined as in the proof of (i), we have

$$\|\rho\| \geq |\rho(Z_k)| = \left| \rho\left(\sum_{n=1}^{k} z_n E_n \right) \right|$$

$$= \left| \sum_{n=1}^{k} z_n y_n \right| = \sum_{n=1}^{k} |y_n|.$$

Hence $\sum_{n=1}^{\infty} |y_n| \leq \|\rho\| < \infty$, and $Y \in l_1$.

When $X = \{x_n\} \in c_0$, $X = \sum_{n=1}^{\infty} x_n E_n$ by the result of Exercise 1.9.20(i); and the continuity and linearity of ρ entails

$$\rho(X) = \sum_{n=1}^{\infty} x_n \rho(E_n) = \sum_{n=1}^{\infty} x_n y_n.$$

Thus ρ is derived from Y as in (i).

(iii)   For each Y in l_1, let TY denote the bounded linear functional on c_0, derived from Y as in (i). It is clear that $T : l_1 \to c_0^{\#}$ is linear. By (i), it is isometric, and by (ii) its range is the whole of $c_0^{\#}$.   ∎

1.9.22.   Let c and l_1 be the Banach spaces defined in Exercises 1.9.19 and 1.9.21.

(i)   Show that, if $\{y_0, y_1, y_2, \ldots\}$ is a complex sequence such that $\sum_{n=0}^{\infty} |y_n| < \infty$, the equation

$$\rho(X) = y_0 \lim_{n \to \infty} x_n + \sum_{n=1}^{\infty} y_n x_n \qquad (X = \{x_n\} \in c)$$

defines a bounded linear functional ρ on c, and $\|\rho\| = \sum_{n=0}^{\infty} |y_n|$.

(ii)   Prove that every bounded linear functional on c arises, as in (i), from such a sequence $\{y_0, y_1, y_2, \ldots\}$. [*Hint.* Use the results of Exercises 1.9.20(ii) and 1.9.21(ii).]

(iii)   Deduce that the Banach dual space $c^{\#}$ is isometrically isomorphic to l_1.

(iv)   Deduce that $c_0^{\#}$ and $c^{\#}$ are isometrically isomorphic, while c_0 and c are not.

Solution. (i)   When $X = \{x_1, x_2, \ldots\} \in c$, let $x = \lim_{n \to \infty} x_n$; so that $|x_n| \le \|X\|$, for all n, and $|x| \le \|X\|$. Hence

$$|y_0 x| + \sum_{n=1}^{\infty} |y_n x_n| \le \|X\| \sum_{n=0}^{\infty} |y_n|.$$

Hence the equation $\rho(X) = y_0 \lim_{n \to \infty} x_n + \sum_{n=1}^{\infty} y_n x_n$ defines a mapping $\rho : c \to \mathbb{C}$, and $|\rho(X)| \le \|X\| \sum_{n=0}^{\infty} |y_n|$. It is apparent that ρ is linear, so it is a bounded linear functional on c, with $\|\rho\| \le \sum_{n=0}^{\infty} |y_n|$.

For $n = 0, 1, 2, \ldots$, let z_n be a complex number such that $|z_n| = 1$ and $y_n z_n = |y_n|$. The sequence $\{z_1, z_2, \ldots, z_k, z_0, z_0, z_0, \ldots\}$ is an element Z_k of the unit ball of c, for $k = 1, 2, \ldots$. Thus

$$\|\rho\| \ge |\rho(Z_k)| = \left| y_0 z_0 + \sum_{n=1}^{k} y_n z_n + \sum_{n=k+1}^{\infty} y_n z_0 \right|$$

$$= \left| |y_0| + \sum_{n=1}^{k} |y_n| + z_0 \sum_{n=k+1}^{\infty} y_n \right|$$

$$\ge \sum_{n=0}^{k} |y_n| - \sum_{n=k+1}^{\infty} |y_n| \qquad (k = 1, 2, \ldots).$$

Thus $\|\rho\| \ge \sum_{n=0}^{\infty} |y_n|$; and the reverse inequality has already been noted.

(ii)   Suppose that ρ is a bounded linear functional on c. Since the restriction $\rho|c_0$ is a bounded linear functional on c_0, it follows from the result of Exercise 1.9.21(ii) that there is a complex sequence $\{y_1, y_2, \ldots\}$ satisfying

$$\sum_{n=1}^{\infty} |y_n| < \infty, \quad \rho(X) = \sum_{n=1}^{\infty} y_n x_n \qquad (X = \{x_n\} \in c_0).$$

If, now, $X = \{x_n\} \in c$, and $x = \lim_{n \to \infty} x_n$, then (as in Exercise 1.9.20(ii)) $X - xU = \{x_n - x\} \in c_0$, and thus

$$\rho(X - xU) = \sum_{n=1}^{\infty} y_n(x_n - x).$$

Thus

$$\rho(X) = x\rho(U) + \sum_{n=1}^{\infty} y_n(x_n - x)$$

$$= x\Big(\rho(U) - \sum_{n=1}^{\infty} y_n\Big) + \sum_{n=1}^{\infty} y_n x_n$$

$$= y_0 \lim_{n\to\infty} x_n + \sum_{n=1}^{\infty} y_n x_n,$$

where $y_0 = \rho(U) - \sum_{n=1}^{\infty} y_n$.

(iii)   It follows from (i) that, for each $Y = \{y_1, y_2, \ldots\}$ in l_1, the equation

$$(TY)(X) = y_1 \lim_{n\to\infty} x_n + \sum_{n=1}^{\infty} y_{n+1} x_n \qquad (X = \{x_n\} \in c)$$

defines a bounded linear functional TY on c. It is apparent that T is a linear mapping from l_1 into $c^{\#}$. From (i), T is isometric, and from (ii) its range is the whole of $c^{\#}$.

(iv)   The assertion follows from (iii), together with the conclusion of Exercise 1.9.21(iii) and the last statement of Exercise 1.9.19.   ∎

1.9.23.   Let l_∞ and l_1 be the Banach spaces defined in Exercises 1.9.19 and 1.9.21, and, for each positive integer k, let e_k (in l_1) be the sequence that has 1 in the kth position and zeros elsewhere. Without using Theorem 1.7.8:

(i)   Prove that, if $Y = \{y_1, y_2, \ldots\} \in l_1$, then $Y = \sum_{k=1}^{\infty} y_k e_k$, the series converging in the norm topology on l_1.

(ii)   Show that, if $X = \{x_1, x_2, \ldots\} \in l_\infty$, the equation

$$\rho(Y) = \sum_{n=1}^{\infty} x_n y_n \qquad (Y = \{y_n\} \in l_1)$$

defines a bounded linear functional ρ on l_1, and $\|\rho\| = \|X\|$.

(iii)   Prove that each bounded linear functional on l_1 arises, as in (ii), from an element X of l_∞.

(iv)   Deduce that the Banach dual space $l_1^{\#}$ is isometrically isomorphic to l_∞.

Solution. (i)   If $Y = \{y_1, y_2, \ldots\} \in l_1$, then $\sum_{k=1}^{n} y_k e_k$ is the sequence $\{y_1, y_2, \ldots, y_n, 0, 0, \ldots\}$, $Y - \sum_{k=1}^{n} y_k e_k$ is the sequence $\{0, 0, \ldots, 0, y_{n+1}, y_{n+2}, \ldots\}$, and thus

$$\left\| Y - \sum_{k=1}^{n} y_k e_k \right\| = \sum_{k=n+1}^{\infty} |y_k| \to 0 \quad \text{as } n \to \infty.$$

Hence $Y = \sum_{k=1}^{\infty} y_k e_k$.

(ii)   If $X = \{x_1, x_2, \ldots\} \in l_\infty$ then, for each $Y = \{y_1, y_2, \ldots\}$ in l_1,

$$\sum_{n=1}^{\infty} |x_n y_n| \le \sum_{n=1}^{\infty} \|X\| \, |y_n| = \|X\| \, \|Y\|.$$

Thus the equation $\rho(Y) = \sum_{n=1}^{\infty} x_n y_n$ defines a mapping $\rho : l_1 \to \mathbb{C}$, and $|\rho(Y)| \le \|X\| \, \|Y\|$. Moreover ρ is linear, and is thus a bounded linear functional, with $\|\rho\| \le \|X\|$. For each k, $|x_k| = |\rho(e_k)| \le \|\rho\| \, \|e_k\| = \|\rho\|$; so

$$\|X\| = \sup\{|x_k| : k \in \mathbb{N}\} \le \|\rho\|,$$

and thus $\|X\| = \|\rho\|$.

(iii)   Let ρ be a bounded linear functional on l_1. Define a complex sequence $\{x_k\}$ by $x_k = \rho(e_k)$; and note that $|x_k| \le \|\rho\|$, whence $\{x_k\}$ is an element X of l_∞. Given any $Y = \{y_k\}$ in l_1, it follows from (i) that

$$\rho(Y) = \rho\left(\sum_{k=1}^{\infty} y_k e_k\right) = \sum_{k=1}^{\infty} y_k \rho(e_k) = \sum_{k=1}^{\infty} x_k y_k.$$

Thus ρ arises, as in (ii), from the element X of l_∞.

(iv)   Given X in l_∞, let TX denote the bounded linear functional on l_1 that is derived from X as in (ii). It is apparent that $T : l_\infty \to l_1^{\#}$ is linear; by (ii), it is isometric, and by (iii) its range is the whole of $l_1^{\#}$.   ∎

1.9.24.   By using the results of Exercises 1.9.21, 1.9.22, and 1.9.23, show that neither of the Banach spaces c_0, c is reflexive. Deduce that neither of the Banach spaces l_1, l_∞ is reflexive.

Solution.   Let U be the isometric isomorphism, from l_1 onto $c^{\#}$, for which, given $Y = \{y_1, y_2, \ldots\}$ in l_1,

$$(UY)(X) = y_1 \lim_{n \to \infty} x_n + \sum_{n=1}^{\infty} y_{n+1} x_n \qquad (X = \{x_1, x_2, \ldots\} \in c).$$

Let e_k (in l_1) be defined as in Exercise 1.9.23, and let $X \to \widehat{X}$ be the natural isometric isomorphism from c into $c^{\#\#}$. Then

$$\widehat{X}(Ue_{k+1}) = (Ue_{k+1})(X) = x_k \qquad (X = \{x_1, x_2, \ldots\} \in c).$$

It follows that $\lim_{k \to \infty} \widehat{X}(Ue_{k+1})$ exists, for each X in c.

From Exercise 1.9.23(ii), the equation

$$\rho_0(Y) = \sum_{n=1}^{\infty} (-1)^{n+1} y_n \qquad (Y = \{y_1, y_2, \ldots\} \in l_1)$$

defines a bounded linear functional ρ_0 on l_1. Thus $(U^{-1})^{\#}\rho_0$ is an element ω_0 of $c^{\#\#}$; and

$$\omega_0(Ue_{k+1}) = \rho_0(U^{-1}Ue_{k+1}) = \rho_0(e_{k+1}) = (-1)^k.$$

Since $\lim_{k \to \infty} \omega_0(Ue_{k+1})$ does not exist, ω_0 is an element of $c^{\#\#}$ which is not of the form $\widehat{X}$, where $X \in c$; and thus c is not reflexive.

A similar argument shows that c_0 is not reflexive. We replace U by the isometric isomorphism U_0 from l_1 onto $c_0^{\#}$, where for each $Y = \{y_1, y_2, \ldots\}$ in l_1,

$$(U_0Y)(X) = \sum_{n=1}^{\infty} y_n x_n \qquad (X = \{x_1, x_2, \ldots\} \in c_0).$$

The argument proceeds as before, with $\lim_{k \to \infty} \widehat{X}(U_0 e_k) = 0$, for each X in c_0.

Since c is not reflexive, we deduce in succession (using Exercises 1.9.12, 1.9.22 and 1.9.23) that the spaces $c^{\#}$, l_1, $l_1^{\#}$, l_∞ are not reflexive. ∎

1.9.25. Give a second proof that none of the Banach spaces c_0, c, l_∞ is reflexive, by using the results of Exercises 1.9.19 and 1.9.11.

Solution. A reflexive Banach space $\mathfrak{X}$ is (naturally) isometrically isomorphic to a Banach dual, namely $(\mathfrak{X}^{\#})^{\#}$. This, together with the final assertion of Exercise 1.9.19, shows that c_0 is not reflexive. Since c_0 is a closed subspace of both c and l_∞, it follows from Exercise 1.9.11 that neither c nor l_∞ is reflexive. ∎

1.9.26.   (i)   Suppose that $\varepsilon > 0$ and $\{x_{m,n} : m, n = 1, 2, \ldots\}$ is a double sequence of complex numbers that satisfies the conditions

$$4\varepsilon < \sum_{n=1}^{\infty} |x_{m,n}| < \infty \qquad (m = 1, 2, \ldots),$$

$$\lim_{m \to \infty} x_{m,n} = 0 \qquad (n = 1, 2, \ldots).$$

Show that there exist integers $0 = n(0) < n(1) < n(2) < \cdots$ and $1 = m(1) < m(2) < m(3) < \cdots$ such that, for $j = 1, 2, \ldots,$

$$\sum_{n=1}^{n(j-1)} |x_{m(j),n}| < \varepsilon,$$

$$\sum_{n=1+n(j-1)}^{n(j)} |x_{m(j),n}| > 3\varepsilon,$$

$$\sum_{n=1+n(j)}^{\infty} |x_{m(j),n}| < \varepsilon.$$

Prove also that there is a sequence $\{y_n\}$ of complex numbers of modulus 1 such that

$$\left| \sum_{n=1}^{\infty} y_n x_{m(j),n} \right| > \varepsilon \qquad (j = 1, 2, \ldots).$$

(ii)   Prove that, if a sequence $\{X_m\}$ of elements of the Banach space l_1 is weakly convergent to 0, then it converges to 0 in the norm topology.

Solution. (i)   Since $4\varepsilon < \sum_{n=1}^{\infty} |x_{1,n}| < \infty$, while $n(0) = 0$ and $m(1) = 1$, the three conditions required when $j = 1$ are all satisfied (the first one vacuously) provided $n(1)$ is chosen sufficiently large.   Now suppose that integers $n(0), \ldots, n(k)$, $m(1), \ldots, m(k)$ have been chosen, and the stated conditions are satisfied when $j = 1, \ldots, k$. Since $\lim_{m \to \infty} x_{m,n} = 0$, for each n, we can choose $m(k+1)$ $(> m(k))$ so that $\sum_{n=1}^{n(k)} |x_{m(k+1),n}| < \varepsilon$.   From this, and since $4\varepsilon < \sum_{n=1}^{\infty} |x_{m(k+1),n}| < \infty$, we have $\sum_{n=1+n(k)}^{n(k+1)} |x_{m(k+1),n}| > 3\varepsilon$ and $\sum_{n=1+n(k+1)}^{\infty} |x_{m(k+1),n}| < \varepsilon$, provided that $n(k + 1)$ is chosen

$(> n(k))$ sufficiently large. This provides an inductive proof of the existence of sequences $\{m(j)\}$, $\{n(j)\}$ with the stated properties.

We can choose complex numbers $y_1, y_2, y_3, \ldots$ of modulus 1 so that

$$y_n x_{m(j),n} = |x_{m(j),n}| \qquad (n(j-1) < n \le n(j); \ j = 1, 2, \ldots).$$

Then, for $j = 1, 2, \ldots$, we obtain

$$\left| \sum_{n=1}^{\infty} y_n x_{m(j),n} \right| \ge \left| \sum_{n=1+n(j-1)}^{n(j)} \right| - \left| \sum_{n=1}^{n(j-1)} \right| - \left| \sum_{n=1+n(j)}^{\infty} \right|$$

$$> \sum_{n=1+n(j-1)}^{n(j)} |x_{m(j),n}| - 2\varepsilon > \varepsilon.$$

(ii)　For each $m = 1, 2, \ldots$, X_m is a sequence $\{x_{m,1}, x_{m,2}, \ldots\}$ such that $\sum_{n=1}^{\infty} |x_{m,n}| < \infty$. If $\{X_m\}$ does not converge to 0 in the norm topology, we can choose $\varepsilon \ (> 0)$ such that $\|X_m\| > 4\varepsilon$ for infinitely many values of m. Upon replacing $\{X_m\}$ by a subsequence, we may suppose that

$$4\varepsilon < \sum_{n=1}^{\infty} |x_{m,n}| < \infty \qquad (m = 1, 2, \ldots).$$

Since $\{X_m\}$ is weakly convergent to 0, while (for each fixed n) the mapping $\{x_1, x_2, \ldots\} \to x_n$ is a bounded linear functional on l_1, it follows that

$$\lim_{m \to \infty} x_{m,n} = 0 \qquad (n = 1, 2, \ldots).$$

From (i), it now follows that there is an element $Y = \{y_n\}$ in the unit ball of l_∞, and a subsequence $\{X_{m(j)}\}$ of $\{X_m\}$, such that $|\sum_{n=1}^{\infty} y_n x_{m(j),n}| > \varepsilon \ (j = 1, 2, \ldots)$. In view of Exercise 1.9.23(ii), this means that there is a bounded linear functional ρ on l_1, such that $|\rho(X_{m(j)})| > \varepsilon$ for all j. This provides a contradiction, since $\{X_{m(j)}\}$ is weakly convergent to 0. Thus $\{X_m\}$ converges to 0 in the norm topology of l_1.　∎

1.9.27.   Suppose that $1 < p < \infty$, $q = p/(p-1)$, and l_p is the Banach space $l_p(\mathbb{N}, \mathbb{C})$ of Example 1.7.3, so that an element of l_p is a complex sequence $X = \{x_1, x_2, \ldots\}$ such that $\sum_{n=1}^{\infty} |x_n|^p \; (= \|X\|^p) < \infty$.

(i)   Show that, for each $Y = \{y_1, y_2, \ldots\}$ in l_q, the equation

$$\rho_Y(X) = \sum_{n=1}^{\infty} y_n x_n \qquad (X = \{x_n\} \in l_p)$$

defines a bounded linear functional ρ_Y on l_p, and $\|\rho_Y\| \le \|Y\|_q$ (the norm of Y in l_q). By considering the sequence $X_0 = \{t_n |y_n|^{q/p}\}$, where $t_1, t_2, \ldots$ are suitable complex numbers of modulus 1, show that $\|\rho_Y\| = \|Y\|_q$.

(ii)   Prove that the mapping $Y \to \rho_Y$ is an isometric isomorphism from l_q onto the Banach dual space $l_p^{\#}$, and deduce that l_p is reflexive (that is, that the *natural* isomorphism of l_p into $l_p^{\#\#}$ is onto).

Solution. (i)   By Hölder's inequality, the series $\sum_{n=1}^{\infty} |y_n x_n|$ converges, with sum not exceeding $\|Y\|_q \|X\|_p$, when $X \in l_p$ and $Y \in l_q$. From this it is clear that, given Y in l_q, ρ_Y (as defined) is a bounded linear functional on l_p, and $\|\rho_Y\| \le \|Y\|_q$. Let X_0 be the sequence $\{t_n |y_n|^{q/p}\}$ $(= \{x_n\})$, where the complex number t_n satisfies $|t_n| = 1$ and $t_n y_n = |y_n|$. Since $\sum_{n=1}^{\infty} |x_n|^p = \sum_{n=1}^{\infty} |y_n|^q$, it follows that $X_0 \in l_p$ and $\|X_0\|_p = \left(\|Y\|_q\right)^{q/p}$. Note also that, since $1 + q/p = q$,

$$\rho_Y(X_0) = \sum_{n=1}^{\infty} y_n x_n = \sum_{n=1}^{\infty} t_n y_n |y_n|^{q/p} = \sum_{n=1}^{\infty} |y_n|^{1+q/p}$$

$$= \sum_{n=1}^{\infty} |y_n|^q = \left(\|Y\|_q\right)^q = \left(\|Y\|_q\right)^{1+q/p} = \|Y\|_q \|X_0\|_p \, .$$

This shows that $\|\rho_Y\| \ge \|Y\|_q$, whence $\|\rho_Y\| = \|Y\|_q$.

(ii)   It is clear from (i) that the mapping $Y \to \rho_Y$ is an isometric isomorphism from l_q into $l_p^{\#}$. We have to prove that its range is the whole of $l_p^{\#}$. For this, suppose that $\rho \in l_p^{\#}$, and define a complex sequence $Y = \{y_1, y_2, \ldots\}$ by $y_k = \rho(E_k)$, where E_k (in l_p) is the sequence with 1 in the kth position and zeros elsewhere. When

$X = \{x_1, x_2, \ldots\} \in l_p$, for each $n = 1, 2, \ldots$ we have

$$X - \sum_{k=1}^{n} x_k E_k = \{0, \ldots, 0, x_{n+1}, x_{n+2}, \ldots\},$$

$$\left\| X - \sum_{k=1}^{n} x_k E_k \right\|_p = \left[\sum_{k=n+1}^{\infty} |x_k|^p \right]^{1/p} \to 0 \quad \text{as } n \to \infty,$$

and thus

$$(1) \quad \rho(X) = \lim_{n \to \infty} \rho\left(\sum_{k=1}^{n} x_k E_k \right) = \lim_{n \to \infty} \sum_{k=1}^{n} y_k x_k = \lim_{n \to \infty} \rho_n(X),$$

where ρ_n is the linear functional associated (as in (i)) with the element $\{y_1, y_2, \ldots y_n, 0, 0, \ldots\}$ of l_q. By the principle of uniform boundedness,

$$\infty > \sup_{n \in \mathbb{N}} \|\rho_n\| = \sup_{n \in \mathbb{N}} \left[\sum_{k=1}^{n} |y_k|^q \right]^{1/q};$$

so $\sum_{k=1}^{\infty} |y_k|^q < \infty$, and $Y \in l_q$. From (1), $\rho(X) = \sum_{k=1}^{\infty} y_k x_k$, so $\rho = \rho_Y$. Hence the isometric isomorphism $Y \to \rho_Y$ has range $l_p^{\#}$.

If Ω is a bounded linear functional on $l_p^{\#}$, the equation $\Omega_0(Y) = \Omega(\rho_Y)$ defines a bounded linear functional on l_q. Since $q > 1$ and $p = q/(q-1)$, we can apply the preceding discussion (with q in place of p) to deduce that there is an element $X_0 = \{x_n\}$ of l_p such that

$$\Omega_0(Y) = \sum_{n=1}^{\infty} x_n y_n \qquad (Y = \{y_n\} \in l_q).$$

Since

$$\Omega(\rho_Y) = \Omega_0(Y) = \sum_{n=1}^{\infty} x_n y_n = \rho_Y(X_0) = \widehat{X}_0(\rho_Y),$$

it follows that $\Omega = \widehat{X}_0$. Hence the natural isometric isomorphism $x \to \hat{x} : l_p \to l_p^{\#\#}$ has range $l_p^{\#\#}$, and l_p is reflexive. $\blacksquare$

1.9.28.   Suppose that $p > 1$, $q = p/(p-1)$, and $Y = \{y_1, y_2, \ldots\}$ is a complex sequence with the following property: whenever $X = \{x_1, x_2, \ldots\} \in l_p$, the sequence $\{y_n x_n\}$ is an element of l_1. Prove that $Y \in l_q$.

Solution.   Let ρ_n be the linear functional on l_p defined by

$$\rho_n(X) = \sum_{k=1}^{n} y_k x_k \qquad (X = \{x_n\} \in l_p).$$

Then $\|\rho_n\| = \left[\sum_{k=1}^{n} |y_k|^q\right]^{1/q}$, from Exercise 1.9.27(i). Also, for each $X (= \{x_n\})$ in l_p, $\lim_{n \to \infty} \rho_n(X)$ exists, since the series $\sum_{n=1}^{\infty} y_n x_n$ is (absolutely) convergent. By the principle of uniform boundedness,

$$\infty > \sup_{n \in \mathbb{N}} \|\rho_n\| = \left[\sum_{n=1}^{\infty} |y_n|^q\right]^{1/q};$$

so $Y \in l_q$.   ∎

1.9.29.   Suppose that $1 < p < \infty$, $q = p/(p-1)$, and L_p is the Banach space associated, as in Example 1.7.5, with a σ-finite measure space $(S, \mathcal{S}, m)$.

(i)   Prove that, for each f in L_q, the equation

$$\rho_f(g) = \int_S f(s)g(s)\, dm(s) \qquad (g \in L_p)$$

defines a bounded linear functional ρ_f on L_p, and $\|\rho_f\| \leq \|f\|_q$ (the norm of f in L_q). By considering a suitable function g_0, of the form $g_0(s) = |f(s)|^{q/p} t(s)$, where $t(s)f(s) = |f(s)|$, show that $\|\rho_f\| = \|f\|_q$.

(ii)   Suppose that $m(S) < \infty$, and let ρ be a bounded linear functional on L_p. Show that, if $X (\subseteq S)$ is measurable, the characteristic function χ_X lies in L_p. Prove that there is an element f of L_1 such that

$$\rho(\chi_X) = \int_S f(s)\chi_X(s)\, dm(s)$$

for every measurable subset X of S. Show that $f \in L_q$, and $\rho = \rho_f$.

(iii)   In the general case (with m σ-finite, but not necessarily finite) prove that the mapping $f \to \rho_f$ is an isometric isomorphism from L_q onto the Banach dual space $L_p^{\#}$, and deduce that L_p is reflexive.

Solution. (i)   When $f \in L_q$ and $g \in L_p$, it results from Hölder's inequality that $fg \in L_1$ and $\|fg\|_1 \le \|f\|_q \|g\|_p$. It follows that, given f in L_q, ρ_f (as defined) is a bounded linear functional on L_p, and $\|\rho_f\| \le \|f\|_q$. Let g_0 be the measurable function defined by $g_0(s) = |f(s)|^{q/p} \overline{\text{sgn}} f(s)$, where (for complex z) $\overline{\text{sgn}}\, z$ is 0 when $z = 0$ and $|z|/z$ when $z \ne 0$. Since $|g_0(s)|^p = |f(s)|^q$, $g_0 \in L_p$ and $\|g_0\|_p = (\|f\|_q)^{q/p}$. Moreover,

$$|\rho_f(g_0)| = \left| \int_S f(s) g_0(s)\, dm(s) \right| = \int_S |f(s)|^{(q/p)+1}\, dm(s)$$

$$= \int_S |f(s)|^q\, dm(s) = (\|f\|_q)^q = (\|f\|_q)^{(q/p)+1}$$

$$= \|f\|_q \|g_0\|_p .$$

Thus $\|\rho_f\| \ge \|f\|_q$, and hence $\|\rho_f\| = \|f\|_q$.

(ii)   Suppose that $m(S) < \infty$, and ρ is a bounded linear functional on L_p. When X ($\subseteq S$) is measurable,

$$\int_S |\chi_X(s)|^p\, dm(s) = m(X) < \infty;$$

so $\chi_X \in L_p$ and $\|\chi_X\|_p = [m(X)]^{1/p}$. If $\{X_n\}$ is an increasing sequence of mearable sets, with union X, then

$$\|\chi_X - \chi_{X_n}\|_p = [m(X \setminus X_n)]^{1/p} \to 0,$$

and thus $\rho(\chi_{X_n}) \to \rho(\chi_X)$. From this, it follows that the mapping $X \to \rho(\chi_X) : \mathcal{S} \to \mathbb{C}$ is a complex measure on $\mathcal{S}$, and vanishes on m-null sets. The Radon–Nikodým derivative is an element f of L_1, and

$$(1) \qquad \rho(\chi_X) = \int_S f(s) \chi_X(s)\, dm(s)$$

for every measurable set X ($\subseteq S$).

For $n = 1, 2, \ldots$, let $X_n = \{s \in S : |f(s)| \le n\}$. and let $f_n = f \chi_{X_n}$. Since $m(S) < \infty$ and f_n is bounded, we have $f_n \in L_q$. The equation $\omega_n(g) = \rho(g \chi_{X_n})$ defines a bounded linear functional ω_n on L_p, with $\|\omega_n\| \le \|\rho\|$. Moreover

$$\omega_n(\chi_X) = \rho(\chi_X \chi_{X_n}) = \rho(\chi_{X \cap X_n}) = \int_S f(s) \chi_{X_n}(s) \chi_X(s)\, dm(s)$$

$$= \int_S f_n(s) \chi_X(s)\, dm(s) = \rho_{f_n}(\chi_X) \qquad (X \in \mathcal{S}).$$

Since simple functions are everywhere dense in L_p, while ω_n and ρ_{f_n} are bounded functionals on L_p, it follows that $\omega_n = \rho_{f_n}$. Hence $\|f_n\|_q = \|\rho_{f_n}\| = \|\omega_n\| \leq \|\rho\|$; that is

$$\left[\int_{X_n} |f(s)|^q \, dm(s)\right]^{1/q} \leq \|\rho\| \qquad (n = 1, 2, \ldots).$$

Since $\{X_n\}$ is an increasing sequence of measurable sets, with union S, it now follows that $f \in L_q$ (and $\|f\|_q \leq \|\rho\|$). We can now rewrite (1) in the form $\rho(\chi_X) = \rho_f(\chi_X)$ ($X \in \mathcal{S}$); and again because simple functions are everywhere dense in L_p, we have $\rho = \rho_f$.

(iii)   From (i), the mapping $\rho \to \rho_f$ is an isometric isomorphism from L_q into $L_p^{\#}$. We have to prove that its range is the whole of $L_p^{\#}$ (and, in the case where $m(S) < \infty$, this has been done in (ii)). Suppose that $\rho \in L_p^{\#}$. Let $\{S_n\}$ be an increasing sequence of measurable sets of finite measure, with union S. The equation

$$\rho_n(g|S_n) = \rho(g\chi_{S_n}) \qquad (g \in L_p)$$

defines a bounded linear functional ρ_n on $L_p(S_n)$, and $\|\rho_n\| \leq \|\rho\|$. From (ii), ρ_n corresponds to an element f_n of $L_q(S_n)$. With f_n defined as 0 on $S \setminus S_n$, we have $f_n \in L_q$, $\|f_n\|_q = \|\rho_n\| \leq \|\rho\|$, and

$$(2) \qquad \rho(g\chi_{S_n}) = \int_S f_n(s)g(s) \, dm(s) \qquad (g \in L_p).$$

When $m > n$,

$$\int_S f_n(s)g(s) \, dm(s) = \rho(g\chi_{S_n}) = \rho(g\chi_{S_n}\chi_{S_m})$$

$$= \int_S f_m(s)g(s)\chi_{S_n}(s) \, dm(s) \qquad (g \in L_p);$$

so $f_n(s) = f_m(s)\chi_{S_n}(s)$ almost everywhere. Hence there is a measurable function f on S, such that $f_n(s) = f(s)\chi_{S_n}(s)$ almost everywhere ($n = 1, 2, \ldots$). Since $\chi_{S_n}(s)$ is increasing to 1, for all s in S, and

$$\int_S |f(s)|^q \chi_{S_n}(s) \, dm(s) = (\|f_n\|_q)^q \leq \|\rho\|^q,$$

it follows that $f \in L_q$ (and $\|f\|_q \leq \|\rho\|$).

From (2), we have

$$\rho(g\chi_{S_n}) = \int_S f(s)g(s)\chi_{S_n}(s)\,dm(s)$$

$$= \rho_f(g\chi_{S_n}) \qquad (g \in L_p;\ n = 1, 2, \ldots).$$

When $n \to \infty$, we obtain $\rho(g) = \rho_f(g)$, since (by the dominated convergence theorem)

$$\|g - g\chi_{S_n}\|_p = \left[\int_S |g(s)[1 - \chi_{S_n}(s)]|^p\,dm(s)\right]^{1/p} \to 0.$$

Thus $\rho = \rho_f$; and the isometric isomorphism has range $L_p^{\#}$.

If Ω is a bounded linear functional on $L_p^{\#}$, the equation $\Omega_0(f) = \Omega(\rho_f)$ defines a bounded linear functional Ω_0 on L_q. Since $q > 1$ and $p = q/(q - 1)$, it follows from the preceding paragraph (with q in place of p) that Ω_0 corresponds to an element g_0 of L_p. We have

$$\Omega(\rho_f) = \Omega_0(f) = \int_S g_0(s)f(s)\,dm(s)$$

$$= \rho_f(g_0) = \hat{g}_0(\rho_f) \qquad (f \in L_q).$$

Thus $\Omega = \hat{g}_0$; so the natural isomorphism $g \to \hat{g} : L_p \to L_p^{\#\#}$ has range $L_p^{\#\#}$, and L_p is reflexive. $\blacksquare$

1.9.30.   Suppose that $1 < p < \infty$, $q = p/(p-1)$, and L_p, L_q are the corresponding Banach spaces associated with a σ-finite measure space $(S, \mathcal{S}, m)$. Show that, if f is a complex-valued function on S, and $fg \in L_1$ for each g in L_p, then $f \in L_q$. [*Hint.* Show that f is the limit, almost everywhere, of a sequence $\{f_n\}$ of functions in L_q such that $|f_n(s)| \le |f(s)|$, and consider the corresponding sequence $\{\rho_n\}$ of bounded linear functionals on L_p.]

Solution.   Let $\{S_n\}$ be an increasing sequence of sets of finite measure, with union S; and define

$$f_n(s) = \begin{cases} f(s) & \text{if } s \in S_n \text{ and } |f(s)| \le n, \\ 0 & \text{otherwise.} \end{cases}$$

If χ_n is the characteristic function of S_n, then $\chi_n \in L_p$. By hypothesis, $f\chi_n \in L_1$ and, in particular, $f\chi_n$ is measurable. Hence f_n is measurable. Since f_n is bounded and vanishes outside a set of finite measure , we have $f_n \in L_q$. Moreover, $|f_n(s)| \le |f(s)|$, $f_n(s) \to f(s)$ almost everywhere, and $\{|f_n(s)|\}$ increases to $|f(s)|$.

Let ρ_n be the linear functional on L_p, corresponding to f_n. Given g in L_p, we have that $f_n g$ tends to fg almost everywhere, $|f_n g| \le |fg| \in L_1$. By the dominated convergence theorem,

$$\rho_n(g) = \int_S f_n(s)g(s)\,dm(s) \to \int_S f(s)g(s)\,dm(s).$$

By the principle of uniform boundedness,

$$\sup_{n \in \mathbb{N}} \left[\int_S |f_n(s)|^q\,dm(s) \right]^{1/q} = \sup_{n \in \mathbb{N}} \|\rho_n\| < \infty.$$

From the monotone convergence theorem,

$$\int_S |f(s)|^q\,dm(s) < \infty;$$

so $f \in L_q$.   ∎

1.9.31.   Show that the Banach space $l_\infty(\mathbb{A})$ is separable if and only if the set $\mathbb{A}$ is finite. Deduce that a separable Banach space may have a non-separable dual.

Solution.   If $\mathbb{A}$ is finite, then $l_\infty(\mathbb{A})$ is finite-dimensional, and is therefore separable.

If $\mathbb{A}$ is infinite, there is an infinite sequence $\{a_1, a_2, \ldots\}$ of distinct elements of $\mathbb{A}$. Given any subset S of $\mathbb{N}$, the characteristic function χ_S of the set $\{a_n : n \in S\}$ is an element of $l_\infty(\mathbb{A})$, and $\|\chi_{S_1} - \chi_{S_2}\|_\infty = 1$ if $S_1 \ne S_2$. If B_S denotes the open ball with center χ_S and radius $\frac{1}{2}$, then $\{B_S\}$ is an uncountable disjoint family of open sets. A dense subset of $l_\infty(\mathbb{A})$ meets each B_S, and is therefore uncountable: so $l_\infty(\mathbb{A})$ is not separable.

From Proposition 1.7.9, $l_1(\mathbb{N})$ is separable; but $l_1(\mathbb{N})^\#$ is isometrically isomorphic to $l_\infty(\mathbb{N})$ (Exercise 1.9.23(iv)), and is therefore not separable, by the result just proved.   ∎

1.9.32.   Show that the Banach space L_∞, associated with a σ-finite measure space $(S, \mathcal{S}, m)$, is separable if and only if S can be expressed as the disjoint union of a finite family of "atoms." (An *atom* is a measurable subset S_0 of S such that $m(S_0) > 0$ and each measurable subset of S_0 has measure 0 or $m(S_0)$.)

Solution. If S is the disjoint union of atoms $S_1, \ldots, S_n$, then every measurable function on S is constant almost everywhere on each S_j; and L_∞ is the linear span of the characteristic functions of the sets $S_1, \ldots, S_n$. In this case, L_∞ is finite-dimensional, and thus separable.

Suppose now that S is not a finite union of atoms. Then S can be expressed as $S_1 \cup S_1'$, where S_1, S_1' are disjoint measurable non-null sets, at least one of which (say S_1') is not a finite disjoint union of atoms. Then $S_1' = S_2 \cup S_2'$, where S_2, S_2' are disjoint measurable non-null sets, at least one of which (say S_2') is not a dijoint union of atoms. Continuing in this way,

$$S = S_1 \cup S_1' = S_1 \cup S_2 \cup S_2' = S_1 \cup S_2 \cup S_3 \cup S_3' = \cdots,$$

where at each stage we have a dijoint union of non-null measurable sets. Hence $\{S_n\}$ is an infinite sequence of disjoint measurable non-null sets.

Since l_∞ is not separable (Exercise 1.9.31) and there is an isometric isomorphism $U : X \to UX$ from l_∞ into L_∞ defined by

$$(UX)(s) = x_n \quad (s \in S_n), \qquad (UX)(s) = 0 \quad (s \notin \bigcup_{n=1}^\infty S_n),$$

where $X = \{x_1, x_2, \ldots\} \in l_\infty$, it follows that the space L_∞ is not separable. ∎

1.9.33. Consider the Banach spaces L_1 and L_∞ associated with a σ-finite measure space $(S, \mathcal{S}, m)$, and the isometric isomorphism (of Theorem 1.7.8) from L_∞ onto $L_1^\#$.

(i) Suppose that $g, g_1, g_2, \ldots \in L_\infty$, and let $\rho, \rho_1, \rho_2, \ldots$ be the corresponding bounded linear functionals on L_1. Show that, if

$$\sup\{\|g_n\|_\infty : n \in \mathbb{N}\} < \infty$$

and $g(s) = \lim_{n \to \infty} g_n(s)$ for almost all s, then the sequence $\{\rho_n\}$ is weak* convergent to ρ.

(ii) Show that, if m is Lebesgue measure on the interval $[0, 1]$, and $g_1, g_2, g_3, \ldots$ (in L_∞) are defined by

$$g_n(s) = (-1)^r \quad (2^{-n} r \le s < 2^{-n}(r+1), \quad r = 0, 1, \ldots, 2^n - 1),$$

then the corresponding sequence $\{\rho_n\}$ of bounded linear functionals on L_1 is weak* convergent to 0.

Solution. (i)   If $f \in L_1$, we have

$$(*) \qquad \rho_n(f) - \rho(f) = \int_S f(s)[g_n(s) - g(s)]\, dm(s).$$

Let $K = \sup\{\|g_n\|_\infty : n \in \mathbb{N}\}$, and observe that

$$\lim_{n \to \infty} f(s)[g_n(s) - g(s)] = 0, \qquad |f(s)[g_n(s) - g(s)]| \le 2K\,|f(s)|$$

almost everywhere, and $2K|f| \in L_1$. From $(*)$ and the dominated convergence theorem, $\rho_n(f) - \rho(f) \to 0$.

(ii)   With ρ_n defined as in (ii), we have $\|\rho_n\| \le 1$. Given f in L_1 and $\varepsilon\ (> 0)$, we can find a positive integer N and a step function f_0 in L_1 such that $\|f - f_0\| < \varepsilon$ and f_0 is constant on each of the intervals $[2^{-N}s, 2^{-N}(s + 1))$, $s = 0, \ldots, 2^N - 1$. When $n > N$ we have $\rho_n(f_0) = 0$, and thus $|\rho_n(f)| = |\rho_n(f - f_0)| < \varepsilon$; so $\rho_n(f) \to 0$ as $n \to \infty$.   ∎

1.9.34.   Suppose that $\sum_{n=1}^\infty x_n$ is a series of elements of a complex Banach space $\mathfrak{X}$ such that, for every strictly increasing sequence $\{n(1), n(2), \ldots\}$ of positive integers, the subseries $\sum_{j=1}^\infty x_{n(j)}$ converges in the weak topology to an element of $\mathfrak{X}$.

(i)   Prove that $\sum_{n=1}^\infty |\rho(x_n)| < \infty$ for each ρ in $\mathfrak{X}^\#$.

(ii)   Show that the equation $S\rho = \{\rho(x_1), \rho(x_2), \ldots\}$ defines a bounded linear operator S from $\mathfrak{X}^\#$ into the Banach space l_1.

(iii)   Prove that, if $A = \{a_1, a_2, \ldots\} \in l_\infty$, the equation

$$\Omega_A(\rho) = \sum_{n=1}^\infty \rho(a_n x_n) \qquad (\rho \in \mathfrak{X}^\#)$$

defines a bounded linear functional Ω_A on $\mathfrak{X}^\#$ and $\|\Omega_A\| \le \|S\|\,\|A\|_\infty$.

(iv)   Show that Ω_A lies in $\widehat{\mathfrak{X}}$ (the natural image of $\mathfrak{X}$ in $\mathfrak{X}^{\#\#}$) whenever A (in l_∞) is a sequence that takes only finitely many distinct values. Deduce that $\Omega_A \in \widehat{\mathfrak{X}}$ for all A in l_∞.

(v)   Prove that, for every $A = \{a_1, a_2, \ldots\}$ in l_∞, the series $\sum_{n=1}^\infty a_n x_n$ is weakly convergent to an element of $\mathfrak{X}$.

(vi)   Show that, if a sequence $\{\rho_n\}$ in $\mathfrak{X}^\#$ is weak* convergent to an element ρ of $\mathfrak{X}^\#$, then $\{S\rho_n\}$ is norm convergent to $S\rho$. [*Hint.* Use (v) and the result of Exercise 1.9.26.]

(vii)   Prove that, if $\varepsilon > 0$, there is a positive integer $n(\varepsilon)$, such that $\sum_{n=n(\varepsilon)}^{\infty} |\rho(x_n)| \le \varepsilon \|\rho\|$ for each ρ in $\mathfrak{X}^\#$. [*Hint.* Upon replacing $\mathfrak{X}$ by the closed linear span of $\{x_n\}$, reduce to the case in which $\mathfrak{X}$ is separable. If the result were false, we could choose $\rho_1, \rho_2, \ldots$ in the unit ball of $\mathfrak{X}^\#$, satisfying $\sum_{n=k}^{\infty} |\rho_k(x_n)| \ge \varepsilon$. Obtain a contradiction by using Exercise 1.9.8 and (vi).]

(viii)   Prove that, for each bounded complex sequence $\{a_1, a_2, \ldots\}$, the series $\sum_{n=1}^{\infty} a_n x_n$ converges in the norm topology on $\mathfrak{X}$. [The assertion, that weak convergence of every subseries entails norm convergence, is known as the Banach–Orlicz theorem.]

Solution.   (i)   If $\rho \in \mathfrak{X}^\#$, then, for every strictly increasing sequence $\{n(1), n(2), \ldots\}$ of integers, we have that the (complex) series $\sum_{j=1}^{\infty} \rho(x_{n(j)})$ converges. Thus $\sum_{n=1}^{\infty} |\rho(x_n)| < \infty$.

(ii)   It is apparent from (i) that S, as defined, is a linear operator from $\mathfrak{X}^\#$ into l_1. To prove that it is bounded, it suffices to show that it has a closed graph. If $\|\rho_j - \rho\| \to 0$ and $\|S\rho_j - C\|_1 \to 0$, where $C = \{c_n\}$ lies in l_1, we have

$$c_n = \lim_{j \to \infty} (S\rho_j)(n) = \lim_{j \to \infty} \rho_j(x_n) = \rho(x_n) = (S\rho)(n)$$

$(n = 1, 2, \ldots)$, and thus $C = S\rho$.

(iii)   If $A = \{a_n\} \in l_\infty$, and $\rho \in \mathfrak{X}^\#$, then

$$\begin{aligned}
\sum_{n=1}^{\infty} |\rho(a_n x_n)| &= \sum_{n=1}^{\infty} |a_n|\,|\rho(x_n)| \\
&\le \|A\|_\infty \sum_{n=1}^{\infty} |\rho(x_n)| \\
&= \|A\|_\infty \|S\rho\|_1 \\
&\le \|A\|_\infty \|S\| \|\rho\| \,.
\end{aligned}$$

Hence the equation $\Omega_A(\rho) = \sum_{n=1}^{\infty} \rho(a_n x_n)$ defines a bounded linear functional Ω_A on $\mathfrak{X}^\#$, and $\|\Omega_A\| \le \|S\| \|A\|_\infty$.

(iv)   If $A = \{a_n\}$ consists of 0's and 1's, and $n(1) < n(2) < \cdots$ are the integers for which a_n is 1, then $\sum_{j=1}^{\infty} x_{n(j)}$ is weakly convergent to an element x of $\mathfrak{X}$, and

$$\Omega_A(\rho) = \sum_{n=1}^{\infty} \rho(a_n x_n) = \sum_{j=1}^{\infty} \rho(x_{n(j)}) = \rho(x) = \hat{x}(\rho) \qquad (\rho \in \mathfrak{X}^\#).$$

Thus $\Omega_A = \hat{x} \in \widehat{\mathfrak{X}}$, in this case. We still have $\Omega_A \in \widehat{\mathfrak{X}}$, when $A = \{a_n\}$ takes only finitely many distinct values; for then, A is a finite linear combination of certain sequences of 0's and 1's.

Now suppose that $A = \{a_n\}$ is a general element of l_∞. If $\varepsilon > 0$, the disk $\{z \in \mathbb{C} : |z| \le \|A\|_\infty\}$ can be covered by a finite number of disks of radius ε. Replacing each a_n by the center a_n' of one of the disks that contain it, we obtain an element $A' = \{a_n'\}$ of l_∞, such that $\|A - A'\| \le \varepsilon$ and A' takes only finitely many distinct values. Then, $\|\Omega_A - \Omega_{A'}\| \le \varepsilon \|S\|$, and $\Omega_{A'} \in \widehat{\mathfrak{X}}$. Since $\widehat{\mathfrak{X}}$ is closed in $\mathfrak{X}^{\#\#}$, it now follows that $\Omega_A \in \widehat{\mathfrak{X}}$.

(v) Given $A = \{a_n\}$ in l_∞, we have $\Omega_A = \hat{x}_A$ for some x_A in $\mathfrak{X}$, and

$$\rho(x_A) = \hat{x}_A(\rho) = \Omega_A(\rho) = \sum_{n=1}^{\infty} \rho(a_n x_n) \qquad (\rho \in \mathfrak{X}^{\#}).$$

Thus $\sum_{n=1}^{\infty} a_n x_n$ is weakly convergent to x_A.

(vi) Suppose that $\{\rho_j\}$ is weak* convergent to ρ. In order to prove that $\|S\rho_j - S\rho\|_1 \to 0$, it suffices (by the result of Exercise 1.9.26) to show that $\omega(S\rho_j - S\rho) \to 0$ for each bounded linear functional ω on l_1. Each such ω arises, as in Exercise 1.9.23, from an element $A = \{a_n\}$ of l_∞. With x_A the element $\sum_{n=1}^{\infty} a_n x_n$ of $\mathfrak{X}$, we have

$$\omega(S\rho_j - S\rho) = \sum_{n=1}^{\infty} a_n \big[(S\rho_j)(n) - (S\rho)(n)\big]$$
$$= \sum_{n=1}^{\infty} \big[\rho_j(a_n x_n) - \rho(a_n x_n)\big] = (\rho_j - \rho)(x_A) \to 0.$$

(vii) If there is no integer $n(\varepsilon)$ with the stated property, we can choose $\rho_1, \rho_2, \ldots$ in the unit ball of $\mathfrak{X}^{\#}$, so that

$$\sum_{n=k}^{\infty} |\rho_k(x_n)| \ge \varepsilon \qquad (k = 1, 2, \ldots).$$

Upon reducing to the case in which $\mathfrak{X}$ is separable (as in the hint), it then follows from Exercise 1.9.8 that some subsequence $\{\rho_{k_j}\}$ of $\{\rho_k\}$ is weak* convergent to an element ρ of $\mathfrak{X}^{\#}$. Since $\sum_{n=1}^{\infty} |\rho(x_n)| < \infty$, we can choose N so that $\sum_{n=N}^{\infty} |\rho(x_n)| < \varepsilon/3$; and by (vi), we can choose $k_j \ (> N)$ such that $\|S\rho_{k_j} - S\rho\|_1 < \varepsilon/3$. Then, we have

$$\varepsilon \le \sum_{n=k_j}^{\infty} |\rho_{k_j}(x_n)| \le \sum_{n=k_j}^{\infty} \left[|\rho(x_n)| + |\rho(x_n) - \rho_{k_j}(x_n)| \right]$$

$$\le \sum_{n=N}^{\infty} |\rho(x_n)| + \sum_{n=1}^{\infty} |\rho(x_n) - \rho_{k_j}(x_n)|$$

$$< \tfrac{1}{3}\varepsilon + \|S\rho - S\rho_{k_j}\|_1 < \tfrac{2}{3}\varepsilon,$$

a contradiction. Hence there is an integer $n(\varepsilon)$ with the stated property.

(viii)   When $m \ge n(\varepsilon)$ we have

$$\left| \rho\Big(\sum_{n=m}^{\infty} a_n x_n \Big) \right| \le \sum_{n=m}^{\infty} |a_n|\, |\rho(x_n)|$$

$$\le \|A\|_{\infty} \sum_{n=n(\varepsilon)}^{\infty} |\rho(x_n)| \le \varepsilon \|A\|_{\infty} \|\rho\|,$$

for each ρ in $\mathfrak{X}^{\#}$. Thus

$$\left\| \sum_{n=m}^{\infty} a_n x_n \right\| \le \varepsilon \|A\|_{\infty} \qquad \text{whenever } m \ge n(\varepsilon);$$

and the series $\sum_{n=1}^{\infty} a_n x_n$ is norm convergent.   ∎

1.9.35.   (i)   Prove that, if the dual space $\mathfrak{X}^{\#}$ of a Banach space $\mathfrak{X}$ is separable, then $\mathfrak{X}$ is separable. [*Hint.* Let $\{\rho_n\}$ be a countable dense subset of the surface $\{\rho \in \mathfrak{X}^{\#} : \|\rho\| = 1\}$ of the unit ball $(\mathfrak{X}^{\#})_1$; for each $n = 1, 2, \ldots$, choose x_n in $(\mathfrak{X})_1$ such that $|\rho_n(x_n)| > \tfrac{1}{2}$. Show that $\mathfrak{X}$ is the closed linear span of $\{x_n\}$.]

(ii)   Show that a reflexive Banach space is separable if and only if its dual space is separable.

(iii)   Give an example of a Banach space that is not reflexive but has a separable dual space (and is, therefore, separable).

Solution.   Choose $\{\rho_n\}$ and $\{x_n\}$ as in the hint. If the closed linear span of $\{x_n\}$ is not the whole of $\mathfrak{X}$, it is contained in the

null space of a non-zero bounded linear functional ρ on $\mathfrak{X}$. We may
assume that $\|\rho\| = 1$, and choose k so that $\|\rho - \rho_k\| < \frac{1}{2}$. Then

$$0 = |\rho(x_k)| = |\rho_k(x_k) - (\rho_k - \rho)(x_k)|$$
$$\geq |\rho_k(x_k)| - \|\rho_k - \rho\| \|x_k\| > \tfrac{1}{2} - \|\rho_k - \rho\| > 0,$$

a contradiction. Hence $\mathfrak{X}$ is the closed linear span of $\{x_n\}$, and is
therefore separable.

(ii)   If $\mathfrak{X}^{\#}$ is separable, so is $\mathfrak{X}$, by (i). If $\mathfrak{X}$ is reflexive and separable, then $(\mathfrak{X}^{\#})^{\#}$ is separable, since it is isometrically isomorphic
to $\mathfrak{X}$; and, by (i), $\mathfrak{X}^{\#}$ is separable.

(iii)   The space c is not reflexive and $c^{\#}$ is (isometrically isomorphic to) l_1 (Exercises 1.9.24 and 1.9.22(iii), and Proposition
1.7.9).   ∎

1.9.36.   Show that a bounded sequence $\{x_n\}$ of elements of a
reflexive Banach space $\mathfrak{X}$ has a subsequence that is weakly convergent
to an element of $\mathfrak{X}$. [*Hint.* Show that it is sufficient to consider the
case in which $\mathfrak{X}$ is separable, by replacing $\mathfrak{X}$ by the closed linear span
of $\{x_n\}$. In the separable case show, by use of Exercise 1.9.35(ii),
that the required result can be deduced from Exercise 1.9.8.]

Solution.   The closed linear span $\mathcal{M}$ of $\{x_n\}$ is a reflexive Banach space (Exercise 1.9.11); and its weak topology (as a Banach
space in its own right) coincides with the relative weak topology as
a subset of $\mathfrak{X}$, since bounded linear functionals on $\mathcal{M}$ extend to $\mathfrak{X}$.
Upon replacing $\mathfrak{X}$ by $\mathcal{M}$, we may suppose that $\mathfrak{X}$ is separable.

With $\mathfrak{X}$ separable and reflexive, $\mathfrak{X}^{\#}$ is a separable space (Exercise 1.9.35(ii)). By Exercise 1.9.8, the bounded sequence $\{\hat{x}_n\}$
in $(\mathfrak{X}^{\#})^{\#}$ has a subsequence $\{\hat{x}_{n_j}\}$ that converges to an element of
$(\mathfrak{X}^{\#})^{\#} (= \widehat{\mathfrak{X}})$ in the weak* topology on $(\mathfrak{X}^{\#})^{\#}$. Let the limit be $\hat{x}$,
where $x \in \mathfrak{X}$. Then $\{x_{n_j}\}$ is weakly convergent to x.   ∎

1.9.37.   Suppose $\mathfrak{X}$ is a separable normed space and $\{x_n : n \in
\mathbb{N}\}$ is a (norm-)dense subset of $(\mathfrak{X})_1$. Define

$$d(\rho, \rho') = \sum_{n=1}^{\infty} 2^{-n} |(\rho - \rho')(x_n)| \qquad (\rho, \rho' \in \mathfrak{X}^{\#}).$$

(i)   Show that d is a metric on $\mathfrak{X}^{\#}$.

(ii) Show that the metric topology induced by d on $(\mathfrak{X}^{\#})_1$ is the weak* topology on $(\mathfrak{X}^{\#})_1$.

(iii) Use the fact that each sequence in a compact metric space has a convergent subsequence to solve Exercise 1.9.8 (and 1.9.36) again.

Solution. (i) Since $0 \leq d(\rho, \rho') \leq \|\rho\| + \|\rho'\|$, d takes (finite) non-negative real values. If $d(\rho, \rho') = 0$, then $\rho(x_n) = \rho'(x_n)$ for all n. As ρ and ρ' are continuous and $\{x_n\}$ is dense in $(\mathfrak{X})_1$, $\rho = \rho'$. Conversely, if $\rho = \rho'$, then $d(\rho, \rho') = 0$. Clearly $d(\rho, \rho') = d(\rho', \rho)$. Finally,

$$
\begin{aligned}
d(\rho, \rho'') &= \sum_{n=1}^{\infty} 2^{-n} \, |(\rho - \rho'')(x_n)| \\
&\leq \sum_{n=1}^{\infty} 2^{-n} \big[|(\rho - \rho')(x_n)| + |(\rho' - \rho'')(x_n)|\big] \\
&= d(\rho, \rho') + d(\rho', \rho'').
\end{aligned}
$$

Hence d is a metric on $\mathfrak{X}^{\#}$.

(ii) The functionals $\hat{x}_n$ on $\mathfrak{X}^{\#}$ are continuous on $(\mathfrak{X}^{\#})_1$ in the metric topology; for if ρ_0 is in $(\mathfrak{X}^{\#})_1$ and a positive ε is given, then with $d(\rho, \rho_0) < 2^{-n}\varepsilon$, we have

$$
|\hat{x}_n(\rho) - \hat{x}_n(\rho_0)| = |(\rho - \rho_0)(x_n)| \leq 2^n d(\rho, \rho_0) < \varepsilon.
$$

Given x in $(\mathfrak{X})_1$, choose $\{x_{n_j}\}$ tending to x in norm. Then $\hat{x}$ is the uniform limit of $\hat{x}_{n_j}$ on $(\mathfrak{X}^{\#})_1$; so that $\hat{x}$ is continuous on $(\mathfrak{X}^{\#})_1$ in the metric topology. A basic weak* open neighborhood of ρ_0 in $(\mathfrak{X}^{\#})_1$ is given by a positive ε and a finite set of elements $y_1, \ldots, y_n$ in $\mathfrak{X}$ as

$$
\{\rho : \rho \in (\mathfrak{X}^{\#})_1, \ |(\rho - \rho_0)(y_j)| \, (= |\hat{y}_j(\rho) - \hat{y}_j(\rho_0)|) < \varepsilon\}.
$$

From what we have proved, this neighborhood, and, hence, each weak* open subset of $(\mathfrak{X}^{\#})_1$ is open in the metric $(d\text{-})$topology on $(\mathfrak{X}^{\#})_1$.

We note, next, that the ball of radius ε (> 0) with center ρ_0 (relative to d) contains a weak* open set of which ρ_0 is a member (so that this ball, and each open set in the d-topology, is weak* open).

Choose n_0 such that $\sum_{n=n_0}^{\infty} 2^{-n} < \varepsilon/4$. Suppose $|(\rho - \rho_0)(x_n)| < \varepsilon/n_0$ for n in $\{1, \ldots, n_0\}$. Then

$$d(\rho, \rho_0) = \sum_{n=1}^{\infty} 2^{-n} |(\rho - \rho_0)(x_n)| \leq \tfrac{1}{2} n_0 \tfrac{\varepsilon}{n_0} + 2 \sum_{n=n_0}^{\infty} 2^{-n} < \varepsilon.$$

Thus $V(\rho_0 : \hat{x}_1, \ldots, \hat{x}_{n_0}; \frac{\varepsilon}{n_0})$ is contained in the open ball of radius ε with center ρ_0. It follows that the metric and weak* topologies on $(\mathfrak{X}^\#)_1$ coincide.

(iii)　From Theorem 1.6.5(i), $(\mathfrak{X}^\#)_1$ is weak* compact. From (ii) (under the assumption that $\mathfrak{X}$ is separable), $(\mathfrak{X}^\#)_1$ is a compact metric space. If $\{\rho_n\}$ is a sequence in $\mathfrak{X}^\#$ and $\|\rho_n\| < k$ for all n, then $k^{-1}\rho_n \in (\mathfrak{X}^\#)_1$ and $\{k^{-1}\rho_n\}$ has a weak* convergent subsequence $\{k^{-1}\rho_{n_j}\}$. It follows that the sequence $\{\rho_{n_j}\}$ $(= \{kk^{-1}\rho_{n_j}\})$ is weak* convergent.　∎

1.9.38.　　Use the Baire category theorem (see the proof of Lemma 1.8.3) to give another proof (direct) of the uniform boundedness principle (Theorem 1.8.9).

Solution.　With the notation of Theorem 1.8.9, let $\mathfrak{X}_n$ be $\{x : x \in \mathfrak{X},\ \|T_a x\| \leq n$ for all a in $\mathbb{A}\}$. Since each T_a is continuous, $\mathfrak{X}_n$ is the intersection of closed subsets of $\mathfrak{X}$. Hence each $\mathfrak{X}_n$ is closed in $\mathfrak{X}$; and by hypothesis, $\bigcup_{n=1}^{\infty} \mathfrak{X}_n = \mathfrak{X}$. From the Baire theorem, some $\mathfrak{X}_{n_0}$ contains an open ball of radius $r\ (> 0)$ with center x_0. Thus, if $\|x\| < r$, $x_0 - x \in \mathfrak{X}_{n_0}$; and $\|T_a(x_0 - x)\| \leq n_0$ for every a in $\mathbb{A}$. It follows that $\|T_a y\| \leq (n_0 + \|T_a x_0\|)r^{-1}$ for all y in $(\mathfrak{X})_1$ and a in $\mathbb{A}$. Thus $\|T_a\| \leq (n_0 + \|T_a x_0\|)r^{-1}$ for all a in $\mathbb{A}$.　∎

1.9.39.　　If $\{T_n\}$ is a sequence of bounded linear transformations of one Banach space $\mathfrak{X}$ into another Banach space $\mathfrak{Y}$ and $\{T_n x\}$ converges for each x in $\mathfrak{X}$, show that T, defined by $Tx = \lim_n T_n x$, is a bounded linear transformation of $\mathfrak{X}$ into $\mathfrak{Y}$.

Solution.　Clearly T is a linear transformation of $\mathfrak{X}$ into $\mathfrak{Y}$. Since $\{T_n x\}$ converges for each x in $\mathfrak{X}$; $\{\|T_n x\|\}$ is a bounded set for each x in $\mathfrak{X}$. From Theorem 1.8.9 (the uniform boundedness principle), $\{\|T_n\|\}$ is bounded (say, by k). Since $\|Tx\| = \lim \|T_n x\| \leq k$ for each x in $(\mathfrak{X})_1$; $\|T\| \leq k$, and T is bounded.　∎

1.9.40.   Suppose $\mathfrak{X}$ and $\mathfrak{Y}$ are Banach spaces and T is a linear transformation of $\mathfrak{X}$ into $\mathfrak{Y}$. With η in the algebraic dual of $\mathfrak{Y}$, let $(T'\eta)(x)$ be $\eta(Tx)$ for each x in $\mathfrak{X}$; and suppose that $T'\rho \in \mathfrak{X}^{\#}$ for each ρ in $\mathfrak{Y}^{\#}$. Show that T is bounded and that $T'|\mathfrak{Y}^{\#} = T^{\#}$. [*Hint.* Consider the graph of T.]

Solution.   We show that T has a closed graph. For this, it will suffice to note that, if $\{x_n\}$ is a sequence in $\mathfrak{X}$ converging to 0 such that $\{Tx_n\}$ converges to y (in $\mathfrak{Y}$), then $y = 0$. With ρ in $\mathfrak{Y}^{\#}$, $T'\rho \in \mathfrak{X}^{\#}$, by assumption; so that;

$$\rho(y) = \lim_n \rho(Tx_n) = \lim_n (T'\rho)(x_n) = 0.$$

From the Hahn–Banach theorem, $y = 0$; and the graph of T is closed. Thus T is bounded; and, from the definition of T' and $T^{\#}$, $T'|\mathfrak{Y}^{\#} = T^{\#}$.   ∎

CHAPTER 2

BASICS OF HILBERT SPACE
AND LINEAR OPERATORS

2.8.   Exercises

2.8.1.   Show that a finite set $\{x_1,\ldots,x_n\}$ of n vectors in a Hilbert space $\mathcal{H}$ is linearly independent if and only if the $n \times n$ matrix that has $\langle x_j, x_k \rangle$ in the (j,k) position is non-singular.

Solution.   Suppose, first, that the matrix is non-singular. If $a_1,\ldots,a_n$ are scalars such that

$$a_1 x_1 + \cdots + a_n x_n = 0,$$

then

$$a_1 \langle x_1, x_k \rangle + \cdots + a_n \langle x_n, x_k \rangle = 0 \qquad (k = 1,\ldots,n);$$

and (since the matrix is non-singular) $a_1 = \cdots = a_n = 0$. Thus $x_1,\ldots,x_n$ are linearly independent.

Next, suppose that the matrix is singular. There exist scalars $b_1,\ldots,b_n$ (not all zero) such that

$$b_1 \langle x_1, x_k \rangle + \cdots + b_n \langle x_n, x_k \rangle = 0 \qquad (k = 1,\ldots,n).$$

With y the vector $b_1 x_1 + \cdots + b_n x_n$, $\langle y, x_k \rangle = 0$ for each k; thus

$$\langle y, y \rangle = \bar{b}_1 \langle y, x_1 \rangle + \cdots + \bar{b}_n \langle y, x_n \rangle = 0,$$

and

$$b_1 x_1 + \cdots + b_n x_n = y = 0.$$

Thus $x_1,\ldots,x_n$ are linearly dependent.   ∎

40

2.8.2.   Show that a Hilbert space is uniformly convex (in the sense defined in Exercise 1.9.13).

Solution.   If x_1, x_2 lie in the unit ball of a Hilbert space,

$$\|x_1 - x_2\|^2 = 2\|x_1\|^2 + 2\|x_2\|^2 - \|x_1 + x_2\|^2$$
$$\leq 4 - 4\left\|\tfrac{1}{2}(x_1 + x_2)\right\|^2.$$

Thus $\|x_1 - x_2\| < \varepsilon$ whenever $\|x_1\| \leq 1$, $\|x_2\| \leq 1$, and

$$\left\|\tfrac{1}{2}(x_1 + x_2)\right\| > [1 - \tfrac{1}{4}\varepsilon^2]^{1/2} = 1 - \delta(\varepsilon). \qquad \blacksquare$$

2.8.3.   Show that, if $\mathcal{H}$ is a real Hilbert space, then $\mathcal{H} \times \mathcal{H}$ becomes a (complex) Hilbert space $\mathcal{H}_{\mathbb{C}}$ when its linear structure, inner product, and norm, are defined by

$$(x,y) + (u,v) = (x+u, y+v),$$
$$(a + ib)(x,y) = (ax - by, bx + ay),$$
$$\langle (x,y), (u,v) \rangle = \langle x, u \rangle + \langle y, v \rangle + i\langle y, u \rangle - i\langle x, v \rangle,$$
$$\|(x,y)\|^2 = \|x\|^2 + \|y\|^2,$$

for all x, y, u, v in $\mathcal{H}$ and a, b in $\mathbb{R}$.

Prove also that the set $\{(x,0) : x \in \mathcal{H}\}$ is a closed real-linear subspace $\mathcal{H}_{\mathbb{R}}$ of $\mathcal{H}_{\mathbb{C}}$, that $\mathcal{H}_{\mathbb{C}} = \{h + ik : h, k \in \mathcal{H}_{\mathbb{R}}\}$, and that the mapping $x \to (x,0)$ is an isometric isomorphism from $\mathcal{H}$ onto (the real Hilbert space) $\mathcal{H}_{\mathbb{R}}$.

Solution.   It is apparent that, with the given structure, $\mathcal{H}_{\mathbb{C}}$ is a complex inner product space, with a positive definite inner product. The associated norm on $\mathcal{H}_{\mathbb{C}}$ satisfies

$$\max\{\|x\|, \|y\|\} \leq \|(x,y)\| \leq \|x\| + \|y\|;$$

so the induced topology and uniform structure on $\mathcal{H}_{\mathbb{C}}$ ($= \mathcal{H} \times \mathcal{H}$) are the products of those on $\mathcal{H}$. Hence $\mathcal{H}_{\mathbb{C}}$ is complete, and is thus a Hilbert space. The remaining assertions are immediate. $\qquad \blacksquare$

2.8.4.   Suppose that $\{x_1, x_2, x_3, \ldots\}$ is an orthonormal basis in a Hilbert space $\mathcal{H}$ and that

$$Y = \left\{ y \in \mathcal{H} : \sum_{n=1}^{\infty} \left(1 + \frac{1}{n}\right)^2 |\langle y, x_n \rangle|^2 \leq 1 \right\}.$$

Prove that Y is a bounded closed convex set that has no element with greatest norm.

Solution.   The equation

$$Tx = \sum_{n=1}^{\infty} \left(1 + \frac{1}{n}\right) \langle x, x_n \rangle x_n \qquad (x \in \mathcal{H})$$

defines a bounded linear operator T acting on $\mathcal{H}$ (see Example 2.4.10), and

$$Y = \{y \in \mathcal{H} : \|Ty\| \leq 1\}.$$

Thus Y is a closed convex set.

If $y \in Y$ and $y \neq 0$, then $\langle y, x_k \rangle \neq 0$ for some k, and

$$\|y\|^2 = \sum_{n=1}^{\infty} |\langle y, x_n \rangle|^2 < \sum_{n=1}^{\infty} \left(1 + \frac{1}{n}\right)^2 |\langle y, x_n \rangle|^2 \leq 1.$$

Thus $\|y\| < 1$ for each y in Y.

However, $\left(1 + \frac{1}{n}\right)^{-1} x_n \in Y$ $(n = 1, 2, \ldots)$, and we conclude that $\sup\{\|y\| : y \in Y\} = 1$.   ∎

2.8.5.   Suppose that Y is a closed convex set in a Hilbert space $\mathcal{H}$, $\mathcal{U}_Y$ is the set of all unitary operators U acting on $\mathcal{H}$ for which $U(Y) = Y$, and

$$Y_0 = \{y \in Y : Uy = y \text{ for each } U \text{ in } \mathcal{U}_Y\}.$$

(i)   Prove that Y_0 is not empty. [*Hint.* Use Proposition 2.2.1.]

(ii)   Show that, if Y_0 consists of a single non-zero vector y_0, then Y is a subset of the hyperplane

$$\{x \in \mathcal{H} : \mathbf{Re}\langle x - y_0, y_0 \rangle = 0\}.$$

Solution.   Since each U in $\mathcal{U}_Y$ maps 0 to 0, Y to Y, and pre-serves distances, it is apparent that $U y_0 = y_0$ $(U \in \mathcal{U}_Y)$, where y_0 is the point of Y that is closest to 0 (see Proposition 2.2.1). This proves (i).

Under the conditions of (ii), it follows from the preceding para-graph that y_0 is the element of Y closest to 0. However, since $2y_0$ is fixed by each U in $\mathcal{U}_Y$, the same is true of the point of Y that is closest to $2y_0$; so this, too, is y_0. Since y_0 is the best approximant in Y to both 0 and $2y_0$, it follows that

$$\mathbf{Re}\langle y - y_0, y_0 \rangle = 0 \qquad (y \in Y)$$

by Proposition 2.2.1.   ■

2.8.6.   Prove that a bounded sequence of vectors in a Hilbert space has a weakly convergent subsequence.

Solution.   Since a Hilbert space is reflexive, by Corollary 2.3.3, the required conclusion is an immediate consequence of the result of Exercise 1.9.36. We provide an alternative and slightly more direct proof, along the lines of the solutions given for Exercises 1.9.8 and 1.9.36, but with some simplification made possible by the Hilbert space structure.

Let $\{x_1, x_2, x_3, \ldots\}$ be a bounded sequence in a Hilbert space $\mathcal{H}$. We may suppose that

$$\|x_n\| \le M \qquad (n = 1, 2, 3, \ldots),$$

whence

$$|\langle x_m, x_n \rangle| \le M^2 \qquad (m, n = 1, 2, 3, \ldots).$$

Since the complex sequence $\{\langle x_1, x_n \rangle\}$ is bounded, it has a con-vergent subsequence $\{\langle x_1, x_{n,1} \rangle\}$. Since the sequence of complex numbers $\{\langle x_2, x_{n,1} \rangle\}$ is bounded, it has a convergent subsequence $\{\langle x_2, x_{n,2} \rangle\}$. Since the complex sequence $\{\langle x_3, x_{n,2} \rangle\}$ is bounded, it has a convergent subsequence $\{\langle x_3, x_{n,3} \rangle\}$. By continuing in this way, for each positive integer k we obtain, inductively, a subsequence $\{x_{n,k} : n = 1, 2, 3, \ldots\}$ of $\{x_n\}$, such that the complex sequence $\{\langle x_k, x_{n,k} \rangle : n = 1, 2, 3, \ldots\}$ converges and $\{x_{n,k+1}\}$ is a subse-quence of $\{x_{n,k}\}$. The diagonal sequence $\{x_{n,n}\}$ is a subsequence of $\{x_n\}$, and coincides, from its kth term onwards, with a subsequence of $\{x_{n,k}\}$. Hence the complex sequence $\{\langle x_k, x_{n,n}\rangle\}$ converges (as $n \to \infty$), for each $k = 1, 2, 3, \ldots$.

We show next that the sequence $\{\langle x, x_{n,n}\rangle\}$ converges, for each x in $\mathcal{H}$. To this end, let E be the projection from $\mathcal{H}$ onto the closed subspace X generated by the vectors $x_1, x_2, x_3, \ldots$. Given any positive real number ε, since $Ex \in X$, there is a linear combination y of finitely many of the vectors x_n, such that

$$\|Ex - y\| < \frac{\varepsilon}{3M}.$$

From the final statement of the preceding paragraph, the sequence $\{\langle y, x_{n,n}\rangle\}$ converges, so there is a positive integer n_0 such that

$$\left|\langle y, x_{m,m}\rangle - \langle y, x_{n,n}\rangle\right| < \tfrac{1}{3}\varepsilon$$

whenever $m, n \geq n_0$. Thus

$$
\begin{aligned}
&\left|\langle x, x_{m,m}\rangle - \langle x, x_{n,n}\rangle\right| \\
&= \left|\langle x, Ex_{m,m}\rangle - \langle x, Ex_{n,n}\rangle\right| \\
&= \left|\langle Ex, x_{m,m} - x_{n,n}\rangle\right| \\
&= \left|\langle Ex - y, x_{m,m} - x_{n,n}\rangle + \langle y, x_{m,m} - x_{n,n}\rangle\right| \\
&\leq \left\|Ex - y\right\|\left\|x_{m,m} - x_{n,n}\right\| + \left|\langle y, x_{m,m}\rangle - \langle y, x_{n,n}\rangle\right| \\
&< \frac{\varepsilon}{3M}2M + \frac{\varepsilon}{3} = \varepsilon,
\end{aligned}
$$

whenever $m, n \geq n_0$. From this we see that the complex sequence $\{\langle x, x_{n,n}\rangle\}$ is a Cauchy sequence, and so converges, for each x in $\mathcal{H}$.

We can now define a function $\varphi : \mathcal{H} \to \mathbb{C}$ by

$$\varphi(x) = \lim_{n \to \infty} \langle x, x_{n,n}\rangle \qquad (x \in \mathcal{H}).$$

Since $\|x_{n,n}\| \leq M$, we have $|\varphi(x)| \leq M\|x\|$; from this, and since the inner product is linear in its first variable, φ is a bounded linear functional on $\mathcal{H}$. By Riesz's representation theorem (2.3.1), there is an element x_0 of $\mathcal{H}$ such that

$$\langle x, x_0\rangle = \varphi(x) = \lim_{n \to \infty} \langle x, x_{n,n}\rangle \qquad (x \in \mathcal{H}).$$

By Corollary 2.3.4, the subsequence $\{x_{n,n}\}$ of $\{x_n\}$ is weakly convergent to x_0. ∎

2.8.7. Suppose that $\mathbb{A}$ is an uncountable set and, for each a in $\mathbb{A}$, m_a is Lebesgue measure on the σ-algebra $\mathcal{S}_a$ of Borel subsets of the interval $[0,1]$ $(= S_a)$. Show that, if $(S, \mathcal{S}, m)$ is the corresponding infinite-product measure space (see [H: p. 158]), then $L_2(S, \mathcal{S}, m)$ is non-separable.

Solution. Let $f : [0,1] \to \mathbb{R}$ be a continuous function such that

$$\int_0^1 f(t)\, dt = 0, \qquad \int_0^1 |f(t)|^2\, dt = 1.$$

Each element of S is a function $s : \mathbb{A} \to [0,1]$. When $a \in \mathbb{A}$, the equation $f_a(s) = f(s(a))$ defines a (continuous) function f_a in $L_2(S, \mathcal{S}, m)$. Since $\{f_a : a \in \mathbb{A}\}$ is an uncountable orthonormal system in $L_2(S, \mathcal{S}, m)$, this space is non-separable. ∎

2.8.8. Suppose that $\mathcal{H}$ is a Hilbert space in which the inner product is denoted by $\langle\ ,\ \rangle$ and that $K \in \mathcal{B}(\mathcal{H})^+$. Show that the equation

$$\langle x, y \rangle_1 = \langle Kx, y \rangle \qquad (x, y \in \mathcal{H})$$

defines an inner product $\langle\ ,\ \rangle_1$ on $\mathcal{H}$. By means of the Cauchy–Schwarz inequality for $\langle\ ,\ \rangle_1$, prove that

$$\|K\| = \min\{a : a \in \mathbb{R},\ \ K \le aI\}.$$

Solution. It is apparent that $\langle\ ,\ \rangle_1$ is an inner product on $\mathcal{H}$. (See the introduction to Section 2.1.)

By the discussion preceding Remark 2.4.9, $K \le \|K\|\,I$. If $a \in \mathbb{R}$ and $K \le aI$, then for all x and y in $\mathcal{H}$,

$$|\langle Kx, y \rangle|^2 = |\langle x, y \rangle_1|^2 \le \langle x, x \rangle_1 \langle y, y \rangle_1$$

$$= \langle Kx, x \rangle \langle Ky, y \rangle \le a^2 \|x\|^2 \|y\|^2.$$

Thus $|\langle Kx, y \rangle| \le a\,\|x\|\,\|y\|$, for all x and y in $\mathcal{H}$; and $\|K\| \le a$. (See Theorem 2.4.1.) ∎

2.8.9. Let $\mathcal{H}$ be a Hilbert space in which the inner product and norm are denoted by $\langle\ ,\ \rangle$ and $\|\ \|$, respectively. Suppose that $\langle\ ,\ \rangle_1$ is another definite inner product on $\mathcal{H}$ and the corresponding norm $\|\ \|_1$ satisfies $\|x\|_1 \le \|x\|$ for each x in $\mathcal{H}$. Prove that there is a positive self-adjoint operator K, acting on $\mathcal{H}$, such that $\|K\| \le 1$, K has null space $\{0\}$, and $\langle x, y \rangle_1 = \langle Kx, y \rangle$ for all x, y in $\mathcal{H}$.

Solution.   Since

$$|\langle x, y \rangle_1| \le \|x\|_1 \|y\|_1 \le \|x\| \|y\|,$$

the equation

$$b(x, y) = \langle x, y \rangle_1$$

defines a bounded conjugate-bilinear functional b on $\mathcal{H} \times \mathcal{H}$. It follows from Theorem 2.4.1 that there is an element K in the unit ball of $\mathcal{B}(\mathcal{H})$ such that

$$\langle x, y \rangle_1 = \langle Kx, y \rangle.$$

Since $\langle Kx, x \rangle = \|x\|_1^2 \ge 0$ (with equality only when $x = 0$), K is a positive self-adjoint operator with null space $\{0\}$.   ∎

2.8.10.   Suppose that $\mathcal{H}$ is a Hilbert space in which the inner product and norm are denoted by $\langle\,,\,\rangle$ and $\|\ \|$, respectively. Let K be a positive element of $\mathcal{B}(\mathcal{H})$, and define an inner product $\langle\,,\,\rangle_1$ on $\mathcal{H}$ by

$$\langle x, y \rangle_1 = \langle Kx, y \rangle \qquad (x, y \in \mathcal{H}).$$

Let $\|x\|_1 = [\langle x, x \rangle_1]^{1/2}$.

(i)   Prove that $\|\ \|_1$ is a norm on $\mathcal{H}$ if and only if K has null space $\{0\}$.

(ii)   Show that, if K has null space $\{0\}$, then the norms $\|\ \|$ and $\|\ \|_1$ give rise to the same topology on $\mathcal{H}$ if and only if K has an inverse in $\mathcal{B}(\mathcal{H})$.

(iii)   Suppose that K has an inverse in $\mathcal{B}(\mathcal{H})$. If A^* denotes the adjoint of an element A of $\mathcal{B}(\mathcal{H})$ relative to the inner product $\langle\,,\,\rangle$, find a formula for the adjoint of A relative to the inner product $\langle\,,\,\rangle_1$.

Solution. (i)   From Proposition 2.1.2, $\|\ \|_1$ is a semi-norm on $\mathcal{H}$. Suppose that $x \in \mathcal{H}$. If $Kx = 0$, then $\|x\|_1 = [\langle Kx, x \rangle]^{1/2} = 0$. Conversely, if $\|x\|_1 = 0$, then $|\langle Kx, y \rangle| = |\langle x, y \rangle_1| \le \|x\|_1 \|y\|_1 = 0$, for each y in $\mathcal{H}$, and thus $Kx = 0$. It follows that $\|\ \|_1$ is a norm if and only if K has null space $\{0\}$.

(ii)   Suppose that K has null space $\{0\}$.

If K has a bounded inverse $K^{-1} : \mathcal{H} \to \mathcal{H}$, we have $\|x\|_1^2 = \langle Kx, x \rangle \le \|K\| \|x\|^2$, and

$$(1) \qquad\qquad \|x\|_1 \le \|K\|^{1/2} \|x\| \qquad (x \in \mathcal{H}).$$

Moreover,

$$\|x\|^2 = \langle K K^{-1}x, x\rangle = \langle K^{-1}x, x\rangle_1 \le \left\|K^{-1}x\right\|_1 \|x\|_1$$
$$\le \|K\|^{1/2}\left\|K^{-1}x\right\|\|x\|_1 \le \|K\|^{1/2}\left\|K^{-1}\right\|\|x\|\|x\|_1 ,$$

and

$$(2) \qquad \|x\| \le \|K\|^{1/2}\left\|K^{-1}\right\|\|x\|_1 \qquad (x \in \mathcal{H}).$$

From (1) and (2), the identity mapping on $\mathcal{H}$ is bicontinuous, as a mapping from $(\mathcal{H}, \|\ \|)$ onto $(\mathcal{H}, \|\ \|_1)$; so the two norms give rise to the same topology on $\mathcal{H}$.

Conversely, suppose that $\|\ \|$ and $\|\ \|_1$ induces the same topology on $\mathcal{H}$. Then the identity mapping is continuous from $\mathcal{H}$ in the $\|\ \|_1$ topology to $\mathcal{H}$ in the $\|\ \|$ topology; so there is a positive real number M such that

$$\|x\| \le M\|x\|_1 \qquad (x \in \mathcal{H}).$$

Since

$$\|x\|\|Kx\| \ge \langle Kx, x\rangle = \|x\|_1^2 \ge M^{-2}\|x\|^2 ,$$

we have

$$\|Kx\| \ge M^{-2}\|x\| \qquad (x \in \mathcal{H}).$$

By Lemma 2.4.8, K has a bounded inverse.

(iii)   When $x, y \in \mathcal{H}$,

$$\langle Ax, y\rangle_1 = \langle K Ax, y\rangle = \langle x, A^*Ky\rangle$$
$$= \langle x, K K^{-1}A^*Ky\rangle$$
$$= \langle Kx, K^{-1}A^*Ky\rangle = \langle x, K^{-1}A^*Ky\rangle_1.$$

Thus A has adjoint $K^{-1}A^*K$ relative to $\langle\ ,\ \rangle_1$.   ∎

2.8.11.   Suppose that T is a bounded self-adjoint operator acting on a Hilbert space $\mathcal{H}$ and k is a positive real number such that $-kI \le T \le kI$. By using the identity

$$4\,\mathbf{Re}\langle Tx, y\rangle = \langle T(x + y), x + y\rangle - \langle T(x - y), x - y\rangle,$$

show that

$$|\mathbf{Re}\langle Tx, y\rangle| \le \tfrac{1}{2}k\{\|x\|^2 + \|y\|^2\}$$

for all x and y in $\mathcal{H}$. Deduce that $\|T\| \le k$ and that

$$\|T\| = \min\{a : a \in \mathbb{R},\ -aI \le T \le aI\}$$
$$= \sup\{|\langle Tx, x\rangle| : x \in \mathcal{H},\ \|x\| = 1\}.$$

Solution. When $x, y \in \mathcal{H}$,

$$\langle T(x+y), x+y \rangle - \langle T(x-y), x-y \rangle = 2\langle Tx, y \rangle + 2\langle Ty, x \rangle$$
$$= 2\langle Tx, y \rangle + 2\langle y, Tx \rangle$$
$$= 4\,\mathbf{Re}\langle Tx, y \rangle.$$

From this, and since $-T, T \leq kI$, we have

$$4|\mathbf{Re}\langle Tx, y \rangle| \leq |\langle T(x+y), x+y \rangle| + |\langle T(x-y), x-y \rangle|$$
$$\leq k\{\|x+y\|^2 + \|x-y\|^2\}$$
$$= 2k\{\|x\|^2 + \|y\|^2\},$$

and

$$|\mathbf{Re}\langle Tx, y \rangle| \leq \tfrac{1}{2}k\{\|x\|^2 + \|y\|^2\}.$$

If $Tx \neq 0$, we can let y be bTx, where b is $\|x\| \,/\, \|Tx\|$. We then obtain

$$b\|Tx\|^2 \leq \tfrac{1}{2}k\{\|x\|^2 + \|x\|^2\}$$
$$= k\|x\|^2 = kb\|x\|\|Tx\|,$$

whence $\|Tx\| \leq k\|x\|$ (and this last is true, also, if $Tx = 0$). Thus $\|T\| \leq k$.

From the discussion preceding Remark 2.4.9, $-\|T\|\,I \leq T \leq \|T\|\,I$. This, together with the argument just given, shows that

$$\|T\| = \min\{a : a \in \mathbb{R}, \ -aI \leq T \leq aI\}.$$

Since $-aI \leq T \leq aI$ if and only if

$$-a \leq \langle Tx, x \rangle \leq a$$

(equivalently, $|\langle Tx, x \rangle| \leq a$) for each unit vector x, it now follows that

$$\|T\| = \sup\{|\langle Tx, x \rangle| : x \in \mathcal{H}, \ \|x\| = 1\}. \qquad \blacksquare$$

2.8.12.   A bounded linear operator A, acting on a Hilbart space $\mathcal{H}$, is said to *attain its bound* if $\|Ax\| = \|A\|$ for some unit vector x in $\mathcal{H}$. Give examples of

(a)   a bounded self-adjoint operator with an orthonormal basis of eigenvectors,

(b)   a bounded self-adjoint operator with no eigenvector,

neither of which attains its bound.

Solution. Let $Y = \{y_1, y_2, y_3, \ldots\}$ be an orthonormal basis in a (separable) Hilbert space $\mathcal{H}$, and define a bounded real-valued function g on Y by $g(y_n) = 1 - n^{-1}$. With T constructed as in Example 2.4.10, we have

$$T = T^*, \quad Ty_n = \left(1 - \frac{1}{n}\right) y_n, \quad \|T\| = \sup_{y \in Y} |g(y)| = 1.$$

Each unit vector x in $\mathcal{H}$ has the form $\sum_{n=1}^{\infty} c_n y_n$, where the coefficients c_n satisfy $\sum_{n=1}^{\infty} |c_n|^2 = 1$. Since $Tx = \sum_{n=1}^{\infty} c_n \left(1 - \frac{1}{n}\right) y_n$, we have $\|Tx\|^2 = \sum_{n=1}^{\infty} |c_n|^2 \left(1 - \frac{1}{n}\right)^2 < 1$. Thus T is a bounded self-adjoint operator with an orthonormal basis of eigenvectors, and does not attain its bound.

Let M_f be the operator described in the final sentence of Example 2.4.11. Then M_f is a bounded self-adjoint (in fact, positive) operator, with no eigenvector, acting on L_2 (with Lebesgue measure, on the interval $[0, 1]$). If x is a unit vector in L_2, then

$$\|M_f x\|^2 = \int_0^1 \left|(M_f x)(s)\right|^2 ds = \int_0^1 s^2 \, |x(s)|^2 \, ds$$

$$< \int_0^1 |x(s)|^2 \, ds = \|x\|^2 = 1.$$

On the other hand, $\|M_f\| = \operatorname{ess\,sup} |f(s)| = 1$, so M_f does not attain its bound. ∎

2.8.13.   Let $\mathcal{H}$ be a Hilbert space.

(i)   Prove that each unit vector x in $\mathcal{H}$ is an extreme point of the unit ball $(\mathcal{H})_1$.

(ii)   Prove that each isometric linear operator V from $\mathcal{H}$ into $\mathcal{H}$ is an extreme point of the unit ball $(\mathcal{B}(\mathcal{H}))_1$.

Solution. (i)   Suppose that $x = (1 - \alpha)y + \alpha z$, where $0 < \alpha < 1$ and $y, z \in (\mathcal{H})_1$. Then

$$1 = \langle x, x \rangle = (1 - \alpha)\langle y, x \rangle + \alpha \langle z, x \rangle.$$

Since $\langle y, x \rangle$ and $\langle z, x \rangle$ lie in the unit disk in $\mathbb{C}$, and 1 is an extreme point of that disk, we have

$$\langle y, x \rangle = 1 \geq \|y\| = \|y\| \, \|x\|.$$

Thus $\|y\| = 1 \,(= \|x\|)$, and $y = x$ by Proposition 2.1.3. This shows that x is an extreme point of $(\mathcal{H})_1$.

(ii)   Suppose that $V = (1 - \alpha)A + \alpha B$, where $0 < \alpha < 1$ and $A, B \in (\mathcal{B}(\mathcal{H}))_1$. For each unit vector x in $\mathcal{H}$,

$$Vx = (1 - \alpha)Ax + \alpha Bx.$$

Since $Ax, Bx \in (\mathcal{H})_1$ and $\|Vx\| = 1$, it now follows from (i) that $Ax = Vx$. Since this holds for each unit vector x in $\mathcal{H}$, $A = V$; and V is an extreme point of $(\mathcal{B}(\mathcal{H}))_1$.   ■

2.8.14.   Show that the projection E from a Hilbert space $\mathcal{H}$ onto a closed subspace $\mathcal{K}$ is an extreme point of the set $(\mathcal{B}(\mathcal{H})^+)_1$ of all positive operators in the unit ball of $\mathcal{B}(\mathcal{H})$.

Solution.   Suppose that $E = (1 - \alpha)A + \alpha B$, where $0 < \alpha < 1$ and $A, B \in (\mathcal{B}(\mathcal{H})^+)_1$. Each vector z in $\mathcal{H}$ can be expressed as $ax + by$ where x and y are unit vectors in the ranges of E and $I - E$, respectively. Since

$$x = Ex = (1 - \alpha)Ax + \alpha Bx,$$

while $Ax, Bx \in (\mathcal{H})_1$, it follows from the result of Exercise 2.8.13(i) that $Ax = Ex$. Since

$$0 = \langle Ey, y \rangle = (1 - \alpha)\langle Ay, y \rangle + \alpha \langle By, y \rangle,$$

while A and B are positive operators, it follows that $\langle Ay, y \rangle = 0$. By the Cauchy–Schwarz inequality for the inner product $(u, v) \rightarrow \langle Au, v \rangle$,

$$|\langle Ay, u \rangle|^2 \le \langle Ay, y \rangle \langle Au, u \rangle = 0$$

for each u in $\mathcal{H}$; and thus $Ay = 0 = Ey$. Since $Ax = Ex$ and $Ay = Ey$, we have $Az = Ez$ (for each z in $\mathcal{H}$). Hence $A = E$, and E is an extreme point of $(\mathcal{B}(\mathcal{H})^+)_1$.   ■

2.8.15.   Determine a necessary and sufficient condition for the operator M_g, defined in Example 2.4.11, to have a bounded inverse.

Solution.   We assert that M_g has a bounded inverse if and only if

(∗)   there is a positive real number δ such that $|g(s)| \geq \delta$ almost everywhere.

Indeed, if (∗) is satisfied, there is a measurable function h such that

$$|h(s)| \leq \delta^{-1}, \quad h(s)g(s) = 1, \qquad \text{for almost all } s.$$

Then, $h \in L_\infty$ and M_g has a bounded inverse M_h.

Conversely, suppose that M_g has a bounded inverse T. For each f in L_2,

$$\int_S |g(s)f(s)|^2 \, dm(s) = \|M_g f\|^2$$
$$\geq \|T\|^{-2} \|T M_g f\|^2$$
$$= \|T\|^{-2} \|f\|^2$$
$$= \|T\|^{-2} \int_S |f(s)|^2 \, dm(s).$$

Thus

$$\int_S \left\{ |g(s)|^2 - \|T\|^{-2} \right\} |f(s)|^2 \, dm(s) \geq 0 \qquad (f \in L_2).$$

This implies that (∗) is satisfied, with $\delta = \|T\|^{-1}$.   ∎

2.8.16.   Suppose that $T = \sum_{a \in \mathbb{A}} \oplus T_a$, where $T_a \in \mathcal{B}(\mathcal{H}_a)$ for each a in $\mathbb{A}$, and $\sup\{\|T_a\| : a \in \mathbb{A}\} < \infty$. Show that T has a bounded inverse if and only if the following two conditions are satisfied:

(i)   each T_a has a bounded inverse,
(ii) $\sup\{\|T_a^{-1}\| : a \in \mathbb{A}\} < \infty$.

Solution.   If (i) and (ii) are satisfied, then T has a bounded inverse, namely $\sum_{a \in \mathbb{A}} \oplus T_a^{-1}$.

Conversely, suppose that T has a bounded inverse. Since T is one–to–one and maps $\sum_{a \in \mathbb{A}} \oplus \mathcal{H}_a$ onto itself, it follows that, for each a in $\mathbb{A}$, T_a is one–to–one and maps $\mathcal{H}_a$ onto itself. If $b \in \mathbb{A}$, $x_b \in \mathcal{H}_b$, and x_a is the zero vector in $\mathcal{H}_a$ when $a \in \mathbb{A} \setminus \{b\}$, then

$$\|T_b x_b\| = \|T\{x_a\}\| \geq \|T^{-1}\|^{-1} \|\{x_a\}\| = \|T^{-1}\|^{-1} \|x_b\|.$$

Since T_b is a one–to–one linear operator from $\mathcal{H}_b$ onto $\mathcal{H}_b$, and $\|T_b x_b\| \geq \|T^{-1}\|^{-1} \|x_b\|$ when $x_b \in \mathcal{H}_b$, it follows that T_b^{-1} is bounded and $\|T_b^{-1}\| \leq \|T^{-1}\|$. Hence each T_a has a bounded inverse, and

$$\sup\{\|T_a^{-1}\| : a \in \mathbb{A}\} \leq \|T^{-1}\|. \qquad \blacksquare$$

2.8.17.   Let $\mathcal{P}$ denote the set of all projections from a Hilbert space $\mathcal{H}$ onto its closed subspaces, and suppose that $F \in \mathcal{P}$, $0 \neq F \neq I$. Prove that the mappings

$$E \to E \wedge F, \qquad E \to E \vee F$$

are not continuous, from $\mathcal{P}$ with the norm topology into $\mathcal{P}$ with the strong-operator topology.

Solution.   Choose unit vectors, y in the range of F and z in the range of $I - F$. When $0 \leq \theta \leq \frac{1}{2}\pi$, let P_θ be the projection whose range is the one-dimensional subspace containing the unit vector $x_\theta = (\cos\theta)y + (\sin\theta)z$. Since $P_\theta x = \langle x, x_\theta \rangle x_\theta$ (see, for example, Proposition 2.5.16) the mapping

$$\theta \to P_\theta : \quad [0, \tfrac{1}{2}\pi] \to \mathcal{P}$$

is norm continuous. However

$$
\begin{aligned}
P_0 \wedge F &= P_0, & P_0 \vee F &= F, \\
P_\theta \wedge F &= 0, & P_\theta \vee F &= F + P_{\frac{1}{2}\pi} & (\theta > 0);
\end{aligned}
$$

and hence the mappings

$$\theta \to P_\theta \wedge F, \qquad \theta \to P_\theta \vee F$$

are not strong-operator continuous at 0. But these mappings are the compositions

$$\theta \to P_\theta \to P_\theta \wedge F, \qquad \theta \to P_\theta \to P_\theta \vee F,$$

the first components of which are norm continuous. Thus

$$P_\theta \to P_\theta \wedge F, \qquad P_\theta \to P_\theta \vee F$$

are not continuous from $\mathcal{P}$ in the norm topology to $\mathcal{P}$ in the strong-operator topology. $\blacksquare$

2.8.18.   Let $\mathcal{S}$ denote the set of all bounded self-adjoint operators acting on a Hilbert space $\mathcal{H}$. If $A, B, C \in \mathcal{S}$, we say that C is a *lower bound* of $\{A, B\}$ if $C \le A$, $C \le B$. We say that C is the *greatest lower bound* of $\{A, B\}$ if it is a lower bound of $\{A, B\}$, and $D \le C$ whenever D is a lower bound of $\{A, B\}$.

(i)   Show that, if $A, B \in \mathcal{S}$, then $\{A, B\}$ has a lower bound in $\mathcal{S}$.

(ii)   Suppose that A, B are non-zero elements of $\mathcal{B}(\mathcal{H})^+$. Show that there is a vector x_0 such that $\langle Ax_0, x_0 \rangle > 0$ and $\langle Bx_0, x_0 \rangle > 0$. Prove that, if P_0 is the projection onto the one-dimensional subspace containing x_0, a and b are suitable positive real numbers, and $T = aP_0 - b(I - P_0)$, then $T \le A$, $T \le B$, $T \not\le 0$. Deduce that 0 is not the greatest lower bound of $\{A, B\}$.

(iii)   Suppose that $A, B \in \mathcal{S}$, and $\{A, B\}$ has a greatest lower bound C in $\mathcal{S}$. By applying the result of (ii) to $\{A - C, B - C\}$, show that either $A \le B$ or $B \le A$.

Solution. (i)   Note that $-\{\|A\| + \|B\|\}I$ is a lower bound of $\{A, B\}$.

(ii)   We can choose a pair of vectors u, v such that $\langle Au, u \rangle > 0$ and $\langle Bv, v \rangle > 0$. We obtain a vector x_0 with the desired property, by taking x_0 to be u or v, unless

$$(1) \qquad\qquad \langle Av, v \rangle = \langle Bu, u \rangle = 0.$$

However, if (1) holds, we have $\langle Av, w \rangle = \langle Aw, v \rangle = 0 = \langle Bu, w \rangle = \langle Bw, u \rangle$, for all w in $\mathcal{H}$, by the Cauchy–Schwarz inequality for the inner products $(x, y) \to \langle Ax, y \rangle$, $(x, y) \to \langle Bx, y \rangle$. Then, it suffices to take $x_0 = u + v$.

Every vector in $\mathcal{H}$ has the form $\alpha x_0 + \beta y$, where α, β are scalars and $\langle x_0, y \rangle = 0$. We have

$$\langle A(\alpha x_0 + \beta y), \alpha x_0 + \beta y \rangle$$
$$= |\alpha|^2 \langle Ax_0, x_0 \rangle + \alpha \bar{\beta} \langle Ax_0, y \rangle + \bar{\alpha} \beta \langle Ay, x_0 \rangle + |\beta|^2 \langle Ay, y \rangle$$
$$\ge |\alpha|^2 \langle Ax_0, x_0 \rangle - 2 |\alpha| |\beta| \|A\| \|x_0\| \|y\|,$$

and

$$\langle T(\alpha x_0 + \beta y), \alpha x_0 + \beta y \rangle = \langle \alpha a x_0 - \beta b y, \alpha x_0 + \beta y \rangle$$
$$= |\alpha|^2 \|x_0\|^2 a - |\beta|^2 \|y\|^2 b.$$

Thus

$$\langle (A - T)(\alpha x_0 + \beta y), \alpha x_0 + \beta y \rangle$$
$$\geq \left[\langle Ax_0, x_0 \rangle - a \|x_0\|^2 \right] |\alpha|^2 - 2 |\alpha| |\beta| \|A\| \|x_0\| \|y\| + b |\beta|^2 \|y\|^2$$
$$= \left[\langle Ax_0, x_0 \rangle - a \|x_0\|^2 \right] s^2 - 2 \|A\| \|x_0\| st + bt^2,$$

where $s = |\alpha|$, $t = |\beta| \|y\|$. Thus $A - T \geq 0$ provided a is small enough to ensure $\langle Ax_0, x_0 \rangle > a \|x_0\|^2$ and b is large enough to ensure $\|A\|^2 \|x_0\|^2 \leq \left[\langle Ax_0, x_0 \rangle - a \|x_0\|^2 \right] b$. Similarly $B - T \geq 0$ when a is sufficiently small and b is sufficiently large.

Since $T \leq A$, $T \leq B$ but $T \not\leq 0$ (for suitably chosen a and b), 0 is not the greatest lower bound of $\{A, B\}$.

(iii)   Under the conditions of (iii), $A - C \geq 0$, $B - C \geq 0$ and $\{A - C, B - C\}$ has greatest lower bound 0. From (ii), at least one of $A - C$, $B - C$ is 0. Hence either $A = C \leq B$ or $B = C \leq A$.   ∎

2.8.19.   Suppose that A and B are mappings from a Hilbert space $\mathcal{H}$ into itself, and $\langle Ax, y \rangle = \langle x, By \rangle$ for all x and y in $\mathcal{H}$. Prove that A and B are bounded linear operators, and $A = B^*$.

Solution.   When $x_1, x_2 \in \mathcal{H}$ and $a_1, a_2 \in \mathbb{C}$, we have

$$\langle A(a_1 x_1 + a_2 x_2) - a_1 Ax_1 - a_2 Ax_2, y \rangle$$
$$= \langle A(a_1 x_1 + a_2 x_2), y \rangle - a_1 \langle Ax_1, y \rangle - a_2 \langle Ax_2, y \rangle$$
$$= \langle a_1 x_1 + a_2 x_2, By \rangle - a_1 \langle x_1, By \rangle - a_2 \langle x_2, By \rangle = 0,$$

for all y in $\mathcal{H}$. Thus $A(a_1 x_1 + a_2 x_2) = a_1 Ax_1 + a_2 Ax_2$; and A (and, similarly, B) is linear.

Since A is an everywhere defined linear operator, and $\langle Ax, y \rangle = \langle x, By \rangle$ for all x and y in $\mathcal{H}$, it follows that A^* has domain $\mathcal{H}$ and coincides with B. By Theorem 2.7.8(ii), A is preclosed, and is therefore closed since its domain is $\mathcal{H}$. By the closed graph theorem (1.8.6), A is bounded, whence so is B $(= A^*)$; and $A = B^*$.

The boundedness of A can be established by the following alternative argument. We have to prove the boundedness of the subset $\{Ax : x \in \mathcal{H}, \|x\| \leq 1\}$ of $\mathcal{H}$. From the principle of uniform boundedness (Corollary 1.8.11), together with Riesz's representation theorem, it suffices to show that

$$\sup\{|\langle Ax, y \rangle| : x \in \mathcal{H}, \|x\| \leq 1\} < \infty,$$

for each y in $\mathcal{H}$. This follows from the relations

$$|\langle Ax, y\rangle| = |\langle x, By\rangle| \leq \|By\| \qquad (x \in \mathcal{H},\ \|x\| \leq 1). \qquad \blacksquare$$

2.8.20.   Suppose that $\mathcal{H}$ is a Hilbert space and $A \in \mathcal{B}(\mathcal{H})$. Prove that the following five conditions are equivalent.

(i) A is continuous as a mapping from the unit ball $(\mathcal{H})_1$ (with the weak topology) into $\mathcal{H}$ (with the norm topology).

(ii)   If $x, x_1, x_2, \ldots \in \mathcal{H}$ and $\{x_n\}$ is weakly convergent to x, then $\{Ax_n\}$ is norm convergent to Ax.

(iii)   Every bounded sequence $\{x_n\}$ in $\mathcal{H}$ has a subsequence $\{x_{n(k)}\}$ such that $\{Ax_{n(k)}\}$ is norm convergent.

(iv)   The set $\{Ax : x \in (\mathcal{H})_1\}$ is relatively compact in the norm topology of $\mathcal{H}$.

(v)   The set $\{Ax : x \in (\mathcal{H})_1\}$ is compact in the norm topology of $\mathcal{H}$.

[An element of $\mathcal{B}(\mathcal{H})$ that has any (and, hence, all) of the above properties is described as a *compact* linear operator.]

Solution.   (i) $\implies$ (ii).   Under the conditions of (ii), $\{x_n\}$ is bounded, by the principle of uniform boundedness; upon replacing x, x_n by cx, cx_n, with c a suitable positive real number, we may suppose that $x, x_1, x_2, \ldots \in (\mathcal{H})_1$. It is then apparent, from (i), that weak convergence of $\{x_n\}$ to x entails norm convergence of $\{Ax_n\}$ to Ax.

(ii) $\implies$ (iii).   This follows from the fact that every bounded sequence in $\mathcal{H}$ has a weakly convergent subsequence (Exercise 2.8.6).

(iii) $\implies$ (iv).   This is apparent from the equivalence, in a metric space, of relative compactness and relative sequential compactness.

(iv) $\implies$ (v).   This follows from the fact that the set $\{Ax : x \in (\mathcal{H})_1\}$ is norm closed in $\mathcal{H}$ (see Exercise 1.9.17(ii) and Corollary 2.3.3).

(v) $\implies$ (i).   On the set $\{Ax : x \in (\mathcal{H})_1\}$, the norm topology is compact, and so coincides with the (coarser, Hausdorff) weak topology. Since A is weakly continuous (Proposition 1.3.3), it is continuous from $(\mathcal{H})_1$ (with the weak topology) into $A((\mathcal{H})_1)$ with the weak (=norm) topology. $\blacksquare$

2.8.21.   Prove that the identity operator, acting on an infinite-dimensional Hilbert space, is not compact (in the sense of Exercise 2.8.20).

Solution. If the identity operator I, acting on a Hilbert space $\mathcal{H}$, is compact, then the unit ball $(\mathcal{H})_1$ $(= I((\mathcal{H})_1))$ is compact in the norm topology, and $\mathcal{H}$ is a finite-dimensional vector space by Theorem 1.2.18.   ∎

2.8.22.   Suppose that $\mathcal{H}$ is a Hilbert space and

$$\mathcal{I} = \{A \in \mathcal{B}(\mathcal{H}) : A \text{ has finite-dimensional range}\}.$$

(i)   Prove that, if $A \in \mathcal{I}$ and $\{y_1, \ldots, y_n\}$ is an orthonormal basis of the range of A, there exist vectors $x_1, \ldots, x_n$ in $\mathcal{H}$ such that

$$Ax = \sum_{j=1}^{n} \langle x, x_j \rangle y_j \qquad (x \in \mathcal{H}).$$

(ii)   Prove that $\mathcal{I}$ is a two-sided ideal in $\mathcal{B}(\mathcal{H})$ and that every non-zero two-sided ideal in $\mathcal{B}(\mathcal{H})$ contains $\mathcal{I}$.

(iii)   Prove that an element A of $\mathcal{B}(\mathcal{H})$ lies in $\mathcal{I}$ if and only if A is continuous as a mapping from $\mathcal{H}$ (with the weak topology) into $\mathcal{H}$ (with the norm topology).

(iv)   Prove that the elements of $\mathcal{I}$ are compact linear operators (in the sense of Exercise 2.8.20).

Solution. (i)   When $x \in \mathcal{H}$, Ax lies in the range of A, so

$$Ax = \sum_{j=1}^{n} \langle Ax, y_j \rangle y_j = \sum_{j=1}^{n} \langle x, A^* y_j \rangle y_j.$$

This proves (i), with x_j taken to be $A^* y_j$.

(ii)   It is apparent that $\mathcal{I}$ is a two-sided ideal in $\mathcal{B}(\mathcal{H})$. Suppose that $\mathcal{C}$ is a non-zero ideal in $\mathcal{B}(\mathcal{H})$, and choose a non-zero C in $\mathcal{C}$. There exist unit vectors x_0, y_0 such that $\langle C y_0, x_0 \rangle \neq 0$. Given any vectors x_1, y_1 in $\mathcal{H}$, we can define A, B in $\mathcal{B}(\mathcal{H})$ by

$$Ax = \langle x, x_0 \rangle y_1, \qquad Bx = \langle x, x_1 \rangle y_0.$$

Elementary calculation shows that

$$\langle C y_0, x_0 \rangle^{-1} ACBx = \langle x, x_1 \rangle y_1 \qquad (x \in \mathcal{H}).$$

Hence C contains each operator $x \to \langle x, x_1 \rangle y_1$, where $x_1, y_1 \in \mathcal{H}$, and so contains all finite sums of such operators. It now follows from (i) that $\mathcal{I} \subseteq C$.

(iii)   If $A \in \mathcal{I}$, we can choose $x_1, \ldots, x_n, y_1, \ldots, y_n$ as in (i). Since

$$\|Ax\| \le \sum_{j=1}^{n} |\langle x, x_j \rangle| \, \|y_j\| = \sum_{j=1}^{n} |\langle x, x_j \rangle|,$$

it follows that A is continuous as a mapping from $\mathcal{H}$ (with the weak topology) into $\mathcal{H}$ (with the norm topology).

Conversely, suppose that A has the continuity property just mentioned. The inverse image, under A, of the open unit ball in $\mathcal{H}$ contains a weak neighborhood of 0 in $\mathcal{H}$, and so contains a basic weak neighborhood of 0 in $\mathcal{H}$,

$$V = \{x \in \mathcal{H} : |\langle x, y_j \rangle| < \varepsilon \ (j = 1, \ldots, n)\},$$

where $y_1, \ldots, y_n \in \mathcal{H}$. If $x \in \mathcal{H}$ and $\langle x, y_j \rangle = 0$, $j = 1, \ldots, n$, then $kx \in V$ and thus $\|kAx\| < 1$, for every scalar k; so $Ax = 0$. The property just established, namely

$$(*) \qquad x \in \mathcal{H} \text{ and } \langle x, y_j \rangle = 0 \ (j = 1, \ldots, n) \implies Ax = 0,$$

remains valid if $\{y_1, \ldots, y_n\}$ is replaced by an orthonormal basis of the subspace spanned by the vectors $y_1, \ldots, y_n$; so we may assume that $\{y_1, \ldots, y_n\}$ is an orthonormal system. Given any x in $\mathcal{H}$, let

$$z = x - \sum_{j=1}^{n} \langle x, y_j \rangle y_j.$$

Since $\langle z, y_j \rangle = 0 \ (j = 1, \ldots, n)$, we have $Az = 0$ by $(*)$, and thus

$$Ax = \sum_{j=1}^{n} \langle x, y_j \rangle Ay_j.$$

It follows that the range of A is contained in the finite-dimensional subspace spanned by $Ay_1, \ldots, Ay_n$, whence $A \in \mathcal{I}$.

(iv)   If $A \in \mathcal{I}$, then A is continuous from $\mathcal{H}$ (with the weak topology) into $\mathcal{H}$ (with the norm topology), by (iii). Hence it satisfies condition (i) of Exercise 2.8.20, and is thus a compact linear operator.   ∎

2.8.23.   Suppose that $\{A_n\}$ is a sequence of compact linear operators acting on a Hilbert space $\mathcal{H}$, $A \in \mathcal{B}(\mathcal{H})$, and $\|A_n - A\| \to 0$. Prove that A is compact. [*Hint.* Use condition (i) of Exercise 2.8.20 as the defining property of a compact linear operator.]

Solution.   If $(\mathcal{H})_1$ has the weak topology and $\mathcal{H}$ has the norm topology, then the sequence $\{A_n|(\mathcal{H})_1\}$ of restrictions consists of continuous mappings from $(\mathcal{H})_1$ to $\mathcal{H}$, and converges uniformly to $A|(\mathcal{H})_1$. Hence A is continuous from $(\mathcal{H})_1$ (with the weak topology) to $\mathcal{H}$ (with the norm topology), and is therefore compact.   ∎

2.8.24.   Suppose that A is a compact linear operator acting on a Hilbert space $\mathcal{H}$ (see Exercise 2.8.20).
  (i)   Prove that the (closed) range space $[A(\mathcal{H})]$ is separable.
  (ii)   Suppose that $[A(\mathcal{H})]$ is infinite-dimensional, and let $\{y_1, y_2, y_3, \ldots\}$ be an orthonormal basis of $[A(\mathcal{H})]$. For each positive integer n, let P_n be the projection from $\mathcal{H}$ onto the subspace spanned by $y_1, \ldots, y_n$. Prove that $\|A - P_n A\| \to 0$ as $n \to \infty$.
  (iii)   Deduce that A lies in the norm closure of the ideal $\mathcal{I}$ (see Exercise 2.8.22), and A^* is compact.

Solution. (i)   The set $A((\mathcal{H})_1)$ is compact in the norm topology, and so has a countable norm-dense subset. Positive rational multiples of vectors in this subset constitute a countable set which is norm dense in $[A(\mathcal{H})]$.
  (ii)   When $1 \le m \le n$, we have

$$\|A - P_n A\| = \|(I - P_n)A\| = \|(I - P_n)(I - P_m)A\|$$
$$\le \|(I - P_m)A\| = \|A - P_m A\|.$$

Accordingly, the sequence $\{\|A - P_n A\|\}$ decreases, and converges to a real number η (≥ 0).
  If $\eta > 0$, we have $\|A - P_n A\| \ge \eta$, and we can choose a unit vector x_n such that

$$\|(I - P_n)A x_n\| > \tfrac{1}{2}\eta \qquad (n = 1, 2, \ldots).$$

The sequence $\{A x_n\}$ has a subsequence $\{A x_{n(k)}\}$ that converges in norm to an element u of $\mathcal{H}$; and $u \in [A(\mathcal{H})]$ since $A x_{n(k)} \in [A(\mathcal{H})]$.

Thus

$$u = \sum_{j=1}^{\infty} \langle u, y_j \rangle y_j = \lim_{n \to \infty} \sum_{j=1}^{n} \langle u, y_j \rangle y_j$$
$$= \lim_{n \to \infty} P_n u$$

(in the norm topology). For each positive integer k,

$$\tfrac{1}{2}\eta < \|(I - P_{n(k)})Ax_{n(k)}\|$$
$$\leq \|(I - P_{n(k)})(Ax_{n(k)} - u)\| + \|u - P_{n(k)}u\|$$
$$\leq \|Ax_{n(k)} - u\| + \|u - P_{n(k)}u\|.$$

Since the right-hand side has limit 0 as $k \to \infty$, while $\eta > 0$, we have a contradiction. Hence $\eta = 0$, and $\|A - P_n A\| \to 0$.

(iii)   Since $P_n A$ lies in $\mathcal{I}$, A $(= \lim P_n A)$ lies in the norm closure of $\mathcal{I}$. Also, $A^* P_n \in \mathcal{I}$, hence $A^* P_n$ is compact (Exercise 2.8.22(iv)), and thus A^* $(= \lim A^* P_n)$ is compact (Exercise 2.8.23).   ■

2.8.25.   Let $\mathcal{K}$ denote the set of all compact linear operators acting on a Hilbert space $\mathcal{H}$. By using the results of the three preceding exercises, show that:
   (i) $\mathcal{K}$ is the norm closure of the ideal $\mathcal{I}$ in $\mathcal{B}(\mathcal{H})$;
   (ii) $\mathcal{K}$ is a norm closed two-sided ideal in $\mathcal{B}(\mathcal{H})$;
   (iii)   each non-zero norm closed two-sided ideal in $\mathcal{B}(\mathcal{H})$ contains $\mathcal{K}$.

Solution. (i)   From Exercises 2.8.23 and 2.8.22(iv), $\mathcal{K}$ is norm closed and contains $\mathcal{I}$; so $\mathcal{K}$ contains the norm closure of $\mathcal{I}$. From Exercise 2.8.24(iii), $\mathcal{K}$ is contained in the norm closure of $\mathcal{I}$.
   (ii)   This follows at once, from (i) and Exercise 2.8.22(ii).
   (iii)   A non-zero norm closed two-sided ideal in $\mathcal{B}(\mathcal{H})$ contains $\mathcal{I}$ (Exercise 2.8.22(ii)) and so contains the norm closure $\mathcal{K}$ of $\mathcal{I}$.   ■

2.8.26.   Suppose that $\{y_1, y_2, y_3, \ldots\}$ is an orthonormal system in a Hilbert space $\mathcal{H}$, and $\{\lambda_1, \lambda_2, \lambda_3, \ldots\}$ is a sequence of real numbers such that $|\lambda_1| \geq |\lambda_2| \geq |\lambda_3| \geq \cdots$. Suppose also that the sequences $\{y_n\}$ and $\{\lambda_n\}$ are *either* both finite and of the same length *or* both infinite, with $\{\lambda_n\}$ converging to 0. Show that the equation

$$Ax = \sum_n \lambda_n \langle x, y_n \rangle y_n \qquad (x \in \mathcal{H})$$

defines a compact self-adjoint operator A on $\mathcal{H}$, with $\|A\| = |\lambda_1|$. [The result of Exercise 2.8.29 below shows that every compact self-adjoint operator has the form just described.]

Solution. By Theorem 2.2.10, there is an orthonormal basis Y of $\mathcal{H}$ that contains the orthonormal system $\{y_1, y_2, y_3, \ldots\}$. Let g be the real-valued function defined on Y by

$$g(y_n) = \lambda_n, \qquad g(y) = 0 \quad (y \in Y \setminus \{y_n\}).$$

From Example 2.4.10, there is a self-adjoint operator A on $\mathcal{H}$, defined by

$$Ax = \sum_{y \in Y} g(y)\langle x, y\rangle y$$
$$= \sum_n \lambda_n \langle x, y_n\rangle y_n.$$

Moreover, $\|A\| = \sup\{|g(y)| : y \in Y\} = |\lambda_1|$.

If both of the sequences are finite, A has finite-dimensional range spanned by $\{y_n\}$, and is therefore compact by Exercise 2.8.22(iv).

If the sequences are infinite (and $\lambda_n \to 0$), we can define compact self-adjoint operators $A_1, A_2, \ldots$ by

$$A_m x = \sum_{n=1}^{m} \lambda_n \langle x, y_n\rangle y_n.$$

Since

$$(A - A_m)x = \sum_{n=m+1}^{\infty} \lambda_n \langle x, y_n\rangle y_n,$$

we have that $\|A - A_m\| = |\lambda_{m+1}| \to 0$. Thus, from Exercise 2.8.23, A is compact. $\blacksquare$

2.8.27.   Suppose that A is the compact self-adjoint operator constructed in Exercise 2.8.26 from an orthonormal system $\{y_1, y_2, y_3, \ldots\}$ in a Hilbert space $\mathcal{H}$ and a real sequence $\{\lambda_1, \lambda_2, \lambda_3, \ldots\}$ that satisfy the conditions set out in that exercise. Extend the orthonormal system to an orthonormal *basis* $\{y_1, y_2, y_3, \ldots\} \cup \{z_a\}$.

(i)   Prove that, if λ is a non-zero scalar that does not appear in the sequence $\{\lambda_n\}$, the operator $A - \lambda I$ has an inverse in $\mathcal{B}(\mathcal{H})$, and

$$(A - \lambda I)^{-1}x = \sum_n \frac{1}{\lambda_n - \lambda}\langle x, y_n\rangle y_n - \lambda^{-1}\sum_a \langle x, z_a\rangle z_a$$

$$= \sum_n \frac{\lambda_n}{\lambda(\lambda_n - \lambda)}\langle x, y_n\rangle y_n - \lambda^{-1}x \qquad (x \in \mathcal{H}).$$

[*Hint.* Consider the matrix of A.]

(ii)   Show that, if λ is a non-zero scalar that appears in the sequence $\{\lambda_n\}$ and $x \in \mathcal{H}$, the equation

$$(A - \lambda I)z = x$$

has a solution z in $\mathcal{H}$ if and only if $\langle x, y_k\rangle = 0$ for each integer k satisfying $\lambda_k = \lambda$. What is the most general solution z, when this condition is satified?

Solution.   Since $\{y_n\} \cup \{z_a\}$ is an orthonormal basis of $\mathcal{H}$ and

$$Ax = \sum_n \lambda_n \langle x, y_n\rangle y_n \qquad (x \in \mathcal{H}),$$

we have $Ay_n = \lambda_n y_n$ and $Az_a = 0$. Accordingly, the matrix of A with respect to the given basis is diagonal, with λ_n at the (n, n) entry and 0 at the (a, a) entry, for each n and a. Thus $A - \lambda I$ has a diagonal matrix, with $\lambda_n - \lambda$ at the (n, n) entry and $-\lambda$ at the (a, a) entry.

(i)   Suppose that $\lambda \neq 0$ and λ does not appear in the sequence $\{\lambda_n\}$ (which is *either* finite *or* infinite and convergent to 0). It follows that the sequence $\{(\lambda_n - \lambda)^{-1}\}$ is bounded. The diagonal matrix with $(\lambda_n - \lambda)^{-1}$ at the (n, n) entry and $-\lambda^{-1}$ at the (a, a) entry, for each n and a, is inverse to the matrix of $A - \lambda I$ and represents a bounded linear operator T (of the type considered in Example 2.4.10) given by

$$Tx = \sum_n \frac{1}{\lambda_n - \lambda}\langle x, y_n\rangle y_n - \lambda^{-1}\sum_a \langle x, z_a\rangle z_a \qquad (x \in \mathcal{H}).$$

As suggested by the above matrix considerations, direct calculation shows that the operators $T(A - \lambda I)$ and $(A - \lambda I)T$ both map each of the basis vectors y_n, z_a onto itself, so $A - \lambda I$ has bounded inverse T. Note, also, that since

$$x = \sum_n \langle x, y_n \rangle y_n + \sum_a \langle x, z_a \rangle z_a,$$

the second term on the right-hand side of the above equation for Tx can be replaced by

$$-\lambda^{-1} \left(x - \sum_n \langle x, y_n \rangle y_n \right)$$

and this gives

$$Tx = \sum_n \frac{\lambda_n}{\lambda(\lambda_n - \lambda)} \langle x, y_n \rangle y_n - \lambda^{-1} x \qquad (x \in \mathcal{H}).$$

(ii)   Now suppose that $\lambda \neq 0$ and λ appears in the sequence $\{\lambda_n\}$. Since that sequence is either finite or infinite and convergent to 0, the (non-empty) set $\mathbb{K} = \{k : \lambda_k = \lambda\}$ is finite, and the set $\{\lambda^{-1}, (\lambda_n - \lambda)^{-1} : (n \notin \mathbb{K})\}$ is bounded. Also, λ is real, and $A - \lambda I$ is self-adjoint. Since

$$\langle (A - \lambda I)z, y_k \rangle = \langle z, Ay_k - \lambda_k y_k \rangle = 0 \qquad (z \in \mathcal{H}, \ k \in \mathbb{K}),$$

the equation $(a - \lambda I)z = x$ cannot have a solution z in $\mathcal{H}$ unless the given vector x in $\mathcal{H}$ satisfies

$$(1) \qquad\qquad \langle x, y_k \rangle = 0 \qquad (k \in \mathbb{K});$$

we suppose, henceforth, that this condition is satisfied. Each vector z in $\mathcal{H}$ has an expansion, $z = \sum c_n y_n + \sum d_a y_a$, in terms of the given orthonormal basis. Since

$$(A - \lambda I)z = \sum (\lambda_n - \lambda)c_n y_n - \sum \lambda d_a z_a,$$

$$x = \sum \langle x, y_n \rangle y_n + \sum \langle x, z_a \rangle z_a,$$

it follows that $(A - \lambda I)z = x$ if and only if the coefficients c_n and d_a satisfy

$$(\lambda_n - \lambda)c_n = \langle x, y_n \rangle, \qquad -\lambda d_a = \langle x, z_a \rangle,$$

for all n and a. These conditions impose no restriction on the coefficients c_k $(k \in \mathbb{K})$, since $\lambda - \lambda_k = \langle x, y_k \rangle = 0$ when $k \in \mathbb{K}$; they determine the remaining coefficients c_n $(n \notin \mathbb{K})$ and d_a in the form

$$(2) \qquad c_n = (\lambda_n - \lambda)^{-1}\langle x, y_n \rangle, \qquad d_a = -\lambda^{-1}\langle x, z_a \rangle.$$

From the boundedness of the set $\{\lambda^{-1}, (\lambda_n - \lambda)^{-1} : n \notin \mathbb{K}\}$, and since

$$\sum |\langle x, y_n \rangle|^2 + \sum |\langle x, z_a \rangle|^2 = \|x\|^2 < \infty,$$

it follows that $\sum |c_n|^2 + \sum |d_a|^2 < \infty$ when c_n $(n \notin \mathbb{K})$ and d_a satisfy (2); so the sum $\sum c_n y_n + \sum d_a z_a$ converges to an element z of $\mathcal{H}$, and $(A - \lambda I)z = x$.

We have now shown that the equation $(A - \lambda I)z = x$ has a solution z in $\mathcal{H}$ if and only if the given vector x in $\mathcal{H}$ satisfies (1); the general solution is then given by $z = \sum c_n y_n + \sum d_a z_a$, where the coefficients c_k $(k \in \mathbb{K})$ are arbitrary and the remaining coefficients are determined by (2). $\blacksquare$

2.8.28. Let A be a bounded self-adjoint operator acting on a Hilbert space $\mathcal{H}$.

(i) Show that each eigenvalue of A is real.

(ii) Show that eigenvectors corresponding to distinct eigenvalues of A are orthogonal.

Solution. (i) Suppose that $Ax = \lambda x$ and $\|x\| = 1$. Then

$$\lambda = \langle \lambda x, x \rangle = \langle Ax, x \rangle = \langle x, Ax \rangle = \overline{\langle Ax, x \rangle} = \bar{\lambda},$$

since A is self-adjoint. Thus λ is real.

(ii) If $Ax = \lambda x$, $Ay = \mu y$, and $\lambda \neq \mu$, then

$$\lambda \langle x, y \rangle = \langle \lambda x, y \rangle = \langle Ax, y \rangle = \langle x, Ay \rangle = \langle x, \mu y \rangle = \bar{\mu}\langle x, y \rangle = \mu \langle x, y \rangle$$

from (i). Since $\lambda \neq \mu$, we have that $\langle x, y \rangle = 0$. $\blacksquare$

2.8.29. Let A be a compact self-adjoint operator acting on a Hilbert space $\mathcal{H}$.

(i) By using the result of Exercise 2.8.11, show that there exist unit vectors $x_1, x_2, x_3, \ldots$ in $\mathcal{H}$ such that the real sequence

$\{\langle Ax_n, x_n \rangle\}$ converges with limit ρ equal to $\|A\|$ or $-\|A\|$. Show that $\|Ax_n - \rho x_n\| \to 0$ as $n \to \infty$.

(ii)   Prove that $Ax = \rho x$ for some unit vector x in $\mathcal{H}$ (so that a non-zero compact self-adjoint operator has a non-zero eigenvalue).

(iii)   Show that, if λ is a non-zero eigenvalue of A, then the null space of $A - \lambda I$ has finite dimension. (We call this finite dimension the *multiplicity* of λ as an eigenvalue of A.)

(iv)   Prove that, if ε is a positive real number, there are only a finite number of different eigenvalues μ of A such that $|\mu| > \varepsilon$. Deduce that the distinct non-zero eigenvalues of A *either* form a finite set *or* form a sequence converging to 0.

(v)   Let $\{\mu_1, \mu_2, \mu_3, \ldots\}$ be the (finite or infinite) sequence of all distinct non-zero eigenvalues of A, arranged so that $|\mu_1| \geq |\mu_2| \geq |\mu_3| \geq \cdots$, and suppose that μ_n has multiplicity $m(n)$. Let $\{\lambda_1, \lambda_2, \lambda_3, \ldots\}$ be the real sequence consisting of μ_1 ($m(1)$ times), followed by μ_2 ($m(2)$ times), followed by μ_3 ($m(3)$ times), and so on. Let $\{y_1, y_2, y_3, \ldots\}$ be a sequence of unit vectors consisting of an orthonormal basis of the null space of $A - \mu_1 I$, followed by an orthonormal basis of the null space of $A - \mu_2 I$, followed by an orthonormal basis of the null space of $A - \mu_3 I$, and so on. Show that $\{y_n\}$ is an orthonormal system, and $Ay_n = \lambda_n y_n$ for each n. Prove that, if A_0 is the compact self-adjoint operator defined by

$$A_0 x = \sum_n \lambda_n \langle x, y_n \rangle y_n \qquad (x \in \mathcal{H})$$

(see Exercise 2.8.26), then $A - A_0$ has no non-zero eigenvalue. Deduce that $A = A_0$.

(vi)   Show that $A \geq 0$ if and only if $\lambda_n \geq 0$ for all n. Deduce that, in this case, A has a (compact) "positive square root" A_1 (that is, $A_1^2 = A$ and $A_1 \geq 0$) such that $\|A_1\|^2 = \|A\|$.

Solution. (i)   From the final result of Exercise 2.8.11,

$$\|A\| = \sup\{|\langle Ax, x \rangle| : x \in \mathcal{H}, \ \|x\| = 1\}.$$

Thus we can choose unit vectors $x_1, x_2, x_3, \ldots$ such that $\{|\langle Ax_n, x_n \rangle|\}$ tends to $\|A\|$. Since $\langle Ax_n, x_n \rangle$ is real, we may assume (upon passing to a subsequence) that the real sequence $\{\langle Ax_n, x_n \rangle\}$ converges, with limit ρ equal to $\|A\|$ or $-\|A\|$. We have

$$
\begin{aligned}
\|Ax_n - \rho x_n\|^2 &= \langle Ax_n - \rho x_n, Ax_n - \rho x_n \rangle \\
&= \|Ax_n\|^2 - 2\rho\langle Ax_n, x_n \rangle + \rho^2 \|x_n\|^2 \\
&\leq \|A\|^2 - 2\rho\langle Ax_n, x_n \rangle + \rho^2 \\
&= 2\rho^2 - 2\rho\langle Ax_n, x_n \rangle \to 0 \qquad \text{as } n \to \infty.
\end{aligned}
$$

(ii)   If $A = 0$, then $\rho = 0$; and $Ax = \rho x$ for any unit vector x in $\mathcal{H}$. Now suppose that $A \neq 0$, whence $\rho \neq 0$. By passing to a subsequence of $\{x_n\}$, we may suppose that $\{Ax_n\}$ converges in norm to an element u of $\mathcal{H}$ (see Exercise 2.8.20). Thus $\rho x_n = Ax_n - (Ax_n - \rho x_n) \to u$, and $x_n \to \rho^{-1}u \; (= x)$. Thus x is a unit vector in $\mathcal{H}$, and

$$
Ax = \lim_{n \to \infty} Ax_n = u = \rho x.
$$

(iii)   If $\lambda \neq 0$, the null space $\mathcal{N}$ of $A - \lambda I$ is invariant under A, and $A|\mathcal{N}$ inherits from A the defining property (Exercise 2.8.20(iii)) of a compact operator. Hence $\lambda^{-1}A|\mathcal{N}$ (the identity operator on $\mathcal{N}$) is compact, and $\mathcal{N}$ is finite dimensional (see Exercise 2.8.21).

(iv)   Suppose there is an infinite sequence $\{\mu_1, \mu_2, \mu_3, \ldots\}$ of distinct eigenvalues of A, such that $|\mu_n| \geq \varepsilon$ for each n. We can choose unit vectors $x_1, x_2, x_3, \ldots$ such that $Ax_n = \mu_n x_n$; and these form an orthonormal sequence, by Exercise 2.8.28(ii). Since

$$
\begin{aligned}
\|Ax_m - Ax_n\|^2 &= \|\mu_m x_m - \mu_n x_n\|^2 \\
&= \mu_m^2 + \mu_n^2 \geq 2\varepsilon^2
\end{aligned}
$$

whenever $m \neq n$, the sequence $\{Ax_n\}$ has no convergent subsequence. This contradicts our assumption that A is compact. Hence there are only finitely many eigenvalues μ of A such that $|\mu| > \varepsilon$.

We can now arrange all the non-zero eigenvalues of A in a sequence (finite or infinite) by taking first all eigenvalues μ such that $|\mu| \geq 1$, then all eigenvalues μ such that $1 > |\mu| \geq \frac{1}{2}$, then all eigenvalues μ such that $\frac{1}{2} > |\mu| \geq \frac{1}{3}$, and so on. The sequence so obtained either terminates or converges to 0.

(v)   Since the multiplicity $m(n)$ of μ_n is finite, and $\{\mu_n\}$ terminates or converges to 0, it follows that $\{\lambda_n\}$ terminates or converges to 0. Each λ_n coincides with some μ_k, and y_n lies in an orthonormal basis of the null space of $A - \mu_k I$; so $Ay_n = \lambda_n y_n$. If $m \neq n$ then either y_m and y_n lie in an orthonormal basis of the null space of $A - \mu_k I$ (and $\langle y_m, y_n \rangle = 0$) for some k, or they are eigenvectors associated with different eigenvalues (and $\langle y_m, y_n \rangle = 0$, by Exercise 2.8.28(ii)). Hence $\{y_n\}$ is an orthonormal system. From Exercise 2.8.26, the equation

$$(1) \qquad A_0 x = \sum_n \lambda_n \langle x, y_n \rangle y_n \qquad (x \in \mathcal{H})$$

defines a compact self-adjoint operator A_0.

We prove that $A = A_0$ by showing that the compact self-adjoint operator $A - A_0$ has no non-zero eigenvalue, and is therefore 0 by (ii). For this, note first that $A_0 y_n = \lambda_n y_n = A y_n$ for each n, whence $(A - A_0) y_n = 0$.

If $A - A_0$ has a non-zero eigenvalue μ, and y is a unit eigenvector corresponding to μ, then $\langle y, y_n \rangle = 0$ for all n, by Exercise 2.8.28(ii) (applied to $A - A_0$). From (1), $A_0 y = 0$, so

$$Ay = (A - A_0)y = \mu y.$$

Since y is an eigenvector of A corresponding to a non-zero eigenvalue of A, it is a linear combination of certain y_n's (by choice of $\{y_n\}$ in (v)), and is therefore not orthogonal to all y_n's—a contradiction. Hence $A - A_0$ has no non-zero eigenvalue, and $A = A_0$.

(vi)   Since $A = A_0$, it follows from (1) that

$$\langle Ax, x \rangle = \sum_n \lambda_n \langle x, y_n \rangle \langle y_n, x \rangle = \sum_n \lambda_n \, |\langle x, y_n \rangle|^2$$

for each x in $\mathcal{H}$; in particular, $\langle Ay_n, y_n \rangle = \lambda_n$. From this, $A \geq 0$ if and only if $\lambda_n \geq 0$ for each n. When this is so, the decreasing non-negative real sequence $\{\lambda_n^{1/2}\}$ (is finite or) converges to 0. By Exercise 2.8.26, the equation

$$(2) \qquad A_1 x = \sum_n \lambda_n^{1/2} \langle x, y_n \rangle y_n$$

defines a compact self-adjoint operator A_1 on $\mathcal{H}$, with $\|A_1\| = \lambda_1^{1/2}$, and $A_1 \geq 0$ by the result just established. Moreover, $\|A_1\|^2 = \lambda_1 = \|A\|$, and $A_1^2 = A$ since $\langle A_1 x, y_n \rangle = \lambda_n^{1/2} \langle x, y_n \rangle$ for each x in $\mathcal{H}$, from (2), and $A_1^2 x = A_1(A_1 x) = \sum_n \lambda_n^{1/2} \langle A_1 x, y_n \rangle y_n = \sum_n \lambda_n \langle x, y_n \rangle y_n = Ax$.   ∎

2.8.30.   Let A be a self-adjoint operator acting on a Hilbert space $\mathcal{H}$, and let x be a unit vector in $\mathcal{H}$ such that $\|Ax\| = \|A\|$.

(i)   Show that x is an eigenvector for A^2 corresponding to the eigenvalue $\|A\|^2$.

(ii)   Show that either $Ax = \|A\| x$ or $Az = -\|A\| z$ for some unit vector z. [*Hint.* Consider the vector $\|A\| x - Ax$.]

(iii)   Under the added assumption that $A \geq 0$, show that x is an eigenvector for A corresponding to the eigenvalue $\|A\|$.

Solution. (i)   By hypothesis

$$\|A\|^2 = \|Ax\|^2 = \langle Ax, Ax \rangle = \langle A^2 x, x \rangle \leq \|A^2 x\| \, \|x\| \leq \|A\|^2.$$

Thus $\|A\|^2 = \|A^2 x\| = \langle A^2 x, x \rangle$ and

$$\langle A^2 x, x \rangle = \|A^2 x\| \, \|x\|.$$

From Proposition 2.1.3, $A^2 x = \lambda x$. Since

$$\lambda = \lambda \langle x, x \rangle = \langle A^2 x, x \rangle = \|A\|^2,$$

(i) follows.

(ii)   Let z_0 be $\|A\| x - Ax$. From (i), $Az_0 = -\|A\| z_0$. If $z_0 \neq 0$, then $\|z_0\|^{-1} z_0$ will serve as z.

(iii)   In this case, $-\|A\|$ is an eigenvalue for A only if $A = 0$.   ∎

2.8.31.   Let E and F be projections acting on a Hilbert space $\mathcal{H}$.

(i)   Show that, if E and F commute,

$$(*) \qquad\qquad E \vee F \leq E + F.$$

(ii)   By means of a two-dimensional example, show that $(*)$ need not hold when E and F do not commute.

Solution. (i)   From Proposition 2.5.3, $E \vee F = E + F - EF$ and $E \wedge F = EF$ when E and F commute. In particular EF is a projection, so that

$$0 \leq EF = E + F - E \vee F;$$

and ($*$) follows.

(ii)   In the 2×2 complex matrix algebra,

$$\begin{pmatrix} a & [a - a^2]^{1/2} \\ [a - a^2]^{1/2} & 1 - a \end{pmatrix}$$

is a projection E_a when $0 \le a \le 1$. Let E be E_1 and F be E_a, where $0 < a < 1$. Then $E \ne F$ so that $E \vee F$ is I. Thus $E + F - E \vee F$ is

$$\begin{pmatrix} a & [a - a^2]^{1/2} \\ [a - a^2]^{1/2} & -a \end{pmatrix}.$$

Since $-a < 0$, ($*$) does not hold (that is, $0 \nleq E + F - E \vee F$).   ∎

2.8.32.   Suppose that $\mathcal{H}$ is an infinite-dimensional Hilbert space. Let $\{y_1, y_2, y_3, \ldots\}$ be an orthonormal sequence in $\mathcal{H}$. By considering the sequence $\{V_n\}$ in $\mathcal{B}(\mathcal{H})$, where

$$V_n x = \langle x, y_n \rangle y_1 \qquad (x \in \mathcal{H}),$$

prove that the adjoint operation is not strong-operator continuous on (the unit ball of) $\mathcal{B}(\mathcal{H})$. Deduce that the strong-operator topology on $\mathcal{B}(\mathcal{H})$ is strictly coarser that the norm topology.

Solution.   For each x in $\mathcal{H}$, $\|V_n\| = 1$ and

$$\|V_n x\| = |\langle x, y_n \rangle| \to 0$$

as $n \to \infty$ (since $\sum_{n=1}^{\infty} |\langle x, y_n \rangle|^2 \le \|x\|^2 < \infty$). Thus the sequence $\{V_n\}$ is strong-operator convergent to 0. Also, $V_n^* x = \langle x, y_1 \rangle y_n$; and thus $\|V_n^* y_1\| = 1$, for all n. It follows that $\{V_n^*\}$ is not strong-operator convergent to 0; so the mapping $T \to T^*$ is not strong-operator continuous on (the unit ball of) $\mathcal{B}(\mathcal{H})$.

Since the adjoint operation is norm continuous, it follows that the norm topology and strong-operator topology do not coincide. We have already noted, in the discussion following Proposition 2.5.8, that the strong-operator topology is coarser than the norm topology; so it is strictly coarser.   ∎

2.8.33. Let $\mathcal{H}$ be an infinite-dimensional Hilbert space. Given any finite-dimensional subspace $F\ (\neq \{0\})$ of $\mathcal{H}$, let $A_F = n(I - P_F)$, where n is the dimension of F and P_F is the projection from $\mathcal{H}$ onto F. Choose any orthonormal system of $2n$ vectors, $\{x_1,\ldots,x_n, y_1,\ldots y_n\}$, the first n of which form an orthonormal basis for F, and define V_F, in $\mathcal{B}(\mathcal{H})$, by

$$V_F x = \frac{1}{n} \sum_{j=1}^{n} \langle x, x_j \rangle y_j.$$

In this way, we obtain nets $\{A_F\}$, $\{V_F\}$, $\{A_F V_F\}$, where the finite-dimensional subspaces F are directed by the inclusion relation. Show that

(i) $\{A_F\}$ is strong-operator convergent to 0;

(ii) $\{V_F\}$ is bounded, norm convergent to 0, and hence strong-operator convergent to 0;

(iii) $\{A_F V_F\}$ is not strong-operator convergent to 0.

Conclude that multiplication is not (jointly) strong-operator continuous from $\mathcal{B}(\mathcal{H}) \times (\mathcal{B}(\mathcal{H}))_1$ into $\mathcal{B}(\mathcal{H})$.

Solution. (i) Given x in $\mathcal{H}$, $A_F x = 0$ whenever $F \supseteq [x]$; so $\{\|A_F x\|\}$ converges to 0, and $\{A_F\}$ is strong-operator convergent to 0.

(ii) If k is a positive integer and F_0 is a k-dimensional subspace of $\mathcal{H}$, $\|V_F\| \le k^{-1}$ whenever $F \supseteq F_0$.

(iii) If x is a unit vector in $\mathcal{H}$, then $\|A_F V_F x\| = 1$ whenever $F \supseteq [x]$; so $\{A_F V_F\}$ is not strong-operator convergent to 0.

Since $(A_F, V_F) \in \mathcal{B}(\mathcal{H}) \times (\mathcal{B}(\mathcal{H}))_1$, and $(A_F, V_F) \to (0, 0)$ in the product strong-operator topology, but $\{A_F V_F\}$ is not strong-operator convergent to 0, it follows that multiplication is not (jointly) strong-operator continuous from $\mathcal{B}(\mathcal{H}) \times (\mathcal{B}(\mathcal{H}))_1$ into $\mathcal{B}(\mathcal{H})$. ∎

2.8.34. Show that, if $\mathcal{H}$ is a separable Hilbert space, there is a countable strong-operator dense subset of $\mathcal{B}(\mathcal{H})$.

Solution. If $\mathcal{H}$ is finite-dimensional, then $\mathcal{B}(\mathcal{H})$ (with the strong-operator topology) is a finite-dimensional, complex, locally convex space; it is therefore homeomorphic to $\mathbb{C}^n$ for some n, and so has a countable dense subset.

If $\mathcal{H}$ is separable and infinite-dimensional, it has an orthonormal basis $\{y_1, y_2, y_3, \ldots\}$. For each n, let P_n be the projection from $\mathcal{H}$ onto $[y_1, \ldots, y_n]$. Since T is the strong-operator limit of $\{P_n T P_n\}$ when $T \in \mathcal{B}(\mathcal{H})$, it follows that $\bigcup_{n=1}^{\infty} P_n \mathcal{B}(\mathcal{H}) P_n$ is strong-operator dense in $\mathcal{B}(\mathcal{H})$. However, $P_n \mathcal{B}(\mathcal{H}) P_n$ (with the strong-operator topology) is an n^2-dimensional locally convex space, and so has a countable dense subset $\mathcal{B}_n$. Thus $\bigcup_{n=1}^{\infty} \mathcal{B}_n$ is a countable strong-operator dense subset of $\mathcal{B}(\mathcal{H})$.   ■

2.8.35.   Show that, if $\mathcal{H}$ is a separable Hilbert space and $\{y_1, y_2, y_3, \ldots\}$ is an orthonormal basis of $\mathcal{H}$, the equation

$$d(S, T) = \sum_{n=1}^{\infty} 2^{-n} \|Sy_n - Ty_n\|$$

defines a translation-invariant metric d on $\mathcal{B}(\mathcal{H})$ (that is,

$$d(S + R, T + R) = d(S, T)$$

for each R in $\mathcal{B}(\mathcal{H})$), and the associated metric topology coincides on bounded subsets of $\mathcal{B}(\mathcal{H})$ with the strong-operator topology.

Solution.   Since $\|Sy_n - Ty_n\| \le \|S - T\|$ for all n, the series $\sum_{n=1}^{\infty} 2^{-n} \|Sy_n - Ty_n\|$ converges. It is apparent that $d(S, T) \ge 0$; if $d(S, T) = 0$, then $Sy_n = Ty_n$ for all n, and $S = T$. The triangle inequality for d follows from the triangle inequality for the norm on $\mathcal{H}$; moreover $d(S, T) = d(T, S) = d(R + S, R + T)$, for all R, S, T in $\mathcal{B}(\mathcal{H})$. Thus d is a translation-invariant metric on $\mathcal{B}(\mathcal{H})$.

Now let $\mathcal{S}$ be a bounded subset of $\mathcal{B}(\mathcal{H})$, and choose M (> 0) so that $\|S\| \le M$ whenever $S \in \mathcal{S}$. Given S_0 in $\mathcal{S}$ and ε (> 0), we can choose a positive integer k such that $k^{-1} + M 2^{1-k} < \varepsilon$. Then the metric ball $\{S \in \mathcal{S} : d(S, S_0) < \varepsilon\}$ contains the strong-operator neighborhood $\{S \in \mathcal{S} : \|Sy_j - S_0 y_j\| < k^{-1} \ (j = 1, \ldots, k)\} \ (= V)$ of S_0, since

$$d(S, S_0) \le \sum_{j=1}^{k} k^{-1} 2^{-j} + \sum_{j=k+1}^{\infty} 2M 2^{-j} < \frac{1}{k} + \frac{M}{2^{k-1}} < \varepsilon$$

when S is contained in V. Conversely, it is apparent that the metric ball $\{S \in \mathcal{S} : d(S, S_0) < 2^{-k}\varepsilon\}$ is contained in the strong-operator neighborhood $\{S \in \mathcal{S} : \|Sy_j - S_0 y_j\| < \varepsilon \ (j = 1, \ldots, k)\}$; and these form a base of strong-operator neighborhoods of S_0 in $\mathcal{S}$ (see the discussion preceding Remark 2.5.9).   ■

2.8.36.   Suppose that $\mathcal{H}_1, \ldots, \mathcal{H}_n, \mathcal{K}$ are Hilbert spaces and $L : \mathcal{H}_1 \times \cdots \times \mathcal{H}_n \to \mathcal{K}$ is a bounded multilinear mapping. Suppose also that for each u in $\mathcal{K}$, the bounded multilinear functional L_u, defined by

$$L_u(x_1, \ldots, x_n) = \langle L(x_1, \ldots, x_n), u \rangle,$$

is a Hilbert–Schmidt functional on $\mathcal{H}_1 \times \cdots \times \mathcal{H}_n$. Prove that the mapping $u \to L_u$ from the conjugate Hilbert space $\overline{\mathcal{K}}$ into the Hilbert space $\mathcal{HSF}$ of Proposition 2.6.2 is linear and has closed graph. Deduce that there is a positive real number d such that

$$\|L_u\|_2 \le d \, \|u\| \qquad (u \in \mathcal{K}),$$

where $\| \ \|_2$ denotes the usual norm on $\mathcal{HSF}$.

Solution.   For each u in $\mathcal{K}$, $L_u \in \mathcal{HSF}$. Since the inner product on $\mathcal{K}$ is conjugate-linear in its second variable, $u \to L_u$ is conjugate-linear as a mapping from $\mathcal{K}$ into $\mathcal{HSF}$, and is therefore linear as a mapping from $\overline{\mathcal{K}}$ into $\mathcal{HSF}$.

Suppose that $\{u(k)\}$ is a sequence that converges to 0 in $\overline{\mathcal{K}}$, while $\{L_{u(k)}\}$ converges (with respect to $\| \ \|_2$) to an element φ of $\mathcal{HSF}$. Given any unit vectors, y_1 in $\mathcal{H}_1$, $\ldots$, y_n in $\mathcal{H}_n$, we can choose orthonormal bases Y_1 of $\mathcal{H}_1$, $\ldots$, Y_n of $\mathcal{H}_n$, such that $y_r \in Y_r$ $(r = 1, \ldots, n)$. It then follows from the definition 2.6(5) of $\| \ \|_2$ that

$$|\varphi(y_1, \ldots, y_n) - L_{u(k)}(y_1, \ldots, y_n)| \le \|\varphi - L_{u(k)}\|_2 \to 0$$

as $k \to \infty$. Thus

$$\varphi(y_1, \ldots, y_n) = \lim_{k \to \infty} L_{u(k)}(y_1, \ldots, y_n)$$
$$= \lim_{k \to \infty} \langle L(y_1, \ldots, y_n), u(k) \rangle = 0,$$

since $\|u(k)\| \to 0$. It follows that $\varphi = 0$; so the linear mapping $u \to L_u : \overline{\mathcal{K}} \to \mathcal{HSF}$ has closed graph, and is therefore a bounded linear mapping.   ∎

2.8.37.   Suppose that $\mathcal{H}$ is a Hilbert space, and let $\mathcal{HSO}$ denote the Hilbert space of all Hilbert–Schmidt operators from $\mathcal{H}$ into $\mathcal{H}$ (see the discussion preceding Proposition 2.6.9).

(i)   Prove that $\|T\| \le \|T\|_2$ for all T in $\mathcal{HSO}$, where $\| \ \|$ and $\| \ \|_2$ denote the usual norms in $\mathcal{B}(\mathcal{H})$ and $\mathcal{HSO}$, respectively.

(ii)   By identifying $\mathcal{HSO}$ with $\overline{\mathcal{H}} \otimes \mathcal{H}$, prove that the ideal

$$\{A \in \mathcal{B}(\mathcal{H}) : A \text{ has finite-dimensional range}\}$$

in $\mathcal{B}(\mathcal{H})$ is a $\|\ \|_2$-dense subset of $\mathcal{HSO}$.

(iii)   Prove that the elements of $\mathcal{HSO}$ are compact linear operators.

Solution. (i)   Given a unit vector y_0 in $\mathcal{H}$, we can find an orthonormal basis Y of $\mathcal{H}$ that contains y_0, and

$$\|T\|_2 = \left[\sum_{y \in Y} \|Ty\|^2 \right]^{1/2} \geq \|Ty_0\|.$$

Since this holds for every unit vector y_0, $\|T\|_2 \geq \|T\|$.

(ii)   We appeal to Proposition 2.6.9. When $x, y \in \mathcal{H}$, the equation $T_{x,y}u = \langle u, x \rangle y$ ($u \in \mathcal{H}$) defines a Hilbert–Schmidt operator $T_{x,y}$ from $\mathcal{H}$ into $\mathcal{H}$; and when $\overline{\mathcal{H}} \otimes \mathcal{H}$ is identified with $\mathcal{HSO}$, $x \otimes y$ corresponds to $T_{x,y}$. Since finite sums of simple tensors form a dense subset of $\overline{\mathcal{H}} \otimes \mathcal{H}$, finite sums of operators $T_{x,y}$ form a dense subset of $\mathcal{HSO}$. In view of the result of Exercise 2.8.22(i), this proves (ii).

(iii)   By (ii), each T in $\mathcal{HSO}$ is the limit (with respect to $\|\ \|_2$) of a sequence $\{T_n\}$ in $\mathcal{HSO}$, where each T_n has finite-dimensional range (and is therefore compact). Since

$$\|T - T_n\| \leq \|T - T_n\|_2 \to 0,$$

T is compact (see Exercise 2.8.23).   ∎

2.8.38.   Suppose that $\mathcal{H}$ is the L_2 space associated with a σ-finite measure space $(S, \mathcal{S}, m)$. When $y, z \in \mathcal{H}$, define $q_{y,z}$ in $L_2(S \times S, \mathcal{S} \times \mathcal{S}, m \times m)$ by $q_{y,z}(s,t) = z(s)\overline{y(t)}$.

(i)   Show that, if Y and Z are orthonormal bases of $\mathcal{H}$, then the set $\{q_{y,z} : y \in Y, z \in Z\}$ is an orthonormal basis of $L_2(S \times S, \mathcal{S} \times \mathcal{S}, m \times m)$.

(ii)   Show that, if $k \in L_2(S \times S, \mathcal{S} \times \mathcal{S}, m \times m)$, the equation

$$(T_k x)(s) = \int_S k(s,t)x(t)\, dm(t) \qquad (x \in \mathcal{H})$$

(where the integral exists for almost all s in S) defines an element T_k of $\mathcal{B}(\mathcal{H})$. Prove also that T_k is a Hilbert–Schmidt operator acting on $\mathcal{H}$, and that $\|T_k\|_2 = \|k\|$.

(iii)   Prove that every Hilbert–Schmidt operator on $\mathcal{H}$ arises, as in (ii), from an element of $L_2(S \times S, \mathcal{S} \times \mathcal{S}, m \times m)$.

Solution. (i)   The sets $\overline{Y} = \{\bar{y} : y \in Y\}$ and Z are orthonormal bases of $\mathcal{H}$, so $\{z \otimes \bar{y} : y \in Y,\ z \in Z\}$ is an orthonormal basis of $\mathcal{H} \otimes \mathcal{H}$. The result of (i) now follows from Example 2.6.11, since (in the notation of that example) $q_{y,z} = p_{z,\bar{y}}$.

(ii)   If $k \in L_2(S \times S, \mathcal{S} \times \mathcal{S}, m \times m)$ and $x, y \in \mathcal{H}$, then

$$\langle k, q_{x,y} \rangle = \iint_{S \times S} k(s,t)x(t)\overline{y(s)}\, dm(s)\, dm(t)$$

$$= \int_S \left(\int_S k(s,t)x(t)\, dm(t) \right) \overline{y(s)}\, dm(s),$$

by Fubini's theorem. Since we can choose y in $\mathcal{H}$ such that $y(s) \neq 0$ for *all* s, it follows that the integral, used in the above definition of $(T_k x)(s)$, exists for almost all s in S. Moreover, $(T_k x)\bar{y} \in L_1$ whenever $x, y \in L_2$. By Exercise 1.9.30, $T_k x \in L_2$ when $x \in L_2$. Note also, from the above equations, that

$$\langle T_k x, y \rangle = \langle k, q_{x,y} \rangle.$$

It is now apparent that T_k is a linear operator from $\mathcal{H}$ into $\mathcal{H}$. Given any unit vector y_0 in $\mathcal{H}$, there is an orthonormal basis Y of $\mathcal{H}$ that contains y_0, and

$$\|T_k y_0\| = \left[\sum_{z \in Y} |\langle T_k y_0, z \rangle|^2 \right]^{1/2}$$

$$\leq \left[\sum_{y,z \in Y} |\langle T_k y, z \rangle|^2 \right]^{1/2}$$

$$= \left[\sum_{y,z \in Y} |\langle k, q_{y,z} \rangle|^2 \right]^{1/2} = \|k\|,$$

from (i). Since $\|T_k y_0\| \leq \|k\|$, for each unit vector y_0 in $\mathcal{H}$, T_k is bounded; and the above equations then show that T_k is a Hilbert–Schmidt operator, and $\|T_k\|_2 = \|k\|$.

(iii)   From (ii), $k \to T_k$ is a norm-preserving linear mapping U from $L_2(S \times S, \mathcal{S} \times \mathcal{S}, m \times m)$ into the Hilbert space $\mathcal{HSO}$ of all Hilbert–Schmidt operators acting on $\mathcal{H}$. The range of U is a closed subspace of $\mathcal{HSO}$. When $y, z \in \mathcal{H}$ and $k = q_{y,z}$, $T_k x = \langle x, y \rangle z$ ($x \in \mathcal{H}$). The range of U contains all finite sums of these operators; that is, it contains the set

$$\{A \in \mathcal{B}(\mathcal{H}) : A \text{ has finite-dimensional range}\}.$$

It now follows from Exercise 2.8.37(ii) that the range of U is the whole of $\mathcal{HSO}$.   ∎

2.8.39.   Suppose that $(S, \mathcal{S}, m)$ is a σ-finite measure space, $k \in L_2(S \times S, \mathcal{S} \times \mathcal{S}, m \times m)$, and T_k is the Hilbert–Schmidt operator defined in Exercise 2.8.38(ii). Show that $T_k^* = T_h$, where $h(s,t) = \overline{k(t,s)}$.

Solution.   When $x, y \in L_2(S, \mathcal{S}, m)$, it follows from the Fubini and Tonelli theorems that

$$
\begin{aligned}
\langle T_k x, y \rangle &= \int_S (T_k x)(s)\overline{y(s)}\, dm(s) \\
&= \int_S \left(\int_S k(s,t)x(t)\, dm(t) \right) \overline{y(s)}\, dm(s) \\
&= \int_S \left(\int_S \overline{h(t,s)y(s)}\, dm(s) \right) x(t)\, dm(t) \\
&= \int_S x(t)\overline{T_h y)(t)}\, dm(t) \\
&= \langle x, T_h y \rangle;
\end{aligned}
$$

so $T_h = T_k^*$.   ∎

2.8.40.   Suppose that $\mathcal{H}$ and $\mathcal{K}$ are Hilbert spaces, $A \in \mathcal{B}(\mathcal{H})$, and $B \in \mathcal{B}(\mathcal{K})$. Prove that $A \otimes I$ has an inverse in $\mathcal{B}(\mathcal{H} \otimes \mathcal{K})$ if and only if A has an inverse in $\mathcal{B}(\mathcal{H})$, and that $A \otimes B$ has an inverse in $\mathcal{B}(\mathcal{H} \otimes \mathcal{K})$ if and only if A has an inverse in $\mathcal{B}(\mathcal{H})$ and B has an inverse in $\mathcal{B}(\mathcal{K})$.

Solution.   If A has an inverse A^{-1} in $\mathcal{B}(\mathcal{H})$, then $A \otimes I$ has an inverse $A^{-1} \otimes I$ in $\mathcal{B}(\mathcal{H} \otimes \mathcal{K})$. If, also, B has an inverse B^{-1} in $\mathcal{B}(\mathcal{K})$, then $A \otimes B$ has an inverse $A^{-1} \otimes B^{-1}$ in $\mathcal{B}(\mathcal{H} \otimes \mathcal{K})$.

Now suppose that $A \otimes I$ has an inverse in $\mathcal{B}(\mathcal{H} \otimes \mathcal{K})$. Under the usual isomorphism between $\mathcal{H} \otimes \mathcal{K}$ and $\sum_{b \in \mathbb{B}} \oplus \mathcal{H}$ (where $\mathbb{B}$ has cardinality $\dim \mathcal{K}$), $A \otimes I$ corresponds to the operator $S = \sum_{b \in \mathbb{B}} \oplus A$, represented by the diagonal matrix with A at each diagonal position. Moreover, S has a bounded inverse T, that can be represented by a matrix $[T_{b,c}]_{b,c \in \mathbb{B}}$ of elements of $\mathcal{B}(\mathcal{H})$. By comparing the diagonal coefficients in the matrices of ST and TS and I $(= ST = TS)$, we deduce that $AT_{b,b} = T_{b,b}A = I$, for all b in $\mathbb{B}$. Thus A has an inverse $T_{b,b}$ in $\mathcal{B}(\mathcal{H})$.

Next, suppose that $A \otimes B$ has an inverse R in $\mathcal{B}(\mathcal{H} \otimes \mathcal{K})$. Then $A \otimes I$ has a left inverse $R(I \otimes B)$ and a right inverse $(I \otimes B)R$ in

$\mathcal{B}(\mathcal{H} \otimes \mathcal{K})$, and so has a (two-sided) inverse in $\mathcal{B}(\mathcal{H} \otimes \mathcal{K})$. From the preceding paragraph, it now follows that A has an inverse in $\mathcal{B}(\mathcal{H})$; and, similarly, B has an inverse in $\mathcal{B}(\mathcal{K})$. ■

2.8.41.   Let $\{y_1, y_2, y_3, \ldots\}$ be an orthonormal basis of a Hilbert space $\mathcal{H}$; and let $[a_{j,k}]$, $[b_{j,k}]$ be the matrices, with respect to this basis, of bounded linear operators A, B acting on $\mathcal{H}$. Prove that $[a_{j,k}b_{j,k}]$ is the matrix of a bounded linear operator. [*Hint.* Let P be the projection from $\mathcal{H} \otimes \mathcal{H}$ onto the subspace $\mathcal{K}$ spanned by the orthonormal system $\{y_1 \otimes y_1, y_2 \otimes y_2, \ldots\}$, and consider the operator T obtained by restricting $P(A \otimes B)P$ to $\mathcal{K}$.]

Solution.   The vectors $\{y_1 \otimes y_1, y_2 \otimes y_2, \ldots\}$ form an orthonormal basis of $\mathcal{K}$. The matrix $[t_{j,k}]$ of T, with respect to this basis, is given by

$$\begin{aligned}
t_{j,k} &= \langle T(y_k \otimes y_k)), y_j \otimes y_j \rangle \\
&= \langle P(A \otimes B)P(y_k \otimes y_k), y_j \otimes y_j \rangle \\
&= \langle (A \otimes B)(y_k \otimes y_k), y_j \otimes y_j \rangle \\
&= \langle Ay_k, y_j \rangle \langle By_k, y_j \rangle = a_{j,k}b_{j,k}.
\end{aligned}$$

Since $[a_{j,k}b_{j,k}]$ is the matrix of the bounded operator T, with respect to the orthonormal basis $\{y_1 \otimes y_1, y_2 \otimes y_2, \ldots\}$, it is the matrix of U^*TU with respect to $\{y_1, y_2, \ldots\}$, where $U : \mathcal{H} \to \mathcal{K}$ is the unitary transformation satisfying $Uy_j = y_j \otimes y_j$ $(j = 1, 2, \ldots)$. ■

2.8.42.   Suppose that A is a closed linear operator with domain dense in a Hilbert space $\mathcal{H}$ and with range in $\mathcal{H}$; and let B be in $\mathcal{B}(\mathcal{H})$. Prove that AB is closed. Show also that, if AB is densely defined and bounded, then $AB \in \mathcal{B}(\mathcal{H})$.

Solution.   Suppose that $\{x_n\}$ is a sequence of vectors in $\mathcal{D}(AB)$, and that $x_n \to x$, $ABx_n \to y$, as $n \to \infty$. Since B is continuous, the sequence $\{Bx_n\}$ in $\mathcal{D}(A)$ converges to Bx, while $A(Bx_n) \to y$. From this, and since A is closed, $Bx \in \mathcal{D}(A)$ and $ABx = y$. Thus $x \in \mathcal{D}(AB)$ and $ABx = y$; so AB is closed.

The last assertion now follows from the fact that, if S is closed, densely defined and bounded, then $\mathcal{D}(S) = \mathcal{H}$. Indeed, each x in $\mathcal{H}$ is the limit of a sequence $\{x_n\}$ in $\mathcal{D}(S)$; and

$$\|Sx_m - Sx_n\| \le \|S\| \, \|x_m - x_n\| \to 0$$

as $\min\{m, n\} \to \infty$, where $\|S\|$ denotes the bound of S as an operator from $\mathcal{D}(S)$ into $\mathcal{H}$. Thus $\{Sx_n\}$ is a Cauchy sequence, and so converges to an element y of $\mathcal{H}$. Since $x_n \to x$, $Sx_n \to y$ and S is closed, it follows that $x \in \mathcal{D}(S)$; so $\mathcal{D}(S) = \mathcal{H}$.   ∎

2.8.43.   Let $\{y_1, y_2, y_3, \ldots\}$ be an orthonormal basis for a Hilbert space $\mathcal{H}$, and let

$$\mathcal{D} = \left\{ x \in \mathcal{H} : \sum_{n=1}^{\infty} n^4 \left| \langle x, y_n \rangle \right|^2 < \infty \right\}, \qquad z = \sum_{n=2}^{\infty} n^{-1} y_n.$$

Define B in $\mathcal{B}(\mathcal{H})$ by $Bx = \langle x, z \rangle z$; and define mappings S and T with domain $\mathcal{D}$ by

$$Sx = \sum_{n=2}^{\infty} n^2 \langle x, y_n \rangle y_n, \qquad Tx = Sx + \langle Sx, z \rangle y_1 \qquad (x \in \mathcal{D}).$$

Show that S and T are closed densely defined operators, but that neither $T - S$ not BS is preclosed.

Solution.   From the discussion of Example 2.7.1, $\mathcal{D}$ is a dense subspace of $\mathcal{H}$, and S is a closed densely defined operator. Thus T, also, is densely defined.

Suppose that x_n is in $\mathcal{D}$, and $x_n \to x$, $Tx_n \to y$, as $n \to \infty$. If P denotes the projection from $\mathcal{H}$ onto the one-dimensional subspace $[y_1]$, then $Sx_n = (I - P)Tx_n \to (I - P)y$. Since $\{x_n\} \subseteq \mathcal{D}$, $x_n \to x$, $\{Sx_n\}$ converges, and S is closed; it follows that $x \in \mathcal{D}$ and that $Sx_n \to Sx$. Thus Tx is defined, and

$$Tx = Sx + \langle Sx, z \rangle y_1 = \lim_{n \to \infty} \{Sx_n + \langle Sx_n, z \rangle y_1\}$$
$$= \lim_{n \to \infty} Tx_n = y;$$

so T is closed.

If $u_n = n^{-1} y_n$ $(n \geq 2)$, then $u_n \to 0$ but $(T - S)u_n = y_1$ $(\neq 0)$ and $BSu_n = z$ $(\neq 0)$; so neither $T - S$ nor BS is preclosed.   ∎

2.8.44.   Suppose that A and B are linear operators with their domains dense in a Hilbert space $\mathcal{H}$ and their ranges in $\mathcal{H}$. Prove that $A^* + B^* \subseteq (A + B)^*$ if $A + B$ is densely defined, and that $B^* A^* \subseteq (AB)^*$ if AB is densely defined.

Solution. Suppose that $A + B$ is densely defined, and that $y \in \mathcal{D}(A^* + B^*)$ $(= \mathcal{D}(A^*) \cap \mathcal{D}(B^*))$. When $x \in \mathcal{D}(A + B)$ $(= \mathcal{D}(A) \cap \mathcal{D}(B))$, we have that

$$
\begin{aligned}
\langle (A + B)x, y \rangle &= \langle Ax, y \rangle + \langle Bx, y \rangle \\
&= \langle x, A^*y \rangle + \langle x, B^*y \rangle \\
&= \langle x, (A^* + B^*)y \rangle.
\end{aligned}
$$

Thus $y \in \mathcal{D}((A + B)^*)$, and $(A + B)^*y = (A^* + B^*)y$; so $A^* + B^* \subseteq (A + B)^*$.

Next, suppose that AB is densely defined. If $y \in \mathcal{D}(B^*A^*)$, then $y \in \mathcal{D}(A^*)$ and $A^*y \in \mathcal{D}(B^*)$. Given x in $\mathcal{D}(AB)$, we have $x \in \mathcal{D}(B)$ and $Bx \in \mathcal{D}(A)$; so

$$
\langle ABx, y \rangle = \langle A(Bx), y \rangle = \langle Bx, A^*y \rangle = \langle x, B^*A^*y \rangle.
$$

Thus $y \in \mathcal{D}((AB)^*)$ and $(AB)^*y = B^*A^*y$; so $B^*A^* \subseteq (AB)^*$. ∎

2.8.45. Suppose that T is a closed linear operator with domain dense in a Hilbert space $\mathcal{H}$ and with range in $\mathcal{H}$. Show that the null space $\{x \in \mathcal{D}(T) : Tx = 0\}$ of T is a closed subspace of $\mathcal{H}$.

Let $N(T)$ and $R(T)$ denote the projections whose ranges are, respectively, the null space of T and the closure of the range of T. Prove that

$$
\begin{aligned}
R(T) = I - N(T^*), \qquad N(T) = I - R(T^*), \\
R(T^*T) = R(T^*), \qquad N(T^*T) = N(T).
\end{aligned}
$$

[For the case in which $T \in \mathcal{B}(\mathcal{H})$, these relations have been established in Proposition 2.5.13.]

Solution. If $x = \lim x_n$, where each x_n lies in the null space of T, then $x_n \to x$, $Tx_n = 0 \to 0$. Since T is closed, $x \in \mathcal{D}(T)$ and $Tx = 0$. Thus x lies in the null space of T (which is, therefore, closed).

A vector y lies in the range of $I - R(T)$ if and only if $\langle Tx, y \rangle = 0$ for all x in $\mathcal{D}(T)$; equivalently, if and only if $y \in \mathcal{D}(T^*)$ and $T^*y = 0$. Thus $I - R(T) = N(T^*)$. Since T^* is densely defined and closed, and $T^{**} = T$, we obtain $I - R(T^*) = N(T)$ upon replacing T by T^*.

If $x \in \mathcal{D}(T^*T)$ and $T^*Tx = 0$, then $x \in \mathcal{D}(T)$ and $Tx \in \mathcal{D}(T^*)$; so

$$\langle Tx, Tx \rangle = \langle x, T^*Tx \rangle = 0.$$

Thus $x \in \mathcal{D}(T)$ and $Tx = 0$. It follows that the null space of T^*T is contained in that of T, and the reverse inclusion is evident; so $N(T^*T) = N(T)$. From our earlier results, and since T^*T is self-adjoint

$$R(T^*T) = I - N(T^*T) = I - N(T) = R(T^*). \qquad \blacksquare$$

2.8.46.   Let T be a closed operator with domain dense in a separable Hilbert space $\mathcal{H}$ and with range in $\mathcal{H}$. Show that the ranges of $R(T)$ and $R(T^*)$ have the same dimension.

Solution.   If both the ranges of $R(T)$ and $R(T^*)$ are infinite dimensional, then each has dimension $\aleph_0$. Suppose that one or both of these ranges have finite dimension; say, $R(T)(\mathcal{H})$ has finite dimension. From Exercise 2.8.45, $N(T)x \in \mathcal{D}(T)$ for each x in $\mathcal{H}$. Thus, if $z \in \mathcal{D}(T)$, then $z - N(T)z \in \mathcal{D}(T)$. Let $\mathcal{D}_0$ be the orthogonal complement of $N(T)(\mathcal{H})$ in $\mathcal{D}(T)$. From the preceding, $\mathcal{D}(T) = \mathcal{D}_0 + N(T)(\mathcal{H})$; and T is a linear isomorphism of $\mathcal{D}_0$ onto $R(T)(\mathcal{H})$. It follows that $\mathcal{D}_0$ has the same (finite) dimension as $R(T)(\mathcal{H})$ and T is bounded on $\mathcal{D}(T)$. Since T is closed and densely defined, we have that $T \in \mathcal{B}(\mathcal{H})$; in particular, $\mathcal{D}_0 = (I - N(T))(H)$ (and $\mathcal{D}(T) = \mathcal{H}$). From Exercise 2.8.45 (or Proposition 2.5.13), $I - N(T) = R(T^*)$. Thus the ranges of $R(T)$ and $R(T^*)$ have the same dimension. $\blacksquare$

2.8.47.   Let $\mathcal{H}$ be a Hilbert space and E be a projection with finite-dimensional range. Let F be a projection such that the dimension of $E(\mathcal{H})$ is less than the dimension of $F(\mathcal{H})$. Show that $(I - E) \wedge F \neq 0$.

Solution.   By assumption, we can choose m linearly independent vectors $\{x_1, \ldots, x_m\}$ in $F(\mathcal{H})$, where m exceeds the dimension of $E(\mathcal{H})$. Then $\{Ex_1, \ldots, Ex_m\}$ is linearly dependent, and there are complex numbers $a_1, \ldots, a_m$, not all 0, such that $\sum_{j=1}^{m} a_j Ex_j = 0$. If $z = \sum_{j=1}^{m} a_j x_j$, then $Ez = 0$, $z \neq 0$, and $z \in F(\mathcal{H})$. Thus $(I - E) \wedge F \neq 0$. $\blacksquare$

2.8.48.   Show that, if T is a linear operator with domain dense in a Hilbert space $\mathcal{H}$ and with range in $\mathcal{H}$, and if $\langle Tz, z \rangle$ is real for each z in $\mathcal{D}(T)$, then T is symmetric.

Solution.   When $z \in \mathcal{D}(T)$,

$$\langle Tz, z \rangle = \overline{\langle Tz, z \rangle} = \langle z, Tz \rangle;$$

and, by polarization,

$$\langle Tx, y \rangle = \langle x, Ty \rangle \qquad (x, y \in \mathcal{D}(T)).$$

Thus, T is symmetric.   ■

2.8.49.   Let $\mathcal{H}$ be the Hilbert space L_2, corresponding to Lebesgue measure on the unit interval $[0, 1]$, and let $\mathcal{D}_0$ be the subspace consisting of all complex-valued functions f that have a continuous derivative f' on $[0, 1]$ and satisfy $f(0) = f(1) = 0$. Let D_0 be the operator with domain $\mathcal{D}_0$ and with range in $\mathcal{H}$ defined by $D_0 f = f'$. Show that iD_0 is a densely defined symmetric operator and that

$$(iD_0)M - M(iD_0) = iI|\mathcal{D}_0,$$

where M is the bounded linear operator defined by

$$(Mf)(s) = sf(s) \qquad (f \in L_2; \ 0 \le s \le 1).$$

Solution.   Each element f of $\mathcal{H}$ can be approximated (in L_2 norm) by a continuous function f_1. In turn, f_1 can be approximated (in the uniform norm, hence in the L_2 norm) by a polynomial f_2. Finally, f_2 can be approximated (in L_2 norm) by an element f_3 of $\mathcal{D}_0$; indeed, it suffices to take $f_3 = gf_2$, where $g : [0, 1] \to [0, 1]$ is continuously differentiable, vanishes at the endpoints 0 and 1, and takes the value 1 except at points very close to $0, 1$.

The preceding argument shows that $\mathcal{D}_0$ is dense in $\mathcal{H}$, so D_0 is a densely defined linear operator. When $f, g \in \mathcal{D}_0$, the function $\bar{g}$ has a continuous derivative $\bar{g}'$, and we have

$$\langle D_0 f, g \rangle = \int_0^1 f'(s)\overline{g(s)}\, ds = \left[f(s)\overline{g(s)} \right]_0^1 - \int_0^1 f(s)\overline{g'(s)}\, ds$$

$$= -\int_0^1 f(s)\overline{g'(s)}\, ds = -\langle f, D_0 g \rangle.$$

Thus $\langle iD_0 f, g \rangle = \langle f, iD_0 g \rangle$, for all f and g in $\mathcal{D}_0$; and iD_0 is symmetric.

When $f \in \mathcal{D}_0$, $Mf \in \mathcal{D}_0$ and

$$(D_0 M f)(s) = \frac{d}{ds}\{s f(s)\} = f(s) + s f'(s) = f(s) + (M D_0 f)(s).$$

Thus $D_0 M f - M D_0 f = f \ (f \in \mathcal{D}_0)$.   ∎

2.8.50.   Let $\mathcal{H}$, $\mathcal{D}_0$, and D_0 be defined as in Exercise 2.8.49, and let $\mathcal{H}_1 = \{f_1 \in \mathcal{H} : \langle f_1, u \rangle = 0\}$, where u is the unit vector in $\mathcal{H}$ defined by $u(s) = 1 \ (0 \le s \le 1)$. When $f \in \mathcal{H}$, define Kf in $\mathcal{H}$ by

$$(Kf)(s) = \int_0^s f(t)\, dt \qquad (0 \le s \le 1).$$

(i)   Prove that $K \in \mathcal{B}(\mathcal{H})$, that K has null space $\{0\}$, and that $K(\mathcal{H}_1) \supseteq \mathcal{D}_0$.

(ii)   Show that the equation

$$D_1 K f_1 = f_1 \qquad (f_1 \in \mathcal{H}_1)$$

defines a closed linear operator D_1 with domain $\mathcal{D}_1 = K(\mathcal{H}_1)$. Prove also that D_1 is the closure of D_0.

(iii)   Show that the equation

$$D_2(Kf + au) = f \qquad (f \in \mathcal{H}, \ a \in \mathbb{C})$$

defines a closed linear operator D_2, with domain $\mathcal{D}_2 = \{Kf + au : f \in \mathcal{H}, \ a \in \mathbb{C}\}$, that extends D_1.

(iv)   Prove that $\langle K f_1, f \rangle + \langle f_1, Kf + au \rangle = 0$ for all f_1 in $\mathcal{H}_1$, f in $\mathcal{H}$, and a in $\mathbb{C}$.

(v)   Show that $D_0^* = D_1^* = -D_2$.

(vi)   Let $\mathcal{D}_3 = \{K f_1 + au : f_1 \in \mathcal{H}_1, \ a \in \mathbb{C}\}$, and let D_3 be the restriction $D_2 | \mathcal{D}_3$. Show that D_3 is a closed densely defined linear operator, that $D_1 \subseteq D_3 = -D_3^* \subseteq D_2$, and that iD_3 is self-adjoint.

Solution. (i)   From Exercise 2.8.38(ii), with

$$k(s,t) = \begin{cases} 1 & (s \ge t), \\ 0 & (s < t), \end{cases}$$

it follows that $K \in \mathcal{B}(\mathcal{H})$. If $f \in \mathcal{H}$ and $Kf = 0$, then $\int_0^s f(t)\,dt = 0$ $(0 \le s \le 1)$, and f is a null function; so K has null space $\{0\}$. If $g \in \mathcal{D}_0$, then $g(0) = g(1) = 0$ and g has a continuous derivative g' on $[0,1]$. Since $g' \in L_2\ (= \mathcal{H})$ and

$$0 = g(1) - g(0) = \int_0^1 g'(s)\,ds = \langle g', u \rangle,$$

it follows that $g' \in \mathcal{H}_1$. Moreover

$$(Kg')(s) = \int_0^s g'(t)\,dt = g(s) - g(0) = g(s) \qquad (0 \le s \le 1),$$

so $g = Kg' \in K(\mathcal{H}_1)$. Thus $\mathcal{D}_0 \subseteq K(\mathcal{H}_1)$.

(ii)   From (i), $K(\mathcal{H}_1)$ is dense in $\mathcal{H}$. Since K is one-to-one, the equation $D_1 K f_1 = f_1$ $(f_1 \in \mathcal{H}_1)$ defines a linear operator D_1 with dense domain $\mathcal{D}_1\ (= K(\mathcal{H}_1))$.

If $\{g_n\}$ is a sequence in $\mathcal{D}_1$, such that $g_n \to g$ and $D_1 g_n \to f$, then $g_n = K f_n$ and $D_1 g_n = f_n$ for some sequence $\{f_n\}$ in $\mathcal{H}_1$. Since $f_n \to f$, $\mathcal{H}_1$ is closed, and K is bounded, we have $f \in \mathcal{H}_1$ and

$$Kf = \lim K f_n = \lim g_n = g.$$

Thus $g \in K(\mathcal{H}_1) = \mathcal{D}_1$, and $D_1 g = f$; so D_1 is closed.

If $g \in \mathcal{D}_0$, then (as noted above) $g = Kg'$ and $g' \in \mathcal{H}_1$. Thus $g \in \mathcal{D}_1$, and $D_1 g = g' = D_0 g$; so $D_0 \subseteq D_1$. Since D_1 is closed, $\overline{D}_0 \subseteq D_1$.

We now prove that $\overline{D}_0 \supseteq D_1$ (whence $\overline{D}_0 = D_1$). To this end, suppose that $g \in \mathcal{D}_1$ and $D_1 g = f$. Then $f \in \mathcal{H}_1$, and $Kf = g$. There is a sequence $\{h_n\}$ of continuous functions on the unit interval such that $\|f - h_n\| \to 0$; and $\langle h_n, u \rangle \to \langle f, u \rangle = 0$. With f_n defined as $h_n - \langle h_n, u \rangle u$, f_n is continuous, $\langle f_n, u \rangle = 0$, and $\|f - f_n\| \to 0$. Let $g_n = K f_n$, so that $g_n \to Kf = g$. Since

$$g_n(s) = \int_0^s f_n(t)\,dt, \qquad \int_0^1 f_n(t)\,dt = \langle f_n, u \rangle = 0,$$

it follows that g_n has a continuous derivative f_n, and satisfies $g_n(0) = g_n(1) = 0$. Thus g_n lies in the domain $\mathcal{D}_0$ of D_0, $g_n \to g$, and $D_0 g_n = f_n \to f = D_1 g$. This shows that each point $(g, D_1 g)$ in the graph $\mathcal{G}(D_1)$ of D_1 is the limit of a sequence $\{(g_n, D_0 g_n)\}$ in $\mathcal{G}(D_0)$; so $D_1 \subseteq \overline{D_0}$.

(iii)   If $f \in \mathcal{H}$, $a \in \mathbb{C}$ and $Kf + au = 0$, then

$$a + \int_0^s f(t)\,dt = au(s) + (Kf)(s) = 0$$

for *almost all* s in $[0,1]$ and hence, by continuity, for *all* s in $[0,1]$. With $s = 0$, we obtain $a = 0$; so $\int_0^s f(t)\,dt = 0$ $(0 \le s \le 1)$, and f is a null function.

In view of the preceding paragraph, the equation

$$D_2(Kf + au) = f \qquad (f \in \mathcal{H}, \ a \in \mathbb{C})$$

defines a linear operator D_2 with domain $\mathcal{D}_2$ as described in (iii). Moreover, $\mathcal{D}_2 \supseteq \mathcal{D}_1$ and $D_2 \supseteq D_1$; in particular, D_2 is densely defined.

Suppose that $g_n \in \mathcal{D}_2$, $g_n \to g$, and $D_2 g_n \to f$. Then g_n has the form $Kf_n + a_n u$, where $f_n \in \mathcal{H}$ and $a_n \in \mathbb{C}$; and $D_2 g_n = f_n$. Thus

$$f_n \to f, \qquad Kf_n \to Kf, \qquad a_n u = g_n - Kf_n \to g - Kf,$$

and therefore $g - Kf = au$ for some scalar a. Thus

$$g = Kf + au \in \mathcal{D}_2, \qquad D_2 g = f;$$

and D_2 is closed.

(iv)   If $f_1 \in \mathcal{H}_1$, $f \in \mathcal{H}$ and $a \in \mathbb{C}$, we have $\langle f_1, u \rangle = 0$, and

$$\langle Kf_1, f \rangle + \langle f_1, Kf + au \rangle = \langle Kf_1, f \rangle + \langle f_1, Kf \rangle$$

$$= \int_0^1 (Kf_1)(s)\overline{f(s)}\,ds + \int_0^1 f_1(t)\overline{(Kf)(t)}\,dt$$

$$= \int_0^1 \overline{f(s)}\left(\int_0^s f_1(t)\,dt\right)ds + \int_0^1 f_1(t)\left(\int_0^t \overline{f(s)}\,ds\right)dt$$

$$= \int_0^1 f_1(t)\left(\int_t^1 \overline{f(s)}\,ds\right)dt + \int_0^1 f_1(t)\left(\int_0^t \overline{f(s)}\,ds\right)dt$$

$$= \int_0^1 f_1(t)\left(\int_0^1 \overline{f(s)}\,ds\right)dt = \langle f_1, u \rangle \langle u, f \rangle = 0.$$

(v) The equation established in (iv) may be rewritten in the form

$$\langle g_1, D_2 g_2 \rangle = -\langle D_1 g_1, g_2 \rangle \qquad (g_1 \in \mathcal{D}_1, \ g_2 \in \mathcal{D}_2);$$

so $D_1^* \supseteq -D_2$. Moreover, $D_0^* = \overline{D_0}^* = D_1^*$. It remains to prove that $\mathcal{D}(D_1^*) \subseteq \mathcal{D}_2$.

Suppose that $g_2 \in \mathcal{D}(D_1^*)$, and let $D_1^* g_2$ be f. For each g_1 in $\mathcal{D}_1$, $\langle D_1 g_1, g_2 \rangle = \langle g_1, f \rangle$; equivalently,

$$\langle f_1, g_2 \rangle = \langle K f_1, f \rangle \qquad (f_1 \in \mathcal{H}_1).$$

From the equation established in (iv), with $a = 0$, it now follows that $\langle f_1, g_2 \rangle = -\langle f_1, K f \rangle$ for each f_1 in $\mathcal{H}_1$; so $g_2 + K f \in \mathcal{H}_1^\perp = [u]$, and $g_2 = -K f + a u$ for some scalar a. Thus $g_2 \in \mathcal{D}_2$, and $\mathcal{D}(D_1^*) \subseteq \mathcal{D}_2$.

(vi) Since $\mathcal{D}_1 \subseteq \mathcal{D}_3 \subseteq \mathcal{D}_2$ and $\mathcal{D}_1 \subseteq \mathcal{D}_2$, it is evident that D_3 $(= D_2 | \mathcal{D}_3)$ is densely defined and $D_1 \subseteq D_3 \subseteq D_2$. If we prove that $D_3 = -D_3^*$, it follows that D_3 is closed and that $i D_3$ is self-adjoint.

Suppose that $g_1, g_2 \in \mathcal{D}_3$, and let $g_j = K f_j + a_j u$, where $f_1, f_2 \in \mathcal{H}_1$ and $a_1, a_2 \in \mathbb{C}$. Since $\langle f_j, u \rangle = 0$, it follows from the equation established in (iv) that

$$\langle D_3 g_1, g_2 \rangle + \langle g_1, D_3 g_2 \rangle = \langle f_1, K f_2 + a_2 u \rangle + \langle K f_1 + a_1 u, f_2 \rangle$$
$$= \langle f_1, K f_2 \rangle + \langle K f_1, f_2 \rangle = 0.$$

Thus $g_2 \in \mathcal{D}(D_3^*)$, and $D_3^* g_2 = -D_3 g_2$; so $D_3^* \supseteq -D_3$.

It remains to prove that $\mathcal{D}(D_3^*) \subseteq \mathcal{D}_3$. To this end, suppose that $g \in \mathcal{D}(D_3^*)$, and $D_3^* g = h$. Given any f_1 in $\mathcal{H}_1$, and any scalar a, $K f_1 + a u \in \mathcal{D}_3$ and $D_3(K f_1 + a u) = f_1$. Thus

$$\langle f_1, g \rangle = \langle D_3(K f_1 + a u), g \rangle = \langle K f_1 + a u, h \rangle.$$

By varying a, it follows that $\langle h, u \rangle = 0$; so $h \in \mathcal{H}_1$, and $\langle f_1, g \rangle = \langle K f_1, h \rangle$. From the equation established in (iv), we now have

$$\langle f_1, g \rangle = -\langle f_1, K h \rangle \qquad (f_1 \in \mathcal{H}_1).$$

Thus $g + K h \in \mathcal{H}_1^\perp = [u]$, and $g = -K h + a u$ for some scalar a. Thus $g \in \mathcal{D}_3$, and $\mathcal{D}(D_3^*) \subseteq \mathcal{D}_3$.   ■

CHAPTER 3

BANACH ALGEBRAS

3.5. Exercises

3.5.1. In Remark 3.1.3 the algebra $\mathfrak{A}_1$ obtained from a normed algebra $\mathfrak{A}$ is defined. Complete the indicated check that $\mathfrak{A}_1$ is a normed algebra, and that $\mathfrak{A}_1$ is a Banach algebra when $\mathfrak{A}$ is.

Solution. Algebraically, $\mathfrak{A}_1$ represents the classic adjunction of a unit to an algebra. We check that: $\|(aI, A)\| = |a| + \|A\|$, defines a norm on $\mathfrak{A}_1$ relative to which it is a normed algebra. Clearly $\|(aI, A)\| = 0$ if and only if $a = 0$ and $A = 0$; thas is, if and only if $(aI, A) = 0$. We have

$$\|c(aI, A)\| = \|(caI, cA)\| = |ca| + \|cA\|$$
$$= |c|\,(|a| + \|A\|) = |c|\,\|(aI, A)\|$$

and

$$\|(aI, A) + (bI, B)\| = \|((a+b)I, A + B)\|$$
$$= |a + b| + \|A + B\|$$
$$\leq |a| + |b| + \|A\| + \|B\|$$
$$= \|(aI, A)\| + \|(bI, B)\|.$$

Thus, $\|(aI, A)\| = |a| + \|A\|$, defines a norm on $\mathfrak{A}_1$.
Since

$$\|(aI, A)(bI, B)\| = \|(abI, aB + bA + AB)\|$$
$$= |ab| + \|aB + bA + AB\|$$
$$\leq |a|\,|b| + |a|\,\|B\| + |b|\,\|A\| + \|A\|\,\|B\|$$
$$= (|a| + \|A\|)(|b| + \|B\|)$$
$$= \|(aI, A)\|\,\|(bI, B)\|,$$

and $\|(I,0)\| = 1$; $\mathfrak{A}_1$ with the norm described is a normed algebra.

Suppose $\mathfrak{A}$ is a Banach algebra and $\{(a_n I, A_n)\}$ is a Cauchy sequence in $\mathfrak{A}_1$. Then

$$\|(a_n I, A_n) - (a_m I, A_m)\| = |a_n - a_m| + \|A_n - A_m\| \to 0 \quad (n, m \to \infty).$$

Since the scalar field is complete, $\{a_n\}$ converges to some a. Since $\mathfrak{A}$ is complete, $\{A_n\}$ converges to some A in $\mathfrak{A}$. Now,

$$\|(a_n I, A_n) - (aI, A)\| = |a_n - a| + \|A_n - A\| \to 0 \quad (n \to \infty);$$

so that $\{(a_n I, A_n)\}$ converges to (aI, A), and $\mathfrak{A}_1$ is a Banach algebra with unit $(I, 0)$. $\blacksquare$

3.5.2. Let $\mathfrak{A}$ be a Banach algebra over $\mathbb{R}$, and let $\mathfrak{A}_{\mathbb{C}}$ be the Banach space "complexification" of $\mathfrak{A}$ described in Exercise 1.9.6. Define the product $(A, B)(C, D)$ to be $(AC - BD, AD + BC)$. Show that:

(i) $\|(A, B)(C, D)\| \le 2\,\|(A, B)\|\,\|(C, D)\|$;

(ii) $\mathfrak{A}_{\mathbb{C}}$ has another norm $\|\|\ \|\|$ relative to which it is a Banach algebra, the identity mapping from $(\mathfrak{A}_{\mathbb{C}}, \|\ \|)$ onto $(\mathfrak{A}_{\mathbb{C}}, \|\|\ \|\|)$ is a homeomorphism, and the mapping $A \to (A, 0)$ is an isometric (real) algebraic isomorphism.

Solution. From Exercise 1.9.6, $\mathfrak{A}_{\mathbb{C}}$, with the norm described there, is a complex Banach space. If we establish (i), the multiplication on $\mathfrak{A}_{\mathbb{C}}$ is jointly continuous, by the inequality analogous to that following 3(1). The argument preceding 3(1) establishes that $\mathfrak{A}_{\mathbb{C}}$ admits a norm "equivalent" to its initial norm relative to which it becomes a Banach algebra. Since the mapping $A \to (A, 0)$ is a real-linear isometric isomorphism of $\mathfrak{A}$ into $\mathfrak{A}_{\mathbb{C}}$ relative to the initial norm (from Exercise 1.9.6), it is a homeomorphism relative to the new norm. Moreover, $(A, 0)(B, 0) = (AB, 0)$; so that this mapping is a (real) algebraic isomorphism as well. It remains to establish (i) and that the mapping $A \to (A, 0)$ is an isometry relative to the new norm.

Note that

$$
\begin{aligned}
\|(A, 0)(B, C)\| &= \|(AB, AC)\| \\
&= \sup\{\|(\cos\theta)AB + (\sin\theta)AC\| : 0 \le \theta \le 2\pi\} \\
&\le \|A\| \sup\{(\cos\theta)B + (\sin\theta)C\| : 0 \le \theta \le 2\pi\} \\
&= \|(A, 0)\|\,\|(B, C)\|.
\end{aligned}
$$

$(*)$

Now

$$\tfrac{1}{2}(\|A\| + \|B\|) \le \max\{\|A\|, \|B\|\} \le \|(A, B)\|;$$

so that

$$
\begin{aligned}
\|(A, B)(C, D)\| &= \|(AC - BD, AD + BC)\| \\
&= \|(A, 0)(C, D) + (B, 0)(-D, C)\| \\
&\le \|A\|\,\|(C, D)\| + \|B\|\,\|(-D, C)\| \\
&= (\|A\| + \|B\|)\,\|(C, D)\| \\
&\le 2\,\|(A, B)\|\,\|(C, D)\| .
\end{aligned}
$$

From the discussion preceding Definition 3.1.1, the new norm of $(A, 0)$ is $\|L_{(A,0)}\|$; and from $(*)$, $\|L_{(A,0)}\| \le \|(A, 0)\|$. However, $L_{(A,0)}(I, 0) = (A, 0)$ and $\|(I, 0)\| = 1$. Thus $\|L_{(A,0)}\| = \|(A, 0)\|$, and the mapping $A \to (A, 0)$ is an isometry relative to the new norm $\|\ \|$ (as well as relative to the norm $\|\ \|$).   ■

3.5.3.   Let $\{\mathfrak{A}_a : a \in \mathbf{A}\}$ be a family of Banach algebras, and let $\mathfrak{A}$ be the set of functions f on $\mathbf{A}$ such that $f(a) \in \mathfrak{A}_a$ and

$$\sup\{\|f(a)\| : a \in \mathbf{A}\}(= \|f\|) < \infty.$$

(i)   Show that $\mathfrak{A}$, provided with the operations of pointwise addition, multiplication, multiplication by scalars, and with the norm described, is a Banach algebra. (It is usually denoted by "$\sum_{a \in \mathbf{A}} \oplus \mathfrak{A}_a$." Compare it with the l_∞ spaces of Example 1.7.1.)

(ii)   With $\mathbb{N}$ in the place of $\mathbf{A}$, show that $\mathfrak{A}_0$, the set of functions f in $\mathfrak{A}$ such that $\lim_{n \to \infty} \|f(n)\| = 0$ is a (norm-)closed proper two-sided ideal in $\mathfrak{A}$.

(iii)   Determine whether or not $\mathfrak{A}_0$ is a maximal (two-sided) ideal in $\mathfrak{A}$.

Solution. (i)   The discussion of Example 1.7.1 applies, with minor modifications, to show that $\mathfrak{A}$, with the given linear space structure and norm, is a Banach space. If f and g are in $\mathfrak{A}$, then

$$
\begin{aligned}
\|fg\| &= \sup\{\|(fg)(a)\| : a \in \mathbf{A}\} \\
&\le \sup\{\|f(a)\|\,\|g(a)\| : a \in \mathbf{A}\} \\
&\le \|f\|\,\|g\| .
\end{aligned}
$$

If $I(a) = I_a$, where I_a is the unit element for $\mathfrak{A}_a$, then I is the multiplicative unit element for $\mathfrak{A}$ and $\|I\| = \sup\{\|I_a\| : a \in \mathbb{A}\} = 1$. Thus $\mathfrak{A}$ is a Banach algebra.

(ii) If f and g are in $\mathfrak{A}_0$ then

$$\lim_n \|\alpha f(n) + g(n)\| \le |\alpha| \lim_n \|f(n)\| + \lim_n \|g(n)\| = 0,$$

for each scalar α; so that $\mathfrak{A}_0$ is a linear subspace of $\mathfrak{A}$. Suppose $f_m \in \mathfrak{A}_0$ and $\{f_m\}$ converges to f_0 in $\mathfrak{A}$. Define $g_m(n)$ to be $\|f_m(n)\|$, for n in $\mathbb{N}$, and $g_m(\infty)$ to be 0. Then g_m is continuous on $(\mathbb{N}, \infty)$ for m in $\{1, 2, \ldots\}$, and g_0 is the uniform limit of $\{g_m\}$ on $(\mathbb{N}, \infty)$; for

$$|g_0(n) - g_m(n)| \le \|f_0(n) - f_m(n)\| \le \|f_0 - f_m\| \to \infty.$$

Thus g_0 is continuous on $(\mathbb{N}, \infty)$ and

$$\lim_{n \to \infty} \|f_0(n)\| = \lim_{n \to \infty} g_0(n) = g_0(\infty) = 0.$$

If follows that $f_0 \in \mathfrak{A}_0$, and $\mathfrak{A}_0$ is (norm) closed in $\mathfrak{A}$. If $f \in \mathfrak{A}_0$ and $g \in \mathfrak{A}$, then

$$\|(fg)(n)\| \le \|f(n)\| \, \|g(n)\| \le \|f(n)\| \, \|g\| \to 0.$$

Similarly $\|(gf)(n)\| \le \|g\| \, \|f(n)\| \to 0$. Thus fg and gf are in $\mathfrak{A}_0$. Since $I(n) = I_n$ and $\|I_n\| = 1$, $I \notin \mathfrak{A}_0$. It follows that $\mathfrak{A}_0$ is a proper (norm-)closed two-sided ideal in $\mathfrak{A}$.

(iii) The ideal $\{f \in \mathfrak{A} : \lim_{n \to \infty} \|f(2n)\| = 0\}$ is always larger than $\mathfrak{A}_0$. ∎

3.5.4. With the notation of Exercise 1.9.19:

(i) Show that l_∞ and c, provided with pointwise multiplication, are (commutative) Banach algebras.

(ii) Show that c_0 is a (norm-)closed proper ideal in l_∞.

(iii) Describe a multiplicative linear functional on c with c_0 as its null space; and conclude that c_0 is a maximal ideal in c.

(iv) Is c_0 a maximal ideal in l_∞? Explain. Reconsider Exercise 3.5.3(iii). [*Hint.* Consider "dimension" in combination with Corollary 3.2.5.]

Solution. (i)   From Exercise 1.9.19(i), c and c_0 are closed linear subspaces of l_∞; and from Example 1.7.1, l_∞ is a Banach space. Thus c and c_0 are Banach spaces. The multiplication on l_∞ is the multiplication it inherits from the algebra of (complex-valued) functions on $\mathbb{N}$; so that l_∞ is a (commutative) algebra with multiplicative unit element $\{1, 1, \ldots\}$. Moreover,

$$\|\{x_n\}\{y_n\}\| = \|\{x_n y_n\}\| = \sup\{|x_n y_n| : n \in \mathbb{N}\}$$
$$\leq \sup\{|x_n| : n \in \mathbb{N}\} \sup\{|y_n| : n \in \mathbb{N}\}$$
$$= \|\{x_n\}\| \, \|\{y_n\}\|$$

and $\|\{1, 1, \ldots\}\| = 1$. Thus l_∞ and c are Banach algebras.

(ii)   If $\{y_n\} \in c_0$ and $\{x_n\} \in l_\infty$, then $\{x_n\}\{y_n\} = \{x_n y_n\}$ and $|x_n y_n| \leq \|\{x_m\}\| \, |y_n| \to 0$ as $n \to \infty$. Thus $\{x_n\}\{y_n\}$ $(= \{y_n\}\{x_n\})$ $\in c_0$ and c_0 is a (norm-) closed ideal in l_∞ (and in c). Since $\{1, 1, \ldots\} \notin c_0$; it follows that c_0 is a proper ideal in l_∞ (and in c).

(iii)   With $\{x_n\}$ in c, let $\rho(\{x_n\})$ be $\lim x_n$. The properties of the limit operation assure us that ρ is a multiplicative linear functional on c. By definition of c_0, ρ has null space precisely c_0 and $\rho(\{1, 1, \ldots\}) = 1$. Thus c_0 is a maximal ideal in c.

(iv)   No! From Corollary 3.2.5, the quotient of a commutative Banach algebra by a maximal ideal is (isomorphic to) $\mathbb{C}$—in particular, this quotient is one-dimensional. Since no non-trivial linear combination of $\{1, 1, \ldots\}$ and $\{1, 0, 1, 0, \ldots\}$ lies in c_0; the images of these elements of l_∞, under the quotient mapping of l_∞ onto l_∞/c_0, are linearly independent. Thus l_∞/c_0 has (linear) dimension greater than 1, and c_0 is not a maximal ideal in l_∞.

If each $\mathfrak{A}_a$, in Exercise 3.5.3, is $\mathbb{C}$, then $\mathfrak{A}$ is l_∞ and $\mathfrak{A}_0$ is c_0. As we have just noted, $\mathfrak{A}_0$ is not a maximal ideal in the algebra $\mathfrak{A}$, in this case.   ■

3.5.5.   Let $\beta(\mathbb{N})$ be the space of non-zero multiplicative linear functionals on $l_\infty(\mathbb{N}, \mathbb{C})$ $(= l_\infty)$ topologized with the weak* topology. Define $\hat{f}(\rho)$ to be $\rho(f)$ for ρ in $\beta(\mathbb{N})$ and f in l_∞.

(i)   Note that $\beta(\mathbb{N})$ is a compact Hausdorff space; and show that the mapping, $f \to \hat{f}$, is an algebraic isometric isomorphism of l_∞ into $C(\beta(\mathbb{N}))$.

(ii)   Show that $\hat{\bar{f}} = \overline{\hat{f}}$. [*Hint.* Consider the spectrum of a real-valued f in l_∞ relative to l_∞.]

(iii) Show that $\hat{1} = 1$, where 1 denotes the constant function "one" in both l_∞ and $C(\beta(\mathbf{N}))$, and that $\hat{l}_\infty$ is (norm) closed in $C(\beta(\mathbf{N}))$. Conclude that $\hat{l}_\infty = C(\beta(\mathbf{N}))$. (The space $\beta(\mathbf{N})$ is called "the β-compactification of $\mathbf{N}$.")

Solution. (i) From Proposition 3.2.20, $\beta(\mathbf{N})$ is a weak* closed subset of the unit ball of $l_\infty^\#$; so that $\beta(\mathbf{N})$ with the weak* topology is a compact Hausdorff space. Moreover, since $\|\rho\| = 1$ for ρ in $\beta(\mathbf{N})$,

$$\|\hat{f}\| = \sup\{|\hat{f}(\rho)| : \rho \in \beta(\mathbf{N})\} = \sup\{|\rho(f)| : \rho \in \beta(\mathbf{N})\}$$
$$\leq \sup\{\|\rho\|\,\|f\| : \rho \in \beta(\mathbf{N})\} = \|f\|.$$

Let $n^\#(f)$ be $f(n)$ for n in $\mathbf{N}$ and f in l_∞. Then $n^\# \in \beta(\mathbf{N})$ and

$$\|f\| = \sup\{|f(n)| : n \in \mathbf{N}\} = \sup\{|n^\#(f)| : n \in \mathbf{N}\}$$
$$= \sup\{|\hat{f}(n^\#)| : n \in \mathbf{N}\} \leq \|\hat{f}\|.$$

Combining these two inequalities, we have $\|f\| = \|\hat{f}\|$.

The mapping, $f \to \hat{f}$, is the composition of the linear isomorphism mapping l_∞ into $l_\infty^{\#\#}$ and the mapping restricting elements of $l_\infty^{\#\#}$ to $\beta(\mathbf{N})$ $(\subseteq l_\infty^\#)$. Thus $f \to \hat{f}$ is linear. If f and g are in l_∞ and $\rho \in \beta(\mathbf{N})$, then

$$(\widehat{fg})(\rho) = \rho(fg) = \rho(f)\rho(g) = \hat{f}(\rho)\hat{g}(\rho) = (\hat{f}\hat{g})(\rho);$$

so that $f \to \hat{f}$ is an algebraic isometric isomorphism of l_∞ into $C(\beta(\mathbf{N}))$. (Each $\hat{f}$ is continuous on $\beta(\mathbf{N})$ by definition of the weak* topology.)

(ii) With f a real-valued function in l_∞ and z in $\mathbb{C} \setminus \mathbb{R}$, $0 < \frac{1}{2}|z - \bar{z}| < |f(n) - z|$ for each n in $\mathbf{N}$. Thus, if $g(n) = [f(n) - z]^{-1}$, then $g \in l_\infty$ and g is an inverse to $f - z1$. It follows that f has real spectrum relative to l_∞. From Remark 3.2.11, $\rho(f)$ $(= \hat{f}(\rho))$ is real for each ρ in $\beta(\mathbf{N})$—that is, $\hat{f}$ is real-valued if f is. Since the mapping, $f \to \hat{f}$, is complex linear, it now follows that $\hat{\bar{f}} = \overline{\hat{f}}$.

(iii) For each ρ in $\beta(\mathbf{N})$, $1 = \rho(1) = \hat{1}(\rho)$; so that $\hat{1} = 1$. Since $f \to \hat{f}$ is an isometry and l_∞ is complete, $\hat{l}_\infty$ is closed in $C(\beta(\mathbf{N}))$. From (ii), $\hat{l}_\infty$ is stable under the operation of complex conjugation. With ρ and ρ' distict elements of $\beta(\mathbf{N})$, there is an f in l_∞ such that

$\hat{f}(\rho) = \rho(f) \neq \rho'(f) = \hat{f}(\rho')$. The hypotheses of Theorem 3.4.14 (the Stone–Weierstrass theorem) are satisfied with $\hat{l}_\infty$ in place of $\mathcal{A}$ and $\beta(\mathbf{N})$ in place of X. We conclude that $\hat{l}_\infty = C(\beta(\mathbf{N}))$, and the mapping $f \to \hat{f}$ is an algebraic, isometric, conjugation-preserving isomorphism of l_∞ onto $C(\beta(\mathbf{N}))$.   ∎

3.5.6.   With the notation of Exercise 3.5.5:
  (i)   Show that $\beta(\mathbf{N})$ is extremely disconnected.
  (ii)   Let $n^\#(f)$ be $f(n)$ for n in $\mathbf{N}$ and f in l_∞; denote by $\mathbf{N}$, again, the subset $\{n^\# : n \in \mathbf{N}\}$ of $\beta(\mathbf{N})$. Show that $\mathbf{N}$ is dense in $\beta(\mathbf{N})$.
  (iii)   Show that the one-point subset $\{n^\#\}$ of $\beta(\mathbf{N})$ is open in $\beta(\mathbf{N})$. [*Hint.* Consider $\hat{\chi}_n$, where $\chi_n(m)$ is 0 unless $n = m$, and $\chi_n(n) = 1$.]
  (iv)   Show that $\rho(f) = 0$ when $\rho \in \beta(n) \setminus \mathbf{N}$ and $f \in c_0$.

Solution. (i)   The mapping $f \to \hat{f}$ of l_∞ onto $C(\beta(\mathbf{N}))$ (see the preceding exercise) is an order isomorphism between the (real) subalgebras of real-valued functions; for the positive functions are precisely the elements with non-negative real spectrum and $f \to \hat{f}$ preserves spectrum (since it is an algebraic isomorphism). Now $l_\infty(\mathbf{N}, \mathbb{R})$ is a boundedly complete lattice; so that the (real) algebra of real-valued functions in $C(\beta(\mathbf{N}))$ is a boundedly complete lattice. From Theorem 3.4.16, $\beta(\mathbf{N})$ is extremely disconnected.

  (ii)   Let $\mathcal{O}$ $(\neq \emptyset)$ be an open subset of $\beta(\mathbf{N})$. Since $\beta(\mathbf{N})$ is a compact Hausdorff space and the mapping $f \to \hat{f}$, is onto, there is an f in l_∞ such that $\hat{f}$ is 1 at some point of $\mathcal{O}$ and 0 on $\beta(\mathbf{N}) \backslash \mathcal{O}$. Now $f \neq 0$ since $\hat{f} \neq 0$. Choose n in $\mathbf{N}$ such that $0 \neq f(n)$ $(= \hat{f}(n^\#))$. Then $n^\# \in \mathcal{O}$; and $\mathbf{N}$ is dense in $\beta(\mathbf{N})$.

  (iii)   It follows, from (ii) and the fact that $\beta(\mathbf{N})$ is a Hausdorff space, that $\mathbf{N} \setminus \{1^\#, \dots, n^\#\}$ is dense in $\beta(\mathbf{N}) \setminus \{1^\#, \dots, n^\#\}$ for each n in $\mathbf{N}$. With χ_n as in the hint, $\hat{\chi}_n(m^\#) = 0$ unless $m = n$, and $\hat{\chi}_n(n^\#) = 1$. Since $\mathbf{N} \setminus \{n^\#\}$ is dense in $\beta(\mathbf{N}) \setminus \{n^\#\}$ and $\hat{\chi}_n$ is continuous on $\beta(\mathbf{N})$ and vanishes on $\mathbf{N} \setminus \{n^\#\}$; $\hat{\chi}_n$ vanishes on the set $\beta(\mathbf{N}) \setminus \{n^\#\}$. The set of points in $\beta(\mathbf{N})$ where $\hat{\chi}_n$ takes values in $(0, 2)$ is an open subset of $\beta(\mathbf{N})$ and consists of $n^\#$ alone.

  (iv)   Suppose $f \in c_0$ and $\rho \in \beta(\mathbf{N}) \backslash \mathbf{N}$. From (iii), $\mathbf{N} \backslash \{1^\#, \dots, n^\#\}$ is dense in $\beta(\mathbf{N}) \setminus \{1^\#, \dots, n^\#\}$. Since $\hat{f}$ is continuous on $\beta(\mathbf{N})$; $\hat{f}(\rho)$ is in the closure of $\{\hat{f}(m^\#) : m > n\}$ $(= \{f(m) : m > n\})$. As $f \in c_0$, if $0 < \varepsilon$, there is an n such that $|f(m)| < \varepsilon$ when $m > n$.

Thus $|\hat{f}(\rho)| \le \varepsilon$ for each positive ε; and $0 = \hat{f}(\rho) = \rho(f)$. ∎

3.5.7.   Let G be a countable (discrete) group and $\{G_n\}$ an ascending sequence of finite subgroups with union G. (Such a group is said to be *locally finite*.) Let ρ be an element of $(l_\infty(G,\mathbb{C})^\#)_1$ such that $\rho(\{1,1,\ldots\}) = 1$; and define ρ_n to be $|G_n|^{-1} \sum_{g \in G_n} T_g^\#(\rho)$, where $|G_n|$ is the number of elements in G_n, and the bounded linear operator T_g, acting on $l_\infty(G,\mathbb{C})$, is given by $[T_g(f)](g_0) = f(g^{-1}g_0)$.

(i)   Show that $\rho_n \in (l_\infty(G,\mathbb{C})^\#)_1$ and that $\rho_n(\{1,1,\ldots\}) = 1$.

(ii)   Show that the weak* closure of $\{\rho_n : n \in \mathbb{N}\}$ contains an element ρ_0 such that $T_g^\# \rho_0 = \rho_0$ for every g in G, and such that $\rho_0(\{1,1,\ldots\}) = 1$. (The functional ρ_0 is called an *invariant mean on* G.) [*Hint.* With f in $l_\infty(G,\mathbb{C})$, define a function φ on $\mathbb{N}$ by $\varphi(n) = \rho_n(f)$. Fix x_0 in $\beta(\mathbb{N}) \setminus \mathbb{N}$; and let $\rho_o(f)$ be $\hat{\varphi}(x_0)$.]

Solution. (i)   If $f \in l_\infty(G,\mathbb{C})$,

$$\|T_g(f)\| = \sup\{|f(g^{-1}g')| : g' \in G\} = \sup\{|f(g)| : g \in G\} = \|f\|;$$

so that T_g is a linear isometry of $l_\infty(G,\mathbb{C})$ onto itself. It follows that $T_g^\#$ is a linear isometry of $l_\infty(G,\mathbb{C})^\#$ onto itself; and

$$\begin{aligned}
\|\rho_n\| &= \Big\| |G_n|^{-1} \sum_{g \in G_n} T_g^\#(\rho) \Big\| \\
&\le |G_n|^{-1} \sum_{g \in G_n} \|T_g^\#(\rho)\| \\
&= |G_n|^{-1} \sum_{g \in G_n} \|\rho\| \\
&= \|\rho\| \le 1.
\end{aligned}$$

Note, too, that $T_g(\{1,1,\ldots\}) = \{1,1,\ldots\}$, for each g in G, so that

$$\rho_n(\{1,1,\ldots\}) = |G_n|^{-1} \sum_{g \in G_n} \rho[T_g(\{1,1,\ldots\})] = \rho(\{1,1,\ldots\}) = 1,$$

for each n.

(ii)   Let φ be as in the hint. Define ψ, corresponding to g in $l_\infty(G,\mathbb{C})$, similarly. Then $(a\varphi + \psi)(n) = \rho_n(af + g)$ for each complex scalar a; so that

$$\rho_0(af + g) = (a\varphi + \psi)\hat{\ }(x_0) = (a\hat{\varphi} + \hat{\psi})(x_0) = a\rho_0(f) + \rho_0(g).$$

Moreover, since $\|\rho_n\| \leq 1$,

$$|\rho_0(f)| = |\hat{\varphi}(x_0)| \leq \|\hat{\varphi}\| = \sup\{|\rho_n(f)| : n \in \mathbb{N}\} \leq \|f\|,$$

so that $\rho_0 \in (l_\infty(G, \mathbb{C})^\#)_1$.

Let $f_1, \ldots, f_k$ in $l_\infty(G, \mathbb{C})$, a positive integer n, and a positive ε, be given. Let $\varphi_1, \ldots, \varphi_k$ correspond, as in the hint, to $f_1, \ldots, f_k$. From the result of Exercise 3.5.6(ii), $\mathbb{N} \setminus \{1^\#, \ldots, n^\#\}$ is dense in $\beta(\mathbb{N}) \setminus \{1^\#, \ldots, n^\#\}$. Thus there is an m in $\mathbb{N}$ such that $m > n$ and

$$\begin{aligned}
\varepsilon > |\hat{\varphi}_j(m) - \hat{\varphi}_j(x_0)| &= |m^\#(\varphi_j) - \rho_0(f_j)| \\
&= |\varphi_j(m) - \rho_0(f_j)| \\
&= |\rho_m(f_j) - \rho_0(f_j)| \qquad (j = 1, \ldots, k).
\end{aligned}$$

Hence ρ_0 is in the weak* closure of $\{\rho_m : m > n\}$. Choosing $\{1, 1, \ldots\}$ for f, we have that $\rho_0(\{1, 1, \ldots\})$ is in the closure of the set of numbers $\{\rho_n(\{1, 1, \ldots\}) : n \in \mathbb{N}\}$. But, from (i), $\rho_n(\{1, 1, \ldots\}) = 1$ for each n. Thus $\rho_0(\{1, 1, \ldots\}) = 1$.

Note that

$$\begin{aligned}
[T_g T_{g'}(f)](g_0) = [T_{g'}(f)](g^{-1} g_0) = f(g'^{-1} g^{-1} g_0) &= f([gg']^{-1} g_0) \\
&= (T_{gg'} f)(g_0).
\end{aligned}$$

Thus $T_g T_{g'} = T_{gg'}$; and $T_{gg'}^\# = (T_g T_{g'})^\# = T_{g'}^\# T_g^\#$.

If $g_0 \in G_n$, then

$$\begin{aligned}
T_{g_0}^\#(\rho_n) = |G_n|^{-1} \sum_{g \in G_n} T_{g_0}^\# T_g^\#(\rho) &= |G_n|^{-1} \sum_{g \in G_n} T_{gg_0}^\#(\rho) \\
&= |G_n|^{-1} \sum_{g \in G_n} T_g^\#(\rho) = \rho_n;
\end{aligned}$$

since $g \to gg_0$ is a one-to-one mapping of G_n onto itself. It follows that $T_{g_0}^\#(\rho_m) = \rho_m$ when $g_0 \in G_n$ and $n \leq m$ (for then $g_0 \in G_n \subseteq G_m$). Now each g in G lies in some G_n. Thus, with f in $l_\infty(G, \mathbb{C})$, $\rho_m(f) = (T_g^\# \rho_m)(f) = \rho_m(T_g f)$, for all $m \geq n$. Since ρ_0 is in the weak* closure of $\{\rho_m : m \geq n\}$; with a positive ε given, there is an m exceeding n such that $|\rho_m(f) - \rho_0(f)| < \varepsilon$ and $|\rho_m(T_g(f)) - \rho_0(T_g(f))| < \varepsilon$. But $\rho_m(T_g(f)) = \rho_m(f)$ for this m; so that $|\rho_0(f) - \rho_0(T_g(f))| < 2\varepsilon$. As this last inequality holds for each positive ε, $\rho_0(f) = \rho_0(T_g(f)) = (T_g^\# \rho_0)(f)$ for each f in $l_\infty(G, \mathbb{C})$. Thus $T_g^\# \rho_0 = \rho_0$ for each g in G.

[Note that this last argument can also proceed by quoting the weak* continuity of $T_g^\#$ on $l_\infty(G, \mathbb{C})^\#$— Proposition 1.6.8—and using the fact that $l_\infty(G, \mathbb{C})^\#$ is a Hausdorff space in the weak* topology. It follows from these observations that the fixed points of $T_g^\#$ form a weak* closed subset of $l_\infty(G, \mathbb{C})^\#$. This set contains $\{\rho_m : m \geq n\}$; and ρ_0 is in the weak* closure of this last set, from which it follows that $T_g^\# \rho_0 = \rho_0$.] ■

3.5.8.   Let A be an element of the Banach algebra $\mathfrak{A}$; and let $\mathcal{O}$ be an open subset of $\mathbb{C}$ containing $\mathrm{sp}_{\mathfrak{A}} A$. Show that there is a positive δ such that $\mathrm{sp}_{\mathfrak{A}} B \subseteq \mathcal{O}$ if $B \in \mathfrak{A}$ and $\|A - B\| < \delta$.

Solution.   Let $\mathbb{D}$ be the closed disk in $\mathbb{C}$ with center 0 and radius $1 + \|A\|$. For each λ in $\mathbb{C} \setminus \mathrm{sp}_{\mathfrak{A}} A$, let $\mathcal{S}_\lambda$ be an open ball in $\mathfrak{A}$, with center $A - \lambda I$ and radius $2r_\lambda$, each of whose elements is invertible in $\mathfrak{A}$. (See Proposition 3.1.6.) Let $\mathcal{O}_\lambda$ be the open disk in $\mathbb{C}$ with center λ and radius r_λ. From Theorem 3.2.3, $\mathrm{sp}_{\mathfrak{A}} A \subseteq \mathbb{D}$. It follows that $\{\mathcal{O}_\lambda : \lambda \in \mathbb{C} \setminus \mathrm{sp}_{\mathfrak{A}} A\}$ is an open covering of $\mathbb{D} \setminus \mathcal{O}$. Let $\{\mathcal{O}_{\lambda(1)}, \ldots, \mathcal{O}_{\lambda(n)}\}$ be a finite subcovering; and let δ be $\min\{r_{\lambda(1)}, \ldots, r_{\lambda(n)}, 1\}$. If $\|A - B\| < \delta$ and $\lambda \in \mathbb{D} \setminus \mathcal{O}$, then $\lambda \in \mathcal{O}_{\lambda(j)}$ for some j; and

$$\|A - \lambda(j)I - (B - \lambda I)\| \leq \|A - B\| + |\lambda - \lambda(j)| < \delta + r_{\lambda(j)} < 2r_{\lambda(j)}.$$

Thus $B - \lambda I \in \mathcal{S}_{\lambda(j)}$, and $B - \lambda I$ is invertible in $\mathfrak{A}$. It follows that $\mathbb{D} \setminus \mathcal{O} \subseteq \mathbb{D} \setminus \mathrm{sp}_{\mathfrak{A}} B$. Now $\mathrm{sp}_{\mathfrak{A}} B \subseteq \mathbb{D}$, since $\|A - B\| < 1$ and $\|B\| < \|A\| + 1$. If $\lambda' \in \mathrm{sp}_{\mathfrak{A}} B$, then $\lambda' \notin \mathbb{D} \setminus \mathrm{sp}_{\mathfrak{A}} B$ so that $\lambda' \notin \mathbb{D} \setminus \mathcal{O}$. But $\lambda' \in \mathbb{D}$, so that $\lambda' \in \mathcal{O}$; and $\mathrm{sp}_{\mathfrak{A}} B \subseteq \mathcal{O}$. ■

3.5.9.   Let $\{A_n\}$ be a sequence of invertible elements in a Banach algebra $\mathfrak{A}$ and suppose $\{A_n\}$ has a limit A that commutes with each A_n.

(i)   Show, by example, that A need not be invertible.

(ii)   Show that A is invertible if $\{r(A_n^{-1})\}$ is bounded. [*Hint.* Use Corollary 3.3.4.]

Solution. (i)   The sequence $\{n^{-1}I\}$, where $n \in \{1, 2, \ldots\}$, has 0 as its limit; 0 commutes with each term; each term is invertible; but 0 is not invertible.

(ii) Since $I - A_n^{-1}A = A_n^{-1}(A_n - A)$ and A_n^{-1} commutes with $A_n - A$, from Corollary 3.3.4 (and Remark 3.2.7) we have

$$r(I - A_n^{-1}A) = r(A_n^{-1}[A_n - A]) \le r(A_n^{-1})r(A_n - A)$$
$$\le r(A_n^{-1})\|A_n - A\| \to 0.$$

Thus, with n large, sp $A_n^{-1}A$ lies in a small disk with center 1; and $A_n^{-1}A$ is invertible. It follows that A is invertible. ∎

3.5.10.   Given an element A of a Banach algebra $\mathfrak{A}$ and a positive number ε, show that:

(i) if A is singular (that is $0 \in$ sp A), there is a positive number δ_0 such that if B in $\mathfrak{A}$ commutes with A and $\|A - B\| < \delta_0$, then there is a λ in sp B with $|\lambda| < \varepsilon$;

(ii) if $\lambda \in$ sp A there is a positive number δ_λ such that if B in $\mathfrak{A}$ commutes with A and $\|A - B\| < \delta_\lambda$, then, for some λ' in sp B, $|\lambda - \lambda'| < \varepsilon$;

(iii) there is a positive δ such that if $\|A - B\| < \delta$ and $AB = BA$, with B in $\mathfrak{A}$, then sp $A \subseteq S_\varepsilon(B)$, where $S_\varepsilon(B) = \{\lambda : |\lambda - \lambda'| < \varepsilon$ for some λ' in sp $B\}$.

Solution. (i) Suppose the contrary. Then there is a sequence $\{B_n\}$ in $\mathfrak{A}$ tending to A, commuting with A, and such that $\varepsilon \le |\lambda|$ for each λ in sp B_n. It follows that each B_n is invertible and, from Proposition 3.2.10, that $r(B_n^{-1}) \le \varepsilon^{-1}$. The conditions of Exercise 3.5.9 apply, and A is invertible—contradicting our assumption.

(ii) With λ in sp A, $A - \lambda I$ is singular. Choose δ_λ (as in (i)) such that if $\|A - \lambda I - B_0\| < \delta_\lambda$ and $(A - \lambda I)B_0 = B_0(A - \lambda I)$, then $|\lambda_0| < \varepsilon$ for some λ_0 in sp B_0. If B in $\mathfrak{A}$ commutes with A and $\|A - B\| < \delta_\lambda$, then $\|A - \lambda I - B_0\| < \delta_\lambda$, where $B_0 = B - \lambda I$. Since B_0 commutes with $A - \lambda I$, there is a λ_0 in sp B_0 such that $|\lambda_0| < \varepsilon$. Now sp $B_0 = ($sp $B) - \lambda$, so that $\lambda_0 = \lambda' - \lambda$ for some λ' in sp B.

(iii) Apply the result of (ii) to A and $\varepsilon/2$ (in place of ε) to yield a positive δ_λ for each λ in sp A. Let $\mathcal{O}_\lambda$ be the open disk of radius $\varepsilon/2$ and center λ in $\mathbb{C}$. From Theorem 3.2.3, sp A is compact. Thus the covering $\{\mathcal{O}_\lambda\}$ of sp A admits a finite subcovering $\{\mathcal{O}_{\lambda(1)}, \ldots, \mathcal{O}_{\lambda(n)}\}$. Let δ be the smallest of $\delta_{\lambda(1)}, \ldots, \delta_{\lambda(n)}$. If $\|A - B\| < \delta$ and $AB = BA$, then, for each j in $\{1, \ldots, n\}$, $|\lambda(j) - \lambda'(j)| < \varepsilon/2$ for some $\lambda'(j)$ in sp B, by choice of $\delta_{\lambda(j)}$, since $\|A - B\| < \delta \le \delta_{\lambda(j)}$. With λ in sp A, $\lambda \in \mathcal{O}_{\lambda(j)}$ for some j in $\{1, \ldots, n\}$. Thus $|\lambda - \lambda(j)| < \varepsilon/2$; and $|\lambda - \lambda'(j)| < \varepsilon$. It follows that sp $A \subseteq S_\varepsilon(B)$. ∎

3.5.11.   Nilpotent and generalized nilpotent elements are defined following the proof of Theorem 3.3.3 (the spectral radius formula). Use Proposition 3.2.10 to show, without the aid of the spectral radius formula, that:

(i)   each nilpotent element in a Banach algebra has spectrum consisting of 0 alone;

(ii)   each generalized nilpotent element in a Banach algebra has spectrum consisting of 0 alone.

Solution.   Let A be an element of a Banach algebra $\mathfrak{A}$.

(i)   If $A^n = 0$ then,

$$\{0\} = \operatorname{sp} 0 = \operatorname{sp}(A^n) = (\operatorname{sp} A)^n,$$

from Proposition 3.2.10. Thus $\operatorname{sp} A = \{0\}$.

(ii)   If $\|A^n\|^{1/n} \to 0$ and $\lambda \in \operatorname{sp} A$, then $\lambda^n \in \operatorname{sp}(A^n)$, from Proposition 3.2.10. From Theorem 3.2.3, $\operatorname{sp} A$ is non-empty and $|\lambda|^n \le \|A^n\|$. Thus $|\lambda| \le \|A^n\|^{1/n} \to 0$, and $\lambda = 0$. It follows that the (non-empty) spectrum of A consists of 0 alone.   ∎

3.5.12.   Let $\{y_1, y_2, \ldots\}$ be an orthonormal basis for the (complex, separable) Hilbert space $\mathcal{H}$. Let A be the linear transformation of $\mathcal{H}$ into $\mathcal{H}$ that maps y_n onto $a_n y_{\pi(n)}$, where $\{a_n\}$ is a bounded sequence of complex numbers and π is a one-to-one mapping of $\{1, 2, \ldots\}$ into itself.

(i)   Determine $\|A\|$.

(ii)   Describe A^k, where k is a positive integer.

(iii)   Determine $\|A^k\|$.

(iv)   Determine $\|A\|$ and $\|A^k\|$ under the additional assuptions that $|a_n| \le |a_m|$ and $\pi(m) \le \pi(n)$ when $m \le n$.

(v)   Show that A is nilpotent when $a_n = 0$ for all n larger than some k and $j < \pi(j)$ for j in $\{1, \ldots, k\}$.

(vi)   Use Exercises 2.8.22(iv) and 2.8.23 to show that A is compact when $\{a_n\}$ tends to 0.

(vii)   Show that A is a compact generalized nilpotent operator on $\mathcal{H}$ with infinite-dimensional range when $a_n = 1/n$ and $\pi(m) = m+1$.

Solution. (i)   If x is a unit vector in $\mathcal{H}$, then $x = \sum_{n=1}^{\infty} b_n y_n$, where $\sum_{n=1}^{\infty} |b_n|^2 = 1$. If $a = \sup\{|a_n|\}$, then

$$\|Ax\|^2 = \left\| \sum_{n=1}^{\infty} b_n a_n y_{\pi(n)} \right\|^2 = \sum_{n=1}^{\infty} |a_n|^2 |b_n|^2 \le a^2.$$

Hence $\|A\| \leq a$. But $|a_n| = \|a_n y_{\pi(n)}\| = \|A y_n\| \leq \|A\|$ for all n. Thus $a \leq \|A\|$; and

$$(*) \qquad \|A\| = \sup\{|a_n| : n = 1, 2, \ldots\}.$$

(ii)   Note that

$$(**) \qquad \begin{aligned} A^k y_n &= A^{k-1}(a_n y_{\pi(n)}) = A^{k-2}(a_n a_{\pi(n)} y_{\pi^2(n)}) \\ &= \cdots = a_n a_{\pi(n)} \cdots a_{\pi^{k-1}(n)} y_{\pi^k(n)}. \end{aligned}$$

Since π^k is a one-to-one mapping of the set $\{1, 2, \ldots\}$ into itself and $\{|a_n a_{\pi(n)} \cdots a_{\pi^{k-1}(n)}|\}$ is bounded by b^k when $\{|a_n|\}$ is bounded by b, A^k is an operator of the same type as A.

(iii)   From (ii), (i) applies to A^k; and from $(*)$

$$(***) \qquad \|A^k\| = \sup\{|a_n a_{\pi(n)} \cdots a_{\pi^{k-1}(n)}| : n = 1, 2, \ldots\}.$$

(iv)   Since $|a_1| \geq |a_2| \geq \cdots$; $|a_1| = \sup\{|a_n|\}$ and $|a_1| = \|A\|$ in this case. In the same way,

$$\begin{aligned} &|a_1| \geq |a_n|, \\ &|a_{\pi(1)}| \geq |a_{\pi(n)}|, \\ &|a_{\pi^2(1)}| \geq |a_{\pi^2(n)}|, \ldots, |a_{\pi^{k-1}(1)}| \geq |a_{\pi^{k-1}(n)}|, \end{aligned}$$

since $1 \leq n$, $\pi(1) \leq \pi(n)$, $\pi^2(1) \leq \pi^2(n)$, $\ldots$, $\pi^{k-1}(1) \leq \pi^{k-1}(n)$; and

$$|a_1 a_{\pi(1)} a_{\pi^2(1)} \cdots a_{\pi^{k-1}(1)}| \geq |a_n a_{\pi(n)} a_{\pi^2(n)} \cdots a_{\pi^{k-n}(n)}|$$

for all n. Thus, in this case, from $(***)$ we have

$$(****) \qquad \|A^k\| = |a_1 a_{\pi(1)} a_{\pi^2(1)} \cdots a_{\pi^{k-1}(1)}|.$$

(v)   Under the present assumptions, at least one of n, $\pi(n)$, $\ldots$, $\pi^k(n)$ exceeds k; for if the contrary were the case, we would have $1 \leq n < \pi(n) < \cdots < \pi^k(n) \leq k$—an absurdity. Thus, from $(**)$ and the assumption that $a_m = 0$ if $k < m$, we have

$$A^{k+1} y_n = a_n a_{\pi(n)} \cdots a_{\pi^k(n)} y_{\pi^{k+1}(n)} = 0 \cdot y_{\pi^{k+1}(n)} = 0$$

for all n. It follows that $A^{k+1} = 0$.

(vi)   Let A_n be defined as A is with 0 in place of $a_{n+1}, a_{n+2}, \ldots$. Then A_n has range $[y_{\pi(1)}, \ldots, y_{\pi(n)}]$. From Exercise 2.8.22(iv), A_n is compact. Now $A - A_n$ is of the same type as A, where $a_1, \ldots, a_n$ are replaced by 0. Thus $\|A - A_n\| = \sup\{|a_j| : j > n\} \to 0$, since $\{a_j\} \to 0$ by assumption. From Exercise 2.8.23, A is compact.

(vii)   The conditions of (iv) and (vi) apply here. Thus A is compact and, from $(****)$, we have

$$\|A^n\| = |a_1 a_{\pi(1)} \cdots a_{\pi^{n-1}(1)}| = \frac{1}{n!}.$$

It follws that

$$\|A^n\|^{1/n} = \frac{1}{(n!)^{1/n}} \to 0.$$

Thus A is a compact generalized nilpotent operator on $\mathcal{H}$ with range containing $\{y_2, y_3, \ldots\}$.   ∎

3.5.13.   As in Exercise 3.5.12, let $\{y_n\}$ be an orthonormal basis for the Hilbert space $\mathcal{H}$ and let Ay_n be $a_n y_{n+1}$, where $a_n = 2^{-k}$ and k is the largest positive integer such that 2^k divides n.

(i)   Show that

$$a_1 a_2 \cdots a_{2^n - 1} = \prod_{j=1}^{n-1} 2^{-j 2^{n-j-1}} \quad (= b_n);$$

and deduce that $b_n \le \|A^{2^n - 1}\|$. [*Hint.* Observe that 2^j has 2^{n-j-1} odd multiples among the numbers $1, 2, \ldots, 2^n - 1$.]

(ii)   Show that A is *not* a generalized nilpotent operator.

(iii)   Let $A_k y_n$ be $a_n y_{n+1}$ except when n is an odd multiple of 2^k, in which case let $A_k y_n$ be 0. Show that each A_k is nilpotent and that $\|A - A_k\| \to 0$.

(iv)   Deduce that the norm limit of nilpotent operators need not be generalized nilpotent. What relation does this conclusion have to the "spectral (semi-)continuity" considerations of Exercises 3.5.8 and 3.5.10?

Solution. (i)   The multiples of 2^j among $1, \ldots, 2^n$ are 2^j, $2(2^j)$, $3(2^j), \ldots, 2^{n-j}(2^j)$. Half, namely 2^{n-j-1}, of these are odd multiples of 2^j and they are among $1, 2, \ldots, 2^n - 1$. By choice of a_h, these odd multiples of 2^j among $1, 2, \ldots, 2^n - 1$ contribute the factor $(2^{-j})^{2^{n-j-1}}$ to the product b_n, from which the formula of (i) follows. From Exercise 3.5.12(iii), we have $b_n \le \|A^{2^n-1}\|$.

(ii)   From (i), writing b for $\sum_{j=1}^{\infty} j/2^j$, we have

$$2^{-b} \le \prod_{j=1}^{n-1} 2^{-j/2^j}$$

$$\le \prod_{j=1}^{n-1} 2^{-j2^{n-j-1}/(2^n-1)} = b_n^{1/(2^n-1)}$$

$$\le \|A^{2^n-1}\|^{1/(2^n-1)}$$

Thus $0 < 2^{-b} \le r(A)$ and A is not a generalized nilpotent.

(iii)   Since each sequence of consecutive positive integers containing 2^{k+1} integers contains an odd multiple of 2^k, it follows from the definition of A_k and Exercise 3.5.12(ii) that $A^{2^{k+1}} = 0$. Now $(A - A_k)y_k = c_n y_{n+1}$, where $c_n = 0$ unless n is an odd multiple of 2^k in which case $c_n = 2^{-k}$. It follows from Exercise 3.5.12(i) that $\|A - A_k\| = 2^{-k} \to 0$.

(iv)   As noted, A is not a generalized nilpotent operator, while each A_k is nilpotent; and $\{A_k\}$ tends to A in norm. From Exercise 3.5.11, each A_k has spectrum consisting of 0 alone; and from Exercise 3.5.8, $0 \in \text{sp } A$. But sp A does not consist of 0 alone since A is not a generalized nilpotent operator. Thus the "semi-continuity" of spectra established in Exercise 3.5.8 is not full continuity in the case of $\{A_k\}$ and A. Full continuity holds in the "commuting" case as can be seen be combining the results of Exercises 3.5.8 and 3.5.10(iii); but of course the operators A_k do not commute with A.   ∎

3.5.14.   With the notation of Exercise 3.5.12, show that:

(i) $\{a_n\}$ tends to 0 if A is compact (compare (vi) of Exercise 3.5.12);

(ii) A is a compact generalized nilpotent operator on $\mathcal{H}$ when $j < \pi(j)$ for all j and $\{a_n\}$ tends to 0 (compare (vii) of Exercise 3.5.12).

Solution. (i)   For each u in $\mathcal{H}$, $\|u\|^2 = \sum_{n=1}^{\infty} |\langle u, y_n \rangle|^2$, from Theorem 2.2.9; so that $\{\langle u, y_n \rangle\}$ tends to 0 and $\{y_n\}$ converges weakly to 0. From Exercise 2.8.20, $\{Ay_n\}$ converges to 0 in norm when A is compact. But $\|Ay_n\| = \|a_n y_{\pi(n)}\| = |a_n|$. Thus $\{a_n\}$ tends to 0 if A is compact.

(ii)   Given a positive ε, we can choose a positive integer k such that $|a_m| < \varepsilon/2$ when $m \geq k$, since $a_n \to 0$. Let a be the largest of $|a_1|, \ldots, |a_{k-1}|, \varepsilon/2$. Then for each n,

$$\left| a_n a_{\pi(n)} \cdots a_{\pi^{k-1}(n)} \right| \leq a^k$$

since either $\pi^j(n) < k$ and $|a_{\pi^j(n)}| \leq a$ or $k \leq \pi^j(n)$ and $|a_{\pi^j(n)}| < \varepsilon/2 \leq a$. Thus, with m larger than k,

$$\left| a_n a_{\pi(n)} \cdots a_{\pi^{k-1}(n)} a_{\pi^k(n)} \cdots a_{\pi^{m-1}(n)} \right|^{1/m} \leq a^{k/m} \left(\frac{\varepsilon}{2} \right)^{\frac{m-k}{m}}$$

for all n, since $j < \pi^j(n)$ for all j and all n. With m large, we have $|a_n a_{\pi(n)} \cdots a_{\pi^{m-1}(n)}| < \varepsilon$ for all n. From Exercise 3.5.12(iii), $\|A^m\|^{1/m} < \varepsilon$ for large m; and A is a generalized nilpotent operator. From Exercise 3.5.12(vi), A is compact.   ∎

3.5.15.   Does each compact operator have a (non-zero) eigenvector? (Example? Proof? See Exercise 2.8.29(ii).)

Solution.   No! Suppose A of Exercise 3.5.12(iv) is such that $a_n \neq 0$ for all n and $a_n \to 0$. Then, from Exercise 3.5.14(ii), A is a compact generalized nilpotent operator. Thus, if $x \, (= \sum_{n=1}^{\infty} b_n y_n)$ is an eigenvector for A,

$$0 = Ax = \sum_{n=1}^{\infty} a_n b_n y_{\pi(n)};$$

and $a_n b_n = 0$ for all n. By assumption, $a_n \neq 0$ so that $b_n = 0$ for all n; and A has no (non-zero) eigenvectors.   ∎

3.5.16.   Can a non-zero self-adjoint operator be a generalized nilpotent operator? (Example? Proof?)

Solution.   No! From Proposition 3.2.15, if A is self-adjoint, at least one of $\|A\|$ or $-\|A\|$ is in the spectrum of A. If, in addition, A is a generalized nilpotent operator, then sp A is $\{0\}$ so that $\|A\| = 0$ and $A = 0$.   ∎

3.5.17. Let A be a compact operator acting on a Hilbert space $\mathcal{H}$, λ be a non-zero complex number, and $\{x_n\}$ be a sequence of vectors in $\mathcal{H}$ such that $\{(A-\lambda I)x_n\}$ converges in norm to a vector y in $\mathcal{H}$.

(i) Show that if $\{\|x_n\|\}$ is unbounded, $(A - \lambda I)x = 0$ for some unit vector x in $\mathcal{H}$.

(ii) Show that if $\{\|x_n\|\}$ is bounded, y is in the range of $A - \lambda I$. [*Hint.* Use Exercise 1.9.17(ii).]

(iii) Show that the range of $A - \lambda I$ is closed if the null space of $A - \lambda I$ is (0).

(iv) Show that if $\lambda \in \operatorname{sp} A$ either λ is an eigenvalue of A or $\bar{\lambda}$ is an eigenvalue of A^*.

Solution. (i) Passing to a subsequence, we may assume that $n \le \|x_n\|$. Let u_n be $x_n/\|x_n\|$. From Exercise 2.8.6, $\{u_n\}$ has a subsequence converging weakly to some x in $\mathcal{H}$. Passing to this subsequence of $\{u_n\}$ and relabeling, we may assume that $\{u_n\}$ converges weakly to x, $u_n = x_n/\|x_n\|$, and $n \le \|x_n\|$. Since A is compact $\{Au_n\}$ converges in norm to Ax. With n large, $\|(A - \lambda I)x_n\| \le \|y\| + 1$ so that

$$\|(A - \lambda I)u_n\| \le \frac{\|y\| + 1}{\|x_n\|} \le \frac{\|y\| + 1}{n}.$$

It follows that $\{(A - \lambda I)u_n\}$ converges in norm to 0. Since $\lambda u_n = Au_n - (A - \lambda I)u_n$, we have that $\{u_n\}$ converges in norm (and hence, weakly) to $\lambda^{-1}Ax$. Thus $x = \lambda^{-1}Ax$ and $(A - \lambda I)x = 0$. Now $\|u_n\| = 1$ for all n, so that $1 = \|\lambda^{-1}Ax\| = \|x\|$.

(ii) From Exercise 1.9.17(ii), $(A - \lambda I)((\mathcal{H})_1)$ is norm closed. If $\|x_n\| \le k$ for all n, then $k^{-1}y$ is the norm limit of $\{(A-\lambda I)(k^{-1}x_n)\}$ $(\subseteq (A - \lambda I)((\mathcal{H})_1))$; so that y is in the range of $A - \lambda I$.

(iii) If $A - \lambda I$ has null space (0) then $\{x_n\}$ is bounded, from (i); and y is in the range of $A - \lambda I$, from (ii). Thus $A - \lambda I$ has closed range.

(iv) If $\lambda \in \operatorname{sp} A$ and λ is not an eigenvalue of A, then $A - \lambda I$ has null space (0). From (iii), $A - \lambda I$ has closed range. If this range is $\mathcal{H}$, then $A - \lambda I$ has a continuous inverse from the Banach inversion theorem (1.8.5)—contradicting the choice of λ in $\operatorname{sp} A$. Thus the set $(A-\lambda I)(\mathcal{H})$ is closed and properly contained in $\mathcal{H}$; and there is a unit vector x orthogonal to the range of $A - \lambda I$. From Proposition 2.5.13,

x is in the null space of $(A - \lambda I)^*$—that is, x is an eigenvector of A^* corresponding to the eigenvalie $\bar{\lambda}$. ∎

    3.5.18.   Let $\mathcal{H}$ be a Hilbert space and A be an operator on $\mathcal{H}$ such that $Ax = x$ for some unit vector x in $\mathcal{H}$.

    (i)   Show that $A^*y = y$ for some unit vector y in $\mathcal{H}$ when the range of A is finite dimensional.

    (ii)   Show that if A is compact, there is a sequence $\{A_n\}$ of operators on $\mathcal{H}$ converging to A in norm, each A_n having finite-dimensional range, and such that $A_n x = x$ for all n. [*Hint.* Use Exercise 2.8.24(ii).]

    (iii)   Show that if A is compact, $A^*y = y$ for some unit vector y in $\mathcal{H}$.

    (iv)   Show that if A is compact and $0 \neq \lambda \in \operatorname{sp} A$, then λ is an eigenvalue of A.

    (v)   Suppose that $Ax_n = \lambda_n x_n$ $(n = 1, 2, \ldots)$, where $\{\lambda_n\}$ is a sequence of distinct non-zero eigenvalues of A and each x_n is a unit vector. Show that $\{x_n\}$ is linearly independent and that $\langle Ay_n, y_n \rangle = \lambda_n$, where $\{y_n\}$ is the orthonormal sequence obtained from $\{x_n\}$ by the Gram–Schmidt orthogonalization process.

    (vi)   Deduce from (iv) that $\operatorname{sp} A$ is *either* a finite set *or* can be arranged as a sequence of complex numbers tending to 0 and that each non-zero element in $\operatorname{sp} A$ is an eigenvalue of A of finite multiplicity when A is compact.

Solution. (i)   From Exercise 2.8.46, the range of A^* is finite dimensional. Thus the span of the ranges of A and A^* is finite dimensional. Let E be the projection with range this span. Then $E(\mathcal{H})$ is invariant under A and A^*, so that $AE = EA$. Now A restricted to $E(\mathcal{H})$ is an operator on a finite-dimensional space; and since $Ax = x \in E(\mathcal{H})$, 1 is in the spectrum of this restriction. The adjoint of the restricted operator is the restriction of A^* to $E(\mathcal{H})$; and 1 is in the spectrum of this restriction of A^*. But the spectrum and the set of eigenvalues of an operator on a finite-dimensional space coincide. Thus there is a unit vector y in $E(\mathcal{H})$ such that $A^*y = y$, under the present hypotheses.

    (ii)   Let y_1 be x in Exercise 2.8.24(ii). Alternatively, since A is compact, from Exercise 2.8.24(iii), A is the norm limit of a sequence $\{B_n\}$ of operators on $\mathcal{H}$ with finite-dimensional range. Let Y be an orthonormal basis for $\mathcal{H}$ containing x. Let A_n be the operator on $\mathcal{H}$

defined by $A_n y = B_n y$ for all y in $Y \setminus \{x\}$, and let $A_n x$ be x. Then A_n has finite-dimensional range; and if z in $\mathcal{H}$ is $\sum_{y \in Y} a_y y$, then

$$\begin{aligned}
\|(A_n - B_n)z\| &= |a_x|\,\|x - B_n x\| \\
&= |a_x|\,\|Ax - B_n x\| \\
&\leq \|A - B_n\|\,\|z\|\,.
\end{aligned}$$

Hence $\|A_n - B_n\| \leq \|A - B_n\|$; and $\|A_n - A\| \leq 2\|A - B_n\| \to 0$.

(iii) Let $\{A_n\}$ be the sequence constructed in (ii). From (i), there is a unit vector y_n such that $A_n^* y_n = y_n$. From Exercise 2.8.6, some subsequence of $\{y_n\}$ converges weakly to a vector y in $\mathcal{H}$. Passing to this subsequence and relabeling, we may assume that $\{y_n\}$ converges weakly to y. Since A^* is compact, $\{A^* y_n\}$ converges in norm to $A^* y$ and

$$\|y_n - A^* y\| \leq \|y_n - A_n^* y_n\| + \|A_n^* y_n - A^* y_n\| + \|A^* y_n - A^* y\| \to 0.$$

Thus $\{y_n\}$ converges in norm (and hence, weakly) to $A^* y$. It follows that $y = A^* y$ and $\|y\| = \|A^* y\| = \lim \|y_n\| = 1$.

(iv) If A is compact and λ is a non-zero eigenvalue for A^*, then 1 is an eigenvalue for $\bar{\lambda}^{-1} A^*$. Now $\bar{\lambda}^{-1} A^*$ is compact so that 1 is an eigenvalue for $\lambda^{-1} A$, from (iii). Thus λ is an eigenvalue for A in this case. From Exercise 3.5.17(iv), if $0 \neq \lambda \in \operatorname{sp} A$ and A is compact, either λ is an eigenvalue of A or $\bar{\lambda}$ is an eigenvalue of A^*. As just noted, λ is an eigenvalue of A in either event.

(v) Since $p(A)x_j = p(\lambda_j)x_j$ for each polynomial p, we have that $p_0(A)(a_1 x_1 + \cdots + a_n x_n) = a_j x_j$, where $p_0(x) = a \prod_{k \neq j}(x - \lambda_k)$ and $a = \left[\prod_{k \neq j}(\lambda_j - \lambda_k) \right]^{-1}$. Thus $a_j = 0$ for all j when $a_1 x_1 + \cdots + a_n x_n = 0$, and $\{x_n\}$ is linearly independent. Let y_1 be x_1 and suppose, after applying the Gram–Schmidt process to $\{x_1, \ldots, x_n\}$ to obtain $\{y_1, \ldots, y_n\}$, we have that $\langle Ay_j, y_j \rangle = \lambda_j$ for j in $\{1, \ldots, n\}$. Then, for some scalar c,

$$y_{n+1} = c\Big(x_{n+1} - \sum_{j=1}^{n} \langle x_{n+1}, y_j \rangle y_j\Big),$$

where $1 = \langle y_{n+1}, y_{n+1} \rangle = c\langle x_{n+1}, y_{n+1} \rangle$. Thus

$$\begin{aligned}
\langle Ay_{n+1}, y_{n+1} \rangle &= c\Big\langle Ax_{n+1} - \sum_{j=1}^{n} \langle x_{n+1}, y_j \rangle Ay_j, y_{n+1} \Big\rangle \\
&= \lambda_{n+1} c\langle x_{n+1}, y_{n+1} \rangle = \lambda_{n+1},
\end{aligned}$$

since A maps $[y_1, \ldots, y_n]$ $(= [x_1, \ldots, x_n])$ into itself.

(vi) With $\{x_n\}$, $\{y_n\}$ and $\{\lambda_n\}$ as in (v), it follows that $\lambda_n = \langle Ay_n, y_n \rangle \to 0$, since $\{y_n\}$ tends weakly to 0. From (iv) the non-zero elements in $\mathrm{sp}\, A$ are eigenvalues and the solutions to Exercise 2.8.29(iii),(iv) apply to complete this solution.   ∎

3.5.19.   Let A be the operator defined in Exercise 3.5.12. Assume that $j < \pi(j)$ for all j and $a_n \to 0$.

(i)   Show that A has no non-zero eigenvalues.

(ii)   Show that A has no eigenvalues if $a_n \neq 0$ for all n.

(iii)   Deduce again (see Exercise 3.5.14(ii)), with the aid of Exercise 3.5.18, that A is a (compact) generalized nilpotent operator.

Solution. (i) Let z be $\sum_{n=1}^{\infty} b_n y_n$, and suppose that $Az = \lambda z$ for some non-zero λ. Then

$$(*) \qquad Az = \sum_{n=1}^{\infty} a_n b_n y_{\pi(n)} = \sum_{n=1}^{\infty} \lambda b_n y_n.$$

Hence $a_n b_n = \lambda b_{\pi(n)}$ and b_n is 0 unless n is $\pi(m)$ for some m. Let k be the smallest value of n for which $b_n \neq 0$. Then $k = \pi(j)$ for some j; and by assumption, $j < k$. Thus $0 = b_j = a_j b_j = \lambda b_{\pi(j)} = \lambda b_k$, and $b_k = \lambda^{-1} 0 = 0$—contradicting the choice of k. It follows that $b_n = 0$ for all n, and A has no non-zero eigenvalues.

(ii) If $Az = 0$, then from $(*)$, $a_n b_n = 0$ for all n. Under the present assumption that $a_n \neq 0$ for all n, it follows that $b_n = 0$ for all n. Thus A has no eigenvalues in this case.

(iii) Since $\{a_n\}$ tends to 0, Exercise 3.5.12(vi) applies; and A is compact. Hence λ is an eigenvalue for A if $0 \neq \lambda \in \mathrm{sp}\, A$, from Exercise 3.5.18(iv). Since A has no non-zero eigenvalues, from (i), $\mathrm{sp}\, A$ consists of 0 alone; and A is a (compact) generalized nilpotent.   ∎

3.5.20.   Let $\mathcal{H}$ be $L_2(0,1)$ relative to Lebesgue measure, and let $(Kf)(s)$ be $\int_0^s f(t)\, dt$ for each f in $\mathcal{H}$.

(i)   Show that K is a Hilbert–Schmidt operator and a compact operator on $\mathcal{H}$.

(ii)   Show that K has no eigenvalues.

(iii)   Deduce that K is a compact generalized nilpotent operator.

Solution. (i) If T_k is defined as in Exercise 2.8.38(ii), where $k(s,t)$ is 1 when $s \geq t$ and $k(s,t)$ is 0 when $s < t$, then K is T_k.

From that exercise, K is a Hilbert–Schmidt operator; and from Exercise 2.8.37(iii), K is compact.

(ii) If $f \in \mathcal{H}$ and $Kf = \lambda f$ with λ different from 0, then λf is the (indefinite) integral of the L_1-function f so that f is absolutely continuous. Thus f must satisfy $(\lambda f)' = f$; and $f(t) = c \exp(\lambda^{-1} t)$ for some constant c and each t in $[0,1]$. But then $Kf = \lambda(f - c)$. Thus $c = 0$ and $f = 0$.

If $Kg = 0$ for some g in $\mathcal{H}$, then the continuous function Kg vanishes almost everywhere—hence, everywhere. Since $\int_0^s g(t)\, dt$ is 0 for all s in $[0,1]$, g is a null function.

(iii) From (ii), K has no eigenvalues. From (i), K is compact. Thus, from Exercise 3.5.18(iv), $\mathrm{sp}\, K = (0)$; and K is a compact generalized nilpotent operator. ∎

3.5.21. With the notation of Example 3.2.17, $\sum_{n=-\infty}^{\infty} \lambda^{-n} e_n$ is an "eigenvector" for U in a formal sense even when $|\lambda| \neq 1$. Pursue the argument of that example to see how it fails to establish that *each* λ in $\mathbb{C}$ is in $\mathrm{sp}(U)$.

Solution. Suppose $|\lambda| \neq 1$. Then

$$\left\| \sum_{k=-n}^{n} \lambda^{-k} e_k \right\| = \left[\sum_{k=-n}^{n} |\lambda|^{-2k} \right]^{1/2}$$

$$= \left[\frac{|\lambda|^{2n} - |\lambda|^{-2(n+1)}}{-|\lambda|^{-2} + 1} \right]^{1/2} \quad (= a_n^{-1}).$$

Let y_n be $a_n \sum_{k=-n}^{n} \lambda^{-k} e_k$ (so that y_n is the replacement for x_n in Example 3.2.17). Then $\|y_n\| = 1$ and

$$\|(U - \lambda I) y_n\|^2 = a_n^2 \left\| \lambda^{-n} e_{n+1} - \lambda^{n+1} e_{-n} \right\|^2$$

$$= (|\lambda|^{-2} - 1) \frac{|\lambda|^{-2n} + |\lambda|^{2(n+1)}}{|\lambda|^{-2(n+1)} - |\lambda|^{2n}}.$$

It follows that $\|(U - \lambda I) y_n\|$ tends to $|1 - |\lambda|^2|^{1/2}$ as n tends to ∞, and $\{y_n\}$ is not a sequence of approximating eigenvectors when $|\lambda| \neq 1$. ∎

3.5.22.   Let X be a compact Hausdorff space and $\mathcal{A}$ be a subalgebra of $C(X)$. Suppose $\|\ \|'$ and $\|\ \|''$ are norms on $\mathcal{A}$ such that $\mathcal{A}$ is complete relative to $\|\ \|'$ and $\mathcal{A}$ has a completion $\mathcal{A}_0$ in $C(X)$ relative to $\|\ \|''$. Let $\|f\|$ be $\sup\{|f(x)| : x \in X\}$ for f in $C(X)$.

(i)   Show that $\|f\| \le \|f\|'$ and $\|g\| \le \|g\|''$ for f in $\mathcal{A}$ and g in $\mathcal{A}_0$.

(ii)   Show that $\mathcal{A}$ is complete relative to the norm that assigns $\|f\|' + \|f\|''$ to f in $\mathcal{A}$.

(iii)   Show that $\|\ \|'$ and $\|\ \|''$ are equivalent norms on $\mathcal{A}$ (that is, the identity mapping is a homeomorphism from $\mathcal{A}$ relative to $\|\ \|'$ to $\mathcal{A}$ relative to $\|\ \|''$) when $\mathcal{A}$ is complete relative to $\|\ \|''$.

(iv)   Without the assumption that $\mathcal{A}$ is complete relative to $\|\ \|''$, show that the identity mapping from $(\mathcal{A}, \|\ \|')$ to $(\mathcal{A}, \|\ \|'')$ is continuous.

Solution. (i)   If $f \in \mathcal{A}$ and λ is in the range of f, then λ is in the spectrum of f relative to $C(X)$. Hence $\lambda \in \mathrm{sp}_{\mathcal{A}}(f)$ and $|\lambda| \le \|f\|'$. Thus $\|f\| \le \|f\|'$ and, similarly, $\|g\| \le \|g\|''$.

(ii)   Suppose $\{f_n\}$ is a Cauchy sequence in $\mathcal{A}$ relative to $\|\ \|' + \|\ \|''$. Then $\{f_n\}$ is a Cauchy sequence relative to both $\|\ \|'$ and $\|\ \|''$. Since $\mathcal{A}$ is complete relative to $\|\ \|'$ and $\mathcal{A}_0$ is complete relative to $\|\ \|''$, $\{f_n\}$ has limits g in $\mathcal{A}$ and h in $\mathcal{A}_0$ relative to $\|\ \|'$ and $\|\ \|''$, respectively. Thus $\|g - f_n\| \le \|g - f_n\|' \to 0$ and $\|h - f_n\| \le \|h - f_n\|'' \to 0$ as $n \to \infty$. Hence $g = h$ and $\{f_n\}$ has g as its limit relative to $\|\ \|' + \|\ \|''$. Thus $\mathcal{A}$ is complete relative to $\|\ \|' + \|\ \|''$.

(iii)   Since $\|f\|' \le \|f\|' + \|f\|''$ and $\|f\|'' \le \|f\|' + \|f\|''$ for each f in $\mathcal{A}$, the identity mapping is continuous from $\mathcal{A}$ relative to $\|\ \|' + \|\ \|''$ to $\mathcal{A}$ relative to each of $\|\ \|'$ and $\|\ \|''$. From (ii) and the Banach inversion theorem (1.8.5), the identity mapping is a homeomorphism from $\mathcal{A}$ relative to $\|\ \|' + \|\ \|''$ to $\mathcal{A}$ relative to each of $\|\ \|'$ and $\|\ \|''$. Thus $\|\ \|'$ and $\|\ \|''$ are equivalent norms.

(iv)   From the argument of (iii), we see that the identity mappings from $(\mathcal{A}, \|\ \|')$ to $(\mathcal{A}, \|\ \|' + \|\ \|'')$ and from $(\mathcal{A}, \|\ \|' + \|\ \|'')$ to $(\mathcal{A}, \|\ \|'')$ are continuous. The composition of these mappings, the identity mapping from $(\mathcal{A}, \|\ \|')$ to $(\mathcal{A}, \|\ \|'')$ is, therefore, continuous. ∎

3.5.23.   Let $\mathcal{B}$ be a commutative Banach algebra, and let $\mathcal{R}$ be the intersection of the maximal ideals of $\mathcal{B}$. (The ideal $\mathcal{R}$ is called the *radical* of $\mathcal{B}$, and $\mathcal{B}$ is said to be semi-simple when $\mathcal{R} = (0)$.) Let $\mathfrak{A}$ be an arbitrary Banach algebra and φ be a homomorphism of $\mathfrak{A}$ into $\mathcal{B}$ mapping the identity of $\mathfrak{A}$ onto that of $\mathcal{B}$. Let $\mathcal{K}$ be the kernel of φ.

(i)   Show that φ maps $\mathcal{K}^-$, the norm closure of $\mathcal{K}$, into $\mathcal{R}$. Deduce that $\mathcal{K}$ is closed when $\mathcal{R} = (0)$.

(ii)   Show that if $\mathcal{B}$ is semi-simple, then $\mathcal{B}$ is isomorphic with a subalgebra $\mathcal{A}_1$ of $C(X)$, where X is the compact Hausdorff space of non-zero multiplicative linear functionals on $\mathcal{B}$ with the weak* topology. [*Hint.* Let $\hat{A}(\rho)$ be $\rho(A)$, when $A \in \mathcal{B}$ and $\rho \in X$.]

(iii)   Show that if $\mathcal{B}$ is semi-simple, then φ is continuous. [*Hint.* Use the mapping of (ii) and this mapping composed with the quotient mapping $\mathfrak{A} \to \mathfrak{A}/\mathcal{K}$ together with Exercise 3.5.22(iv).]

Solution. (i)   If $\mathcal{M}$ is a maximal ideal in $\mathcal{B}$, then $\varphi^{-1}(\mathcal{M}_0)$ is a maximal ideal in $\mathfrak{A}$, where $\mathcal{M}_0 = \mathcal{M} \cap \mathcal{B}_0$ and $\mathcal{B}_0 = \varphi(\mathfrak{A})$; for $\mathcal{M}$ is the kernel of a non-zero multiplicative linear functional ρ on $\mathcal{B}$ and $\mathcal{M}_0$ is the kernel of its restriction to $\mathcal{B}_0$ (and $\varphi^{-1}(\mathcal{M}_0)$ is the kernel of $\rho \circ \varphi$). From Proposition 3.1.8, $\varphi^{-1}(\mathcal{M}_0)$ is norm closed. Thus $\varphi^{-1}(\mathcal{M}_0)$ contains $\mathcal{K}^-$, and $\varphi(\mathcal{K}^-) \subseteq \mathcal{M}$. It follows that $\varphi(\mathcal{K}^-)$ lies in the intersection $\mathcal{R}$ of all maximal ideals in $\mathcal{B}$. If $\mathcal{R} = (0)$, then $\varphi(\mathcal{K}^-) = 0$; and $\mathcal{K} = \mathcal{K}^-$.

(ii)   With A in $\mathcal{B}$ and ρ in X, let $\hat{A}(\rho)$ be $\rho(A)$. The mapping $A \to \hat{A}$ is a homomorphism of $\mathcal{B}$ into $C(X)$. If $\mathcal{B}$ is semi-simple $(\mathcal{R} = (0))$ and $0 \neq A \in \mathcal{B}$, there is a maximal ideal $\mathcal{M}$ in $\mathcal{B}$ such that $A \notin \mathcal{M}$. Let ρ be the (unique, non-zero) multiplicative linear functional on $\mathcal{B}$ with $\mathcal{M}$ as null space. Then $\hat{A}(\rho) = \rho(A) \neq 0$, so that $\hat{A} \neq 0$ and the mapping $A \to \hat{A}$ is an isomorphism of $\mathcal{B}$ onto a subalgebra $\mathcal{A}_1$ of $C(X)$.

(iii)   If $\mathcal{B}$ is semi-simple, $\mathcal{R} = (0)$; and from (i), $\mathcal{K}$ is norm-closed. Thus $\mathfrak{A}/\mathcal{K}$ is a Banach algebra in its quotient norm, from Proposition 3.1.8. Let ξ be the quotient mapping of $\mathfrak{A}$ onto $\mathfrak{A}/\mathcal{K}$, and let $\widetilde{\varphi}(\xi(A))$ be $\widehat{\varphi(A)}$. Then $\widetilde{\varphi}$ is an isomorphism of $\mathfrak{A}/\mathcal{K}$ onto a subalgebra $\mathcal{A}$ of $C(X)$. We impose two norms on $\mathcal{A}$ (other than the norm $\| \ \|$ it inherits from $C(X)$). The first, $\| \ \|'$, is that transported to it from $\mathfrak{A}/\mathcal{K}$ by the isomorphism $\widetilde{\varphi}$; the second, $\| \ \|''$, is that inherited from $\mathcal{A}_1$ (in the notation of (ii)) in the norm transported to $\mathcal{A}_1$ from $\mathcal{B}$ by the isomorphism $\hat{\ }$. Then $(\mathcal{A}, \| \ \|')$ and $(\mathcal{A}_1, \| \ \|'')$ are Banach algebras (in essence, $\mathfrak{A}/\mathcal{K}$ and $\mathcal{B}$, respectively). The closure $\mathcal{A}_0$ of

$\mathcal{A}$ in $(\mathcal{A}_1, \| \ \|'')$ is a "completion" of $\mathcal{A}$ in $C(X)$. Exercise 3.5.22(iv) applies, and the identity mapping ι' from $(\mathcal{A}, \| \ \|')$ to $(\mathcal{A}, \| \ \|'')$ is continuous. If η is the mapping of $(\mathcal{A}, \| \ \|'')$ into $\mathcal{B}$ inverse to $\widehat{\ }$ (on the image of $(\mathcal{A}, \| \ \|'')$ in $\mathcal{B}$), then η is continuous. The mapping φ can be displayed, now, as the composition,

$$\mathfrak{A} \xrightarrow{\xi} \mathfrak{A}/\mathcal{K} \xrightarrow{\widetilde{\varphi}} (\mathcal{A}, \| \ \|') \xrightarrow{\iota'} (\mathcal{A}, \| \ \|'') \xrightarrow{\eta} \mathcal{B},$$

each of whose components is continuous.   ■

3.5.24.   Establish the following inclusions for a pair of commuting elements A and B of an arbitrary Banach algebra $\mathfrak{A}$:

$$\mathrm{sp}_{\mathfrak{A}}(AB) \subseteq \mathrm{sp}_{\mathfrak{A}}(A)\,\mathrm{sp}_{\mathfrak{A}}(B), \qquad \mathrm{sp}_{\mathfrak{A}}(A+B) \subseteq \mathrm{sp}_{\mathfrak{A}}(A) + \mathrm{sp}_{\mathfrak{A}}(B).$$

[*Hint.* Apply Proposition 3.2.10 to a maximal abelian subalgebra of $\mathfrak{A}$.]

Solution.   An application of Zorn's lemma establishes the existence of a subalgebra $\mathfrak{A}_0$ of $\mathfrak{A}$ containing A and B and maximal with respect to the property of being abelian. By (norm) continuity of multiplication in $\mathfrak{A}$, the norm closure of $\mathfrak{A}_0$ is an abelian subalgebra of $\mathfrak{A}$. Since $\mathfrak{A}_0$ is contained in this norm closure, by its maximal property, $\mathfrak{A}_0$ coincides with its norm closure. Thus $\mathfrak{A}_0$ is a Banach subalgebra of $\mathfrak{A}$. (Clearly $I \in \mathfrak{A}_0$.) If $T \in \mathfrak{A}_0$ and T has an inverse T^{-1} in $\mathfrak{A}$, then for each C in $\mathfrak{A}_0$,

$$CT^{-1} = T^{-1}TCT^{-1} = T^{-1}CTT^{-1} = T^{-1}C.$$

As $\mathfrak{A}_0$ is maximal abelian in $\mathfrak{A}$ and T^{-1} commutes with each C in $\mathfrak{A}_0$, $T^{-1} \in \mathfrak{A}_0$. It follows, now, that $T - \lambda I$ has an inverse in $\mathfrak{A}$ if and only if it has an inverse in $\mathfrak{A}_0$; so that $\mathrm{sp}_{\mathfrak{A}_0}(T) = \mathrm{sp}_{\mathfrak{A}}(T)$ for each T in $\mathfrak{A}_0$. Combining this conclusion with Proposition 3.2.10, we have

$$\mathrm{sp}_{\mathfrak{A}}(AB) = \mathrm{sp}_{\mathfrak{A}_0}(AB) \subseteq \mathrm{sp}_{\mathfrak{A}_0}(A)\,\mathrm{sp}_{\mathfrak{A}_0}(B) = \mathrm{sp}_{\mathfrak{A}}(A)\,\mathrm{sp}_{\mathfrak{A}}(B)$$

and

$$\begin{aligned}
\mathrm{sp}_{\mathfrak{A}}(A+B) &= \mathrm{sp}_{\mathfrak{A}_0}(A+B)\\
&\subseteq \mathrm{sp}_{\mathfrak{A}_0}(A) + \mathrm{sp}_{\mathfrak{A}_0}(B)\\
&= \mathrm{sp}_{\mathfrak{A}}(A) + \mathrm{sp}_{\mathfrak{A}}(B). \qquad ■
\end{aligned}$$

3.5.25. Let A be a normal operator acting on a Hilbert space $\mathcal{H}$.

(i) Show that $A + aI$ is a normal operator for each scalar a and that A^{-1} is normal when A is invertible.

(ii) Is the sum of two normal operators necessarily normal?

(iii) Use Theorem 2.4.2(iv) to show that $r_{B(\mathcal{H})}(A) = \|A\|$.

Solution. (i) Since

$$(A + aI)^*(A + aI) = (A^* + \bar{a}I)(A + aI) = A^*A + aA^* + \bar{a}A + \bar{a}aI$$
$$= AA^* + aA^* + \bar{a}A + a\bar{a}I = (A + aI)(A^* + \bar{a}I)$$
$$= (A + aI)(A + aI)^*,$$

$A + aI$ is normal. If A is invertible, then

$$(A^{-1})^*A^{-1} = (A^*)^{-1}A^{-1} = (AA^*)^{-1} = (A^*A)^{-1}$$
$$= A^{-1}(A^*)^{-1} = A^{-1}(A^{-1})^*;$$

and A^{-1} is normal.

(ii) No! If T has matrix representation $\begin{bmatrix} 0 & 1 \\ 0 & 0 \end{bmatrix}$ relative to an orthonormal basis for two-dimensional Hilbert space, then

$$TT^* = \begin{bmatrix} 0 & 1 \\ 0 & 0 \end{bmatrix}\begin{bmatrix} 0 & 0 \\ 1 & 0 \end{bmatrix} = \begin{bmatrix} 1 & 0 \\ 0 & 0 \end{bmatrix} \neq \begin{bmatrix} 0 & 0 \\ 0 & 1 \end{bmatrix} = \begin{bmatrix} 0 & 0 \\ 1 & 0 \end{bmatrix}\begin{bmatrix} 0 & 1 \\ 0 & 0 \end{bmatrix} = T^*T,$$

so that T is not normal. But each operator T on a Hilbert space is the sum of two normal operators $\frac{1}{2}(T + T^*)$ and $\frac{1}{2}(T - T^*)$.

(iii) From Theorem 2.4.2(iv), $\|A^*A\| = \|A\|^2$. As A^*A is self-adjoint, $r_{B(\mathcal{H})}(A^*A) = \|A^*A\| \ (= \|A\|^2)$, from Proposition 3.2.15. Since A and A^* commute, $r_{B(\mathcal{H})}(A^*A) \leq r_{B(\mathcal{H})}(A^*)r_{B(\mathcal{H})}(A)$, from Exercise 3.5.24. But $r_{B(\mathcal{H})}(A^*) = r_{B(\mathcal{H})}(A)$, from Theorem 3.2.14(v). Thus $r_{B(\mathcal{H})}(A)^2 = \|A\|^2$; and $r_{B(\mathcal{H})}(A) = \|A\|$. ∎

3.5.26. Let A_n be a bounded operator on a Hilbert space $\mathcal{H}_n$ for n in $\{1, 2, \ldots\}$, and suppose that $\{\|A_n\|\}$ is bounded. Let A be $\sum \oplus A_n$ and $\mathcal{H}$ be $\sum \oplus \mathcal{H}_n$. Show that:

(i) (∗) $$\left[\bigcup_{n=1}^{\infty} \mathrm{sp}_{B(\mathcal{H}_n)}(A_n)\right]^{-} \subseteq \mathrm{sp}_{B(\mathcal{H})}(A);$$

(ii) the inclusion (∗) becomes equality when each A_n is normal.

Solution. (i) Note, first, that if $\sum \oplus T_n$ has an inverse B in $\mathcal{B}(\mathcal{H})$ with $[B_{jk}]$ as its matrix representation (see subsection 2.6, *Matrix representations*) then $T_n B_{nn} = B_{nn} T_n = I_n$ for each n, where I_n is the unit element in $\mathcal{B}(\mathcal{H}_n)$; so that each T_n has an inverse. Thus, if $\lambda \notin \mathrm{sp}_{\mathcal{B}(\mathcal{H})}(A)$, then $\lambda \notin \mathrm{sp}_{\mathcal{B}(\mathcal{H}_n)}(A_n)$ for all n; and since $\mathrm{sp}_{\mathcal{B}(\mathcal{H})}(A)$ is closed, $(*)$ holds.

(ii) If each A_n is normal and $\lambda \notin \left[\bigcup_{n=1}^{\infty} \mathrm{sp}_{\mathcal{B}(\mathcal{H}_n)}(A_n) \right]^-$, then λ is at positive distance d from $\left[\bigcup_{n=1}^{\infty} \mathrm{sp}_{\mathcal{B}(\mathcal{H}_n)}(A_n) \right]^-$. By Exercise 3.5.25 and Proposition 3.2.10,

$$\|(A_n - \lambda I)^{-1}\| = r_{\mathcal{B}(\mathcal{H}_n)}([A_n - \lambda I_n]^{-1}) \leq d^{-1}.$$

Thus $\sum \oplus (A_n - \lambda I_n)^{-1}$ is an element B of $\mathcal{B}(\mathcal{H})$ that is inverse to $\sum \oplus (A_n - \lambda I_n) \ (= A - \lambda I)$. It follows that $\lambda \notin \mathrm{sp}_{\mathcal{B}(\mathcal{H})}(A)$ so that, in this case,

$$(**) \qquad \mathrm{sp}_{\mathcal{B}(\mathcal{H})}(A) \subseteq \left[\bigcup_{n=1}^{\infty} \mathrm{sp}_{\mathcal{B}(\mathcal{H}_n)}(A_n) \right]^-.$$

Combining $(*)$ and $(**)$, we have

$$(***) \qquad \mathrm{sp}_{\mathcal{B}(\mathcal{H})}(A) = \left[\bigcup_{n=1}^{\infty} \mathrm{sp}_{\mathcal{B}(\mathcal{H}_n)}(A_n) \right]^-$$

when A_n are normal operators such that $\{\|A_n\|\}$ is bounded. $\quad\blacksquare$

3.5.27. Let $\mathcal{H}_n$ be an n-dimensional Hilbert space, $\{e_1, \ldots, e_n\}$ be an orthonormal basis for $\mathcal{H}_n$, N_n be the linear operator on $\mathcal{H}_n$ determined by $N_n e_1 = 0$ and $N_n e_j = e_{j-1}$ if $1 < j \leq n$, and A_n be $I - N_n$.

(i) Show that $\|A_n\| \leq 2$ and that $\mathrm{sp}_{\mathcal{B}(\mathcal{H}_n)}(A_n) = \{1\}$ for each n in $\{1, 2, \ldots\}$.

(ii) Show that A_n is invertible in $\mathcal{B}(\mathcal{H}_n)$ with inverse $I + N_n + N_n^2 + \cdots + N_n^{n-1}$ $(= B_n)$ and that $\|B_n\| \geq \sqrt{n}$.

(iii) Show that $0 \in \mathrm{sp}_{\mathcal{B}(\mathcal{H})}(A)$, where $A = \sum \oplus A_n$ and $\mathcal{H} = \sum \oplus \mathcal{H}_n$, and deduce that the inclusion $(*)$ of Exercise 3.5.26 is strict in the present case.

Solution. (i) Let y be $\sum_{j=1}^{n} a_j e_j$ and $\|y\|$ be 1. Then

$$\left\| N_n \left(\sum_{j=1}^{n} a_j e_j \right) \right\|^2 = \left\| \sum_{j=2}^{n} a_j e_{j-1} \right\|^2$$

$$= \sum_{j=2}^{n} |a_j|^2 \leq \sum_{j=1}^{n} |a_j|^2 = \|y\|^2 = 1.$$

Thus $\|N_n\| \leq 1$ and $\|A_n\| \leq 2$. Note that $N_n^k e_j = 0$ if $j \leq k$ and $N_n^k e_j = e_{j-k}$ if $k < j$, for k in $\{1,\ldots,n\}$. It follows that $N_n^n = 0$. Thus N_n is a nilpotent operator and $\mathrm{sp}_{\mathcal{B}(\mathcal{H})}(N_n) = \{0\}$, from Exercise 3.5.11(i). Hence $\mathrm{sp}_{\mathcal{B}(\mathcal{H})}(A_n) = \{1\}$ for each n.

(ii) Note that $A_n B_n = B_n A_n = I - N_n^n = I$, from (i); so that B_n is the inverse of A_n in $\mathcal{B}(\mathcal{H})$. Note, too, that

$$\|B_n e_n\|^2 = \|e_n + e_{n-1} + \cdots + e_1\|^2 = n;$$

so that $\sqrt{n} = \|B e_n\| \leq \|B_n\|$.

(iii) If A has an inverse B in $\mathcal{B}(\mathcal{H})$ and $[B_{jk}]$ is the matrix representation of B, then B_{nn} is the inverse of A_n for each n. Thus $B_{n,n} = B_n$ for each n, from (ii). Since $\{\|B_n\|\}$ is not bounded (from (ii)), there is no such inverse B in $\mathcal{B}(\mathcal{H})$, from Proposition 2.6.13(ii); and $0 \in \mathrm{sp}_{\mathcal{B}(\mathcal{H})}(A)$. From (i), $\left[\bigcup_{n=1}^{\infty} \mathrm{sp}_{\mathcal{B}(\mathcal{H}_n)}(A_n) \right]^- = \{1\}$. Thus the inclusion $(*)$ of Exercise 3.5.26 is strict in this case. ■

3.5.28.   Suppose the singular element A in the Banach algebra $\mathfrak{A}$ is a norm limit of the sequence $\{A_n\}$ of regular elements in $\mathfrak{A}$ and $\mathfrak{A}_0$ is a Banach subalgebra of $\mathfrak{A}$. Show that:

(i) $\{\|A_n^{-1}\|\}$ tends to ∞;

(ii)   if $B_n = (A_n^{-1} - I)/\|A_n^{-1} - I\|$ (when $A_n^{-1} \neq I$), then $\{AB_n\}$ and $\{B_n A\}$ tend to 0; (When a sequence such as $\{B_n\}$, not tending to 0, with $\{AB_n\}$ and $\{B_n A\}$ tending to 0, exists, we say that A is a *(two-sided) topological divisor of zero*.) [*Hint.* Note that $A(A_n^{-1} - I) = (A - A_n)(A_n^{-1} - I) - A_n + I.$]

(iii)   if B is a topological divisor of 0 in $\mathfrak{A}$, then B has no inverse in $\mathfrak{A}$;

(iv)   if $B \in \mathfrak{A}_0$ and λ is a boundary point of $\mathrm{sp}_{\mathfrak{A}_0}(B)$, then $\lambda \in \mathrm{sp}_{\mathfrak{A}}(B)$;

(v)   if $B \in \mathfrak{A}_0$ and either one of $\mathrm{sp}_{\mathfrak{A}_0}(B)$ or $\mathrm{sp}_{\mathfrak{A}}(B)$ is real, then $\mathrm{sp}_{\mathfrak{A}_0}(B) = \mathrm{sp}_{\mathfrak{A}}(B)$.

Solution. (i)   If $\{\|A_n^{-1}\|\}$ does not tend to ∞, some subsequence $\{\|A_{n(1)}^{-1}\|, \|A_{n(2)}^{-1}\|, \ldots\}$ is bounded (say, by k); and

$$\|AA_{n(j)}^{-1} - I\| = \|(A - A_{n(j)})A_{n(j)}^{-1}\|$$
$$\leq \|A - A_{n(j)}\|\|A_{n(j)}^{-1}\| \leq k\|A - A_{n(j)}\| \to 0.$$

Thus, for large j, $AA_{n(j)}^{-1}$ is regular (from Lemma 3.1.5) and A is regular, contradicting our hypothesis. Thus $\{\|A_n^{-1}\|\}$ tends to ∞.

(ii)   From (i), $\{\|A_n^{-1} - I\|\}$ tends to ∞, so that

$$AB_n = (A - A_n)B_n - \frac{A_n - I}{\|A_n^{-1} - I\|} \to 0.$$

Similarly,

$$B_n A = B_n(A - A_n) - \frac{A_n - I}{\|A_n^{-1} - I\|} \to 0.$$

(iii)   If B is a (right) topological divisor of zero in $\mathfrak{A}$, there is a sequence $\{C_n\}$ in $\mathfrak{A}$ such that $\{C_n\}$ does not tend to 0 and $BC_n \to 0$. For each T in $\mathfrak{A}$, $\|TBC_n\| \leq \|T\|\|BC_n\| \to 0$. Thus $TBC_n \neq C_n$ for some n, and $TB \neq I$. It follows that B has no left inverse in $\mathfrak{A}$.

(iv)   If λ is a boundary point of $\mathrm{sp}_{\mathfrak{A}_0}(B)$, then $\lambda \in \mathrm{sp}_{\mathfrak{A}_0}(B)$ since $\mathrm{sp}_{\mathfrak{A}_0}(B)$ is closed. Thus $B - \lambda I$ is singular in $\mathfrak{A}_0$. At the same time, there is a sequence $\{\lambda_n\}$ tending to λ such that $\lambda_n \notin \mathrm{sp}_{\mathfrak{A}_0}(B)$. Thus $B - \lambda I$ is the limit of a sequence $\{B - \lambda_n I\}$ of regular elements in $\mathfrak{A}_0$. From (ii), $B - \lambda I$ is a topological divisor of 0 in $\mathfrak{A}_0$ and, hence, in $\mathfrak{A}$. From (iii), $\lambda \in \mathrm{sp}_{\mathfrak{A}}(B)$.

(v)   If $\mathrm{sp}_{\mathfrak{A}_0}(B)$ is real, each λ in $\mathrm{sp}_{\mathfrak{A}_0}(B)$ is a boundary point; and $\mathrm{sp}_{\mathfrak{A}_0}(B) \subseteq \mathrm{sp}_{\mathfrak{A}}(B)$. Since $\mathfrak{A}_0$ is a subalgebra of $\mathfrak{A}$, $\mathrm{sp}_{\mathfrak{A}}(B) \subseteq \mathrm{sp}_{\mathfrak{A}_0}(B)$. Hence $\mathrm{sp}_{\mathfrak{A}_0}(B) = \mathrm{sp}_{\mathfrak{A}}(B)$ when $\mathrm{sp}_{\mathfrak{A}_0}(B) \subseteq \mathbb{R}$.

Suppose, now, that $\lambda \in \mathrm{sp}_{\mathfrak{A}_0}(B)$ and $\lambda = a + ib$ with b not 0 (say, $0 < b$). Then $\{a + it : 0 \leq t\} \cap \mathrm{sp}_{\mathfrak{A}_0}(B)$ is a non-empty compact subset S of $\mathbb{C}$, and there is a t_0 such that $a + it_0 \in S$ and $t \leq t_0$ if $a + it \in S$. Since $a + ib \in S$, $0 < b \leq t_0$; and $a + it_0 \notin \mathbb{R}$. Moreover, $a + it_0$ is a boundary point of $\mathrm{sp}_{\mathfrak{A}_0}(B)$. From (iv), $a + it_0 \in \mathrm{sp}_{\mathfrak{A}}(B)$. Thus, if $\mathrm{sp}_{\mathfrak{A}}(B) \subseteq \mathbb{R}$, then $\mathrm{sp}_{\mathfrak{A}_0}(B) \subseteq \mathbb{R}$; and $\mathrm{sp}_{\mathfrak{A}_0}(B) = \mathrm{sp}_{\mathfrak{A}}(B)$. ∎

3.5.29.   Let A be an element of a Banach algebra $\mathfrak{A}$, and let f be in $\mathcal{H}(A)$. Suppose $f(z) = 0$ for each z in $\mathrm{sp}(A)$.

(i)   Show that $f(A) = 0$ if $\mathrm{sp}(A)$ is infinite and connected.

(ii) Find an example where $\mathrm{sp}(A)$ is connected but not infinite and $f(A) \neq 0$.

(iii) Find an example where $\mathrm{sp}(A)$ is infinite but not connected and $f(A) \neq 0$.

Solution. (i) Since $f \in \mathcal{H}(A)$, there is an open subset $\mathcal{O}$ of $\mathbb{C}$ containing $\mathrm{sp}(A)$ on which f is holomorphic. From the assumption that $\mathrm{sp}(A)$ is connected, some one connected component $\mathcal{O}_0$ of $\mathcal{O}$ contains $\mathrm{sp}(A)$. If $z \in \mathcal{O}_0$ and D is an open disk with center z contained in $\mathcal{O}$, then $D \cup \mathcal{O}_0$ is a connected subset of $\mathcal{O}$ containing $\mathrm{sp}(A)$; so that $D \subseteq \mathcal{O}_0$. Thus $\mathcal{O}_0$ is open. Since $\mathrm{sp}(A)$ is infinite and compact, $\mathrm{sp}(A)$ has an accumulation point. From the identity theorem for analytic functions, f vanishes throughout $\mathcal{O}_0$, since f vanishes on a subset $\mathrm{sp}(A)$ with an accumulation point and $\mathcal{O}_0$ is a connected open set on which f is holomorphic. Thus $f = 0 \cdot f$ on $\mathcal{O}_0$; and $f(A) = 0 \cdot f(A) = 0$, from Theorem 3.3.5.

(ii) Let $\mathcal{H}$ be a two-dimensional Hilbert space and A be the (nilpotent) operator whose matrix is $\begin{bmatrix} 0 & 1 \\ 0 & 0 \end{bmatrix}$ relative to some basis for $\mathcal{H}$. Let $f(z)$ be z for each z in $\mathbb{C}$. Then $f(A) = A \neq 0$, but $\mathrm{sp}(A) = \{0\}$ and $f(0) = 0$.

(iii) Let $\mathcal{H}_1$ and $\mathcal{H}_2$ be l_2 and $\{e_n\}$ be an orthonormal basis for l_2. Let A_1 be a compact generalized nilpotent operator on $\mathcal{H}_1$ with null space (0) (for example, we can take the operator A of Exercise 3.5.12(vii) for A_1), and let $A_2 e_n$ be $n^{-1}(n + 1)e_n$. From Example 3.2.12, there is an A_2 in $\mathcal{B}(\mathcal{H}_2)$ such that

$$\mathrm{sp}(A) = \{1, n^{-1}(n + 1) : n = 1, 2, \ldots\}$$

acting on $\{e_n\}$ as indicated. Let $\mathcal{H}$ be $\mathcal{H}_1 \oplus \mathcal{H}_2$ and A be $A_1 \oplus A_2$. Then $\mathrm{sp}_{\mathcal{B}(\mathcal{H})}(A) = \mathrm{sp}_{\mathcal{B}(\mathcal{H}_1)}(A_1) \cup \mathrm{sp}_{\mathcal{B}(\mathcal{H}_2)}(A_2)$; for if λ is not in the spectra of A_1 or A_2, then $(A_1 - \lambda I_1)^{-1} \oplus (A_2 - \lambda I_2)^{-1}$ is in $\mathcal{B}(\mathcal{H})$ and is inverse to $A - \lambda I$, where I_1 and I_2 are the identity operators on $\mathcal{H}_1$ and $\mathcal{H}_2$, respectively. On the other hand, if B is an inverse to $A - \mu I$ in $\mathcal{B}(\mathcal{H})$, then $B = B_1 \oplus B_2$, where B_1 and B_2 are inverses to $A_1 - \mu I_1$ and $A_2 - \mu I_2$, respectively; so that $\mu \notin \mathrm{sp}_{\mathcal{B}(\mathcal{H}_1)}(A_1) \cup \mathrm{sp}_{\mathcal{B}(\mathcal{H}_2)}(A_2)$. Thus $\mathrm{sp}_{\mathcal{B}(\mathcal{H})}(A)$ consists of $\{0, 1, n^{-1}(n + 1) : n = 1, 2, \ldots\}$.

Let $f(z)$ be z for z in the disk D of radius $\frac{1}{2}$ and center 0; and let $f(z)$ be 0 when $\frac{3}{4} < \mathbf{Re}\, z$. Let $g(z)$ be 1 for z in D and 0 when $\frac{3}{4} < \mathbf{Re}\, z$. Then f and g are in $\mathcal{H}(A)$ and $f = zg$. From Theorem 3.3.5, $f(A) = Ag(A)$; and from the comments preceding

Corollary 3.3.7, $g(A)$ is a non-zero idempotent. If x is a unit vector in the range of $g(A)$, then $f(A)x = Ag(A)x = Ax$. If $x = x_1 \oplus x_2$, then $Ax = A_1 x_1 \oplus A_2 x_2$; and $Ax \neq 0$, since A_1 and A_2 have null spaces (0). Thus $f(A) \neq 0$. ∎

3.5.30. Let $\mathcal{H}$ be a Hilbert space and λ be an isolated point of sp A with sp $A \setminus \{\lambda\}$ non-empty and A in $\mathcal{B}(\mathcal{H})$. Let D be an open disk with center λ and $\mathcal{O}$ be an open set disjoint from D containing sp $A \setminus \{\lambda\}$. Does the idempotent $g(A)$, obtained from the function g that takes the value 1 on D and 0 on $\mathcal{O}$, necessarily project onto a subspace of eigenvectors for A corresponding to λ?

Solution. No! If A is as constructed in the solution to Exercise 3.5.29(iii), then 0 is an isolated point of sp(A); but $g(A)$ does not project onto a space of vectors annihilated by A. From that solution, $g(A) \neq 0$ and A has null space (0). ∎

3.5.31. Let $\mathfrak{A}_1$ and $\mathfrak{A}_2$ be Banach algebras with units I_1 and I_2, respectively. Let φ be a homomorphism or anti-homomorphism (that is, $\varphi(AB) = \varphi(B)\varphi(A)$) of $\mathfrak{A}_1$ into $\mathfrak{A}_2$ such that $\varphi(I_1) = I_2$.
  (i) Show that $\mathrm{sp}_{\mathfrak{A}_2}(\varphi(A)) \subseteq \mathrm{sp}_{\mathfrak{A}_1}(A)$ for each A in $\mathfrak{A}_1$.
  (ii) Show, by example, that the inclusion of spectra described in (i) can be proper.

Solution. (i) Suppose $\lambda \notin \mathrm{sp}_{\mathfrak{A}_1}(A)$. Then $A - \lambda I_1$ has an inverse B and

$$I_2 = \varphi(I_1) = \varphi(B[A - \lambda I_1]) = \varphi([A - \lambda I_1]B).$$

Hence $\varphi(B)\varphi(A - \lambda I_1) = \varphi(A - \lambda I_1)\varphi(B) = I_2$, and $\lambda \notin \mathrm{sp}_{\mathfrak{A}_2}(\varphi(A))$.
  (ii) Let $\mathfrak{A}_1$ be $C([0,1])$ and $\mathfrak{A}_2$ be $C([0,\frac{1}{2}])$. With f in $\mathfrak{A}_1$, let $\varphi(f)$ be the restriction of f to $[0, \frac{1}{2}]$. Then φ is a homomorphism of $\mathfrak{A}_1$ onto $\mathfrak{A}_2$. If $f(\lambda) = \lambda$ for all λ in $[0,1]$, then $f \in \mathfrak{A}_1$ and $\mathrm{sp}_{\mathfrak{A}_1}(f) = [0,1]$. In this case,

$$\mathrm{sp}_{\mathfrak{A}_2}(\varphi(f)) = [0, \tfrac{1}{2}] \subset [0,1] = \mathrm{sp}_{\mathfrak{A}_1}(f). \qquad ∎$$

3.5.32. Let $\mathcal{M}$ be a closed left ideal in a Banach algebra $\mathfrak{A}$, and let $\mathcal{N}$ be a closed right ideal in $\mathfrak{A}$. Let $\mathfrak{X}$ and $\mathfrak{Y}$ be the quotient Banach spaces $\mathfrak{A}/\mathcal{M}$ and $\mathfrak{A}/\mathcal{N}$, respectively.

(i)   Show that $\varphi(A)$ and $\psi(A)$ are bounded linear operators on $\mathfrak{X}$ and $\mathfrak{Y}$, respectively, for each A in $\mathfrak{A}$, where

$$\varphi(A)(B + \mathcal{M}) = AB + \mathcal{M}, \qquad \psi(A)(B + \mathcal{N}) = BA + \mathcal{N}.$$

(ii)   Show that φ is a bounded homomorphism and ψ is a bounded anti-homomorphism of $\mathfrak{A}$ into $\mathcal{B}(\mathfrak{X})$ and $\mathfrak{A}$ into $\mathcal{B}(\mathfrak{Y})$, respectively, and that $\varphi(I)$ and $\psi(I)$ are the respective identity operators on $\mathfrak{X}$ and $\mathfrak{Y}$.

Solution. (i)   Note, first, that $\varphi(A)$ and $\psi(A)$ are well-defined; for if $B + \mathcal{M} = C + \mathcal{M}$, then $B - C \in \mathcal{M}$, and if $B + \mathcal{N} = C + \mathcal{N}$, then $B - C \in \mathcal{N}$. In the first case, $A(B - C) \in \mathcal{M}$, since $\mathcal{M}$ is a left ideal, so that

$$\varphi(A)(B + \mathcal{M}) = AB + \mathcal{M} = AC + \mathcal{M} = \varphi(A)(C + \mathcal{M});$$

in the second case $(B - C)A \in \mathcal{N}$, since $\mathcal{N}$ is a right ideal, so that

$$\psi(A)(B + \mathcal{N}) = BA + \mathcal{N} = CA + \mathcal{N} = \psi(A)(C + \mathcal{N}).$$

Clearly, φ and ψ are linear.

If $\|B + \mathcal{M}\| < 1$, then for some M in $\mathcal{M}$, $\|B + M\| < 1$. Thus

$$\|\varphi(A)(B + \mathcal{M})\| = \|A(B + M) + \mathcal{M}\| \le \|A(B + M)\|$$
$$\le \|A\|\,\|B + M\| \le \|A\|.$$

It follows that $\|\varphi(A)\| \le \|A\|$. Similarly, $\|\psi(A)\| \le \|A\|$.

(ii)   We conclude from the equality,

$$\varphi(aA + B)(C + \mathcal{M}) = (aA + B)C + \mathcal{M} = aAC + BC + \mathcal{M}$$
$$= a\varphi(A)(C + \mathcal{M}) + \varphi(B)(C + \mathcal{M})$$
$$= \big(a\varphi(A) + \varphi(B)\big)(C + \mathcal{M}),$$

that $\varphi(aA + B) = a\varphi(A) + \varphi(B)$, so that φ (and, similarly, ψ) is linear. From (i), $\|\varphi\| \le 1$ and $\|\psi\| \le 1$. Moreover,

$$\varphi(I)(A + \mathcal{M}) = A + \mathcal{M}, \qquad \psi(I)(A + \mathcal{N}) = A + \mathcal{N}$$

for each A in $\mathfrak{A}$, so that $\varphi(I)$ and $\psi(I)$ are the identity operators. Also

$$\Phi(AB)(C + \mathcal{M}) = ABC + \mathcal{M} = \varphi(A)(BC + \mathcal{M})$$
$$= \varphi(A)[\varphi(B)(C + \mathcal{M})] = [\varphi(A)\varphi(B)](C + \mathcal{M}).$$

Thus the mapping φ is a homomorphism (similarly, the mapping ψ is an anti-homomorphism).   ■

3.5.33.   With $\mathbb{Z}$ the additive group of integers and f, g complex-valued functions on $\mathbb{Z}$, let $f * g$ be the function whose domain consists of those integers s for which $\sum_{t \in \mathbb{Z}} f(t)g(s - t)$ converges and whose value at s is the sum. (We call $f * g$ the *convolution* of f and g.) Show that:

(i) $f * g = g * f$;

(ii)   if $f \in l_1(\mathbb{Z})$ and $g \in l_p(\mathbb{Z})$ (where $1 \le p$), then $f * g \in l_p(\mathbb{Z})$ and

$$\|f * g\|_p \le \|f\|_1 \cdot \|g\|_p\,;$$

(iii)   if $f \in l_1(\mathbb{Z})$, g and h are in $l_p(\mathbb{Z})$, and $a \in \mathbb{C}$, then $f*(a{\cdot}g+h)$ and $a \cdot f * g + f * h$ are in $l_p(\mathbb{Z})$ and

$$f * (a \cdot g + h) = a \cdot f * g + f * h;$$

(iv)   if $f, g \in l_1(\mathbb{Z})$ and $h \in l_p(\mathbb{Z})$, then $(f * g) * h$ and $f * (g * h)$ are in $l_p(\mathbb{Z})$ and

$$(f * g) * h = f * (g * h).$$

(v)   Conclude that $l_1(\mathbb{Z})$, provided with the mappings $(f, g) \rightarrow f * g$ and $f \rightarrow \|f\|_1$, is a commutative Banach algebra with unit e, where $e(0) = 1$ and $e(t) = 0$ if $t \ne 0$.

Solution.   What follows is essentially routine modification of the proof of Proposition 3.2.22.

(i)   If s is in the domain of $f * g$, then

$$(f * g)(s) = \sum_{t \in \mathbb{Z}} f(t)g(s - t) = \sum_{t \in \mathbb{Z}} f(s + t)g(-t) = \sum_{t \in \mathbb{Z}} g(t)f(s - t).$$

Thus s is in the domain of $g * f$ and $(f * g)(s) = (g * f)(s)$. By symmetry, the domain of $g * f$ is contained in the domain of $f * g$. Hence $f * g = g * f$.

(ii)   If $h_s(t) = g(s - t)$ and $\mu(S) = \sum_{t \in S} |f(t)|$ for each subset S of $\mathbb{Z}$, then $h_s \in l_p(\mathbb{Z})$ and μ is a finite measure on $\mathbb{Z}$. Apply the Hölder inequality to h_s and the constant function 1 relative to μ:

$$\int_{\mathbb{Z}} |g(s - t)| \cdot 1 \, d\mu(t) \le \left(\sum_{t \in \mathbb{Z}} |h_s(t)|^p \, |f(t)| \right)^{1/p} \|f\|_1^{\frac{p-1}{p}}$$

and, since $\int_{\mathbb{Z}} |g(s-t)|\, d\mu(t) = \sum_{t\in\mathbb{Z}} |f(t)|\, |g(s-t)|$, $(f*g)(s)$ converges for each s. Moreover,

$$\left(\sum_{t\in\mathbb{Z}} |g(s-t)|\,|f(t)|\right)^p \le \|f\|_1^{p-1} \sum_{t\in\mathbb{Z}} |g(s-t)|^p\,|f(t)|.$$

Since

$$\sum_{t\in\mathbb{Z}}\sum_{s\in\mathbb{Z}} |g(s-t)|^p\,|f(t)| = \sum_{t\in\mathbb{Z}} |f(t)|\left(\sum_{s\in\mathbb{Z}} |g(s-t)|^p\right) = \|f\|_1\,\|g\|_p^p,$$

we may interchange the order of summation of the first sum. Thus

$$\sum_{s\in\mathbb{Z}} |(f*g)(s)|^p = \sum_{s\in\mathbb{Z}}\left|\sum_{t\in\mathbb{Z}} f(t)g(s-t)\right|^p$$

$$\le \|f\|_1^{p-1} \sum_{s\in\mathbb{Z}}\sum_{t\in\mathbb{Z}} |g(s-t)|^p\,|f(t)|$$

$$= \|f\|_1^{p-1} \sum_{t\in\mathbb{Z}}\sum_{s\in\mathbb{Z}} |g(s-t)|^p\,|f(t)| = \|f\|_1^p\,\|g\|_p^p.$$

Thus $\|f*g\|_p \le \|f\|_1\,\|g\|_p$.

(iii) From (ii), $f*(a\cdot g+h)$ and $a\cdot f*g+f*h$ are in $l_p(\mathbb{Z})$. In particular $\sum_{t\in\mathbb{Z}} f(t)[ag(s-t)+h(s-t)]$ and $a\sum_{t\in\mathbb{Z}} f(t)g(s-t)+\sum_{t\in\mathbb{Z}} f(t)h(s-t)$ converges for each s in $\mathbb{Z}$. Thus $f*(a\cdot g+h) = a\cdot f*g+f*h$.

(iv) From (ii), $(f*g)*h$ and $f*(g*h)$ are in $l_p(\mathbb{Z})$. Now

$$[f*(g*h)](s) = \sum_{t\in\mathbb{Z}} f(t)\left(\sum_{r\in\mathbb{Z}} g(r)h(s-t-r)\right),$$

$$[(f*g)*h](s) = \sum_{r\in\mathbb{Z}}\left(\sum_{t\in\mathbb{Z}} f(t)g(r-t)\right)h(s-r);$$

so that the sums appearing in each of these equalities converge for each s in $\mathbb{Z}$. Since the same considerations apply to $|f|$, $|g|$, and $|h|$, the sums in question converge absolutely. Hence interchanging the order of summation in these sums is permissible; and

$$[(f*g)*h](s) = \sum_{t\in\mathbb{Z}} f(t)\left(\sum_{r\in\mathbb{Z}} g(r-t)h(s-r)\right)$$

$$= \sum_{t\in\mathbb{Z}} f(t)\left(\sum_{r\in\mathbb{Z}} g(r)h(s-r-t)\right)$$

$$= [f*(g*h)](s).$$

Thus $(f * g) * h = f * (g * h)$.

(v)   From (i)–(iv), the L_1-norm and convolution multiplication provide $l_1(\mathbb{Z})$ with the structure of a commutative Banach algebra. Since

$$(e * f)(s) = \sum_{t \in \mathbb{Z}} e(t)f(s - t) = f(s)$$

for each s in $\mathbb{Z}$ and each function f, e is a unit for this Banach algebra.   ∎

3.5.34.   (i)   Show that for each z in $\mathbb{T}_1$ the equation

$$\xi_z(t) = z^t \qquad (t \in \mathbb{Z})$$

defines a homomorphism ξ_z of the additive group $\mathbb{Z}$ into $\mathbb{T}_1$. (We call such a homomorphism a *character* of $\mathbb{Z}$.)

(ii)   Show that the set $\widehat{\mathbb{Z}}$ of characters of $\mathbb{Z}$ provided with the multiplication $(\xi \cdot \xi')(t) = \xi(t) \cdot \xi'(t)$ for t in $\mathbb{Z}$ is a group. (We call $\widehat{\mathbb{Z}}$ the *dual group* of $\mathbb{Z}$.)

(iii)   Show that the mapping $z \to \xi_z$ is an isomorphism of $\mathbb{T}_1$ onto $\widehat{\mathbb{Z}}$.

Solution. (i)   Since $\xi_z(t + s) = z^{t+s} = z^t z^s = \xi_z(t)\xi_z(s)$, ξ_z is a homomorphism of $\mathbb{Z}$ into $\mathbb{T}_1$.

(ii)   With ξ_1 corresponding to 1 in $\mathbb{T}_1$, as in (i), ξ_1 is a unit element for $\widehat{\mathbb{Z}}$. The multiplication defined on $\widehat{\mathbb{Z}}$ is (pointwise) multiplication of complex-valued functions and is associative. If ξ is a character of $\mathbb{Z}$, so is its complex conjugate $\bar{\xi}$, and $\xi \cdot \bar{\xi} = \xi_1$. Thus $\widehat{\mathbb{Z}}$ is a group.

(iii)   Since

$$\xi_{zz'}(t) = (zz')^t = z^t z'^t = \xi_z(t)\xi_{z'}(t)$$

for each t in $\mathbb{Z}$, $\xi_{zz'} = \xi_z \xi_{z'}$; and the mapping $z \to \xi_z$ is a homomorphism of $\mathbb{T}_1$ into $\widehat{\mathbb{Z}}$. If ξ is a character of $\mathbb{Z}$ and z is $\xi(1)$, then $\xi(t) = z^t$ for each integer t since ξ is a homomorphism. Thus the mapping $z \to \xi_z$ maps $\mathbb{T}_1$ *onto* $\widehat{\mathbb{Z}}$. If $\xi_z = \xi_1$ for some z in $\mathbb{T}_1$, then $z^t = 1$ for each integer t. In particular, $z = z^1 = 1$, and the kernel of the mapping $z \to \xi_z$ is $\{1\}$. It follows that $z \to \xi_z$ is an isomorphism of $\mathbb{T}_1$ onto $\widehat{\mathbb{Z}}$.   ∎

3.5.35.   Using the notation of Exercises 3.5.33 and 3.5.34, let $\xi\ (= \xi_z)$ be a character of $\mathbb{Z}$. Show that

(i)   the formula

$$(*) \qquad \rho_z(f) = \sum_{t\in\mathbb{Z}} f(t)\xi_z(t)\ (= \hat{f}(z)) \qquad (f \in l_1(\mathbb{Z}))$$

defines a non-zero multiplicative linear functional on $l_1(\mathbb{Z})$;

(ii)   each such functional on $l_1(\mathbb{Z})$ corresponds to a (unique) point z in $\mathbb{T}_1$ by means of $(*)$;

(iii)   the function $\hat{f}$ defined in $(*)$ is continuous on $\mathbb{T}_1$, and the mapping $z \to \rho_z$ of $\mathbb{T}_1$ onto the set $\mathcal{M}(\mathbb{Z})$ of non-zero multiplicative linear functionals on $l_1(\mathbb{Z})$ is a homeomorphism of $\mathbb{T}_1$ onto $\mathcal{M}(\mathbb{Z})$ with its weak* topology.

Solution. (i)   Since $f \in l_1(\mathbb{Z})$ and $|\xi_z(t)| = 1$ for each t, the sum in $(*)$ converges. Clearly, ρ_z is linear, and

$$\rho_z(e) = \sum_{t\in\mathbb{Z}} e(t)\xi_z(t) = \xi_z(0) = 1.$$

With f and g in $l_1(\mathbb{Z})$, we have

$$(**) \quad \rho_z(f * g) = \sum_{t\in\mathbb{Z}}(f * g)(t)\xi_z(t) = \sum_{t\in\mathbb{Z}}\left(\sum_{r\in\mathbb{Z}} f(r)g(t - r)\right)\xi_z(t).$$

Now

$$\sum_{t\in\mathbb{Z}}\sum_{r\in\mathbb{Z}}|f(r)|\,|g(t - r)|\,|\xi_z(t)| = \sum_{t\in\mathbb{Z}}\sum_{r\in\mathbb{Z}}|f|\,(r)\,|g|\,(t - r)$$

$$= \sum_{t\in\mathbb{Z}}(|f| * |g|)(t)$$

$$= \||f| * |g|\|_1 ;$$

so that an interchange in the order of summation in $(**)$ is permissible. Thus

$$\rho_z(f * g) = \sum_{r \in \mathbb{Z}} f(r) \Big(\sum_{t \in \mathbb{Z}} g(t - r) \xi_z(t) \Big)$$

$$= \sum_{r \in \mathbb{Z}} f(r) \xi_z(r) \Big(\sum_{t \in \mathbb{Z}} g(t - r) \xi_z(t - r) \Big)$$

$$= \sum_{r \in \mathbb{Z}} f(r) \xi_z(r) \Big(\sum_{t \in \mathbb{Z}} g(t) \xi_z(t) \Big)$$

$$= \sum_{r \in \mathbb{Z}} f(r) \xi_z(r) \rho_z(g)$$

$$= \rho_z(f) \rho_z(g),$$

and ρ_z is multiplicative.

(ii) Suppose ρ is a non-zero multiplicative linear functional on $l_1(\mathbb{Z})$. Let e_t be the function that takes the value 1 at t and 0 at other elements of $\mathbb{Z}$. Note that $(e_t * e_{t'})(s) = \sum_{r \in \mathbb{Z}} e_t(r) e_{t'}(s - r) = e_{t'}(s - t)$ and $e_{t'}(s - t)$ is 1 when $s = t + t'$ and 0 otherwise. Thus $e_t * e_{t'} = e_{t+t'}$. Let $\xi(t)$ be $\rho(e_t)$. Then

$$\xi(t + t') = \rho(e_{t+t'}) = \rho(e_t * e_{t'}) = \rho(e_t) \rho(e_{t'}) = \xi(t) \xi(t')$$

and ξ is a character of $\mathbb{Z}$. From Exercise 3.5.34(iii), $\xi = \xi_z$, where $z = \xi(1)$. If $f \in l_1(\mathbb{Z})$, then $f = \sum_{t \in \mathbb{Z}} f(t) e_t$ where convergence is relative to the norm in $l_1(\mathbb{Z})$. From Proposition 3.2.20, ρ is bounded. Thus

$$\rho(f) = \sum_{t \in \mathbb{Z}} f(t) \rho(e_t) = \sum_{t \in \mathbb{Z}} f(t) \xi_z(t) = \hat{f}(z),$$

and $\rho = \rho_z$.

(iii) Suppose $f \in l_1(\mathbb{Z})$ and $z' \in \mathbb{T}_1$. Given a positive ε, we can choose an integer n such that $\sum_{|t| \geq n} |f(t)| < \varepsilon/4$. Let a be $1 + \sum_{|t| < n} |f(t)|$. There is a positive δ such that if $z \in \mathbb{T}_1$ and $|z - z'| < \delta$, then

$$|z^t - z'^t| = |\xi_z(t) - \xi_{z'}(t)| < \frac{\varepsilon}{2a} \qquad (t \in \mathbb{Z}, \ |t| < n).$$

For such a z, we have

$$|\hat{f}(z) - \hat{f}(z')| = |\rho_z(f) - \rho_{z'}(f)| \leq \sum_{t \in \mathbb{Z}} |f(t)| \, |\xi_z(t) - \xi_{z'}(t)|$$

$$\leq \sum_{|t| < n} |f(t)| \, |\xi_z(t) - \xi_{z'}(t)| + 2 \sum_{|t| \geq n} |f(t)|$$

$$\leq \frac{a\varepsilon}{2a} + \frac{\varepsilon}{2} = \varepsilon.$$

Thus $\hat{f}$ is continuous. At the same time, $z \to \rho_z$ is continuous from $\mathbb{T}_1$ (in its usual metric topology) to $\mathcal{M}(\mathbb{Z})$ in its weak* topology. Since $\mathbb{T}_1$ is compact and $z \to \rho_z$ is a (continuous) one-to-one mapping onto $\mathcal{M}(\mathbb{Z})$ (from (i) and (ii)), it is a homeomorphism of $\mathbb{T}_1$ onto $\mathcal{M}(\mathbb{Z})$. ∎

3.5.36. Let $\eta(s)$ be $\exp is$ for s in $[-\pi, \pi)$, and let $m(S)$ be the Lebesgue measure of $\eta^{-1}(S)$ divided by 2π when S is a subset of $\mathbb{T}_1$ such that $\eta^{-1}(S)$ is Lebesgue measurable.

(i) With the notation of Exercise 3.5.35, show that the mapping $f \to \hat{f}$ of $l_1(\mathbb{Z})$ into $c(\mathbb{T}_1)$ is linear and satisfies

$$\sum_{t \in \mathbb{Z}} |f(t)|^2 = \int_{\mathbb{T}_1} |\hat{f}(z)|^2 \, dm(z).$$

(ii) Show that the mapping $f \to \hat{f}$ of (i) extends to a unitary transformation of $l_2(\mathbb{Z})$ onto $L_2(\mathbb{T}_1, m)$.
[Compare the results of (i) and (ii) with the Plancherel theorem (3.2.31).]

Solution. (i) As noted in Remark 2.2.16 (and Remark 3.4.15), the family $\{(2\pi)^{-1/2}\chi_t : t \in \mathbb{Z}\}$ is an orthonormal basis for $L_2(-\pi, \pi)$, where $\chi_t(s) = \exp its$. If $y_t(z) = z^t$ for z in $\mathbb{T}_1$, then $(y_t \circ \eta)(s) = \exp its = \chi_t(s)$ for each s in $[-\pi, \pi)$; so that $y_t \circ \eta = \chi_t$ and

$$\langle y_t, y_{t'} \rangle = \int_{\mathbb{T}_1} y_t(z)\overline{y_{t'}(z)} \, dm(z) = \frac{1}{2\pi} \int_{[-\pi, \pi)} (y_t \circ \eta)(s)\overline{(y_{t'} \circ \eta)(s)} \, ds$$
$$= \langle (2\pi)^{-1/2}\chi_t, (2\pi)^{-1/2}\chi_{t'} \rangle.$$

Thus $\{y_t : t \in \mathbb{Z}\}$ is an orthonormal basis for $L_2(\mathbb{T}_1, m)$.
If $f \in l_1(\mathbb{Z})$ $(\subseteq l_2(\mathbb{Z}))$, then

$$\hat{f}(z) = \sum_{t \in \mathbb{Z}} f(t)\xi_z(t) = \sum_{t \in \mathbb{Z}} f(t)z^t = \sum_{t \in \mathbb{Z}} f(t)y_t(z)$$

for each z in $\mathbb{T}_1$. In particular, the mapping $f \to \hat{f}$ is linear. Now,

$$\left| \sum_{t=-n}^{n} f(t)y_t \right| \le \sum_{t=-n}^{n} |f(t)| \le \|f\|_1,$$

and the constant functions are in $L_2(\mathbb{T}_1, m)$. It follows from the dominated convergence theorem that $\hat{f} = \sum_{t \in \mathbb{Z}} f(t) y_t$ in $L_2(\mathbb{T}_1, m)$. Thus

$$\sum_{t \in \mathbb{Z}} |f(t)|^2 = \|\hat{f}\|_2 = \int_{\mathbb{T}_1} |\hat{f}(z)|^2 \, dm(z).$$

(ii)　Since $f \to \hat{f}$ is isometric and $l_1(\mathbb{Z})$ is $(L_2\text{-})$ dense in $l_2(\mathbb{Z})$, the linear mapping $f \to \hat{f}$ extends to an isometric linear mapping T of $l_2(\mathbb{Z})$ into $L_2(\mathbb{T}_1, m)$. Suppose $g \in L_2(\mathbb{T}_1, m)$ and g is orthogonal to the range of T. Let e_t be the function on $\mathbb{Z}$ that takes the values 1 at t and 0 elsewhere. Then

$$\hat{e}_t(z) = \rho_z(e_t) = \xi_z(t) = z^t = y_t(z),$$

so that $y_t = \hat{e}_t = Te_t$. Thus $0 = \langle Te_t, g \rangle = \langle y_t, g \rangle$ for all t in $\mathbb{Z}$. Since $\{y_t : t \in \mathbb{Z}\}$ is an orthonormal basis for $L_2(\mathbb{T}_1, m)$, $g = 0$. It follows that T is a unitary transformation of $l_2(\mathbb{Z})$ onto $L_2(\mathbb{T}_1, m)$ extending the mapping $f \to \hat{f}$.　　■

3.5.37.　　With the measure m on $\mathbb{T}_1$ introduced in Exercise 3.5.36, define the convolution of m-measurable functions f and g on $\mathbb{T}_1$ as the function $f * g$ whose domain consists of those w in $\mathbb{T}_1$ for which the integral $\int_{\mathbb{T}_1} f(z) g(z^{-1} w) \, dm(z)$ converges and whose value at w is this integral. Show that (i)–(v) of Proposition 3.2.22 hold for this convolution (where $\mathbb{T}_1$ replaces $\mathbb{R}$ in those assertions).

Solution.　　What follows is a routine translation of the proof of Proposition 3.2.22. All integrations are over $\mathbb{T}_1$ unless otherwise indicated. Note that m is invariant under the transformation $z \to \bar{z}$ $(= z^{-1})$ and $z \to zw$. Thus

$$\int h(\bar{z}w) \, dm(z) = \int h(zw) \, dm(z) = \int h(z) \, dm(z).$$

(i)　With f and g m-measurable,

$$(f * g)(w) = \int f(z) g(\bar{z}w) \, dm(z) = \int f(wz) g(\bar{z}) \, dm(z)$$
$$= \int g(z) f(\bar{z}w) \, dm(z) = (g * f)(w)$$

(ii)   Let $h_w(z)$ be $g(w\bar{z})$ and $\mu(S)$ be $\int_S |f(z)|\, dm(z)$ for each measurable subset S of $\mathbb{T}_1$, then $h_w \in L_p(\mathbb{T}_1)$ and μ is a finite measure on $\mathbb{T}_1$. Applying the Hölder inequality to h_w and the constant function 1 relative to μ, we have

$$\int |g(\bar{z}w)| \cdot 1 \, d\mu(z) \le \left(\int |h_w(z)|^p |f(z)|\, dm(z) \right)^{1/p} \|f\|_1^{(p-1)/p}$$

and

$$\left(\int |g(\bar{z}w)|\, |f(z)|\, dm(z) \right)^p \le \|f\|_1^{p-1} \int |g(\bar{z}w)|^p |f(z)|\, dm(z).$$

Now $(w, z) \to |g(\bar{z}w)|^p f(z)$ is in $L_1(\mathbb{T}_1 \times \mathbb{T}_1)$ since

$$\int \left(\int |g(\bar{z}w)|^p |f(z)|\, dm(w) \right) dm(z)$$

$$= \int |f(z)| \left(\int |g(\bar{z}w)|^p \, dm(w) \right) dm(z)$$

$$= \|g\|_p^p \int |f(z)|\, dm(z) = \|f\|_1 \|g\|_p^p \, .$$

Thus, using Fubini's theorem,

$$\int |(f * g)(w)|^p \, dm(w)$$

$$= \int \left| \int f(z) g(\bar{z}w)\, dm(z) \right|^p dm(w)$$

$$\le \|f\|_1^{p-1} \int \left(\int |g(\bar{z}w)|^p |f(z)|\, dm(z) \right) dm(w)$$

$$= \|f\|_1^{p-1} \int \left(\int |g(\bar{z}w)|^p |f(z)|\, dm(w) \right) dm(z)$$

$$= \|f\|_1^p \|g\|_p^p$$

from which it follows that $f * g \in L_p(\mathbb{T}_1)$ and that

$$\|f * g\|_p \le \|f\|_1 \|g\|_p \, .$$

(iii)   From (ii), $f * (a \cdot g + h)$ and $a \cdot f * g + f * h$ are in $L_p(\mathbb{T}_1)$. In particular,

$$\int f(z)[ag(\bar{z}w) + h(\bar{z}w)]\, dm(z)$$

and

$$a \int f(z)g(\bar{z}w)\,dm(z) + \int f(z)h(\bar{z}w)\,dm(z)$$

converge for almost all w, and (iii) follows.

(iv) From (ii), $(f*g)*h$ and $f*(g*h)$ are in $L_p(\mathbb{T}_1)$. Now

$$(f*(g*h))(w) = \int f(z)\left(\int g(v)h(\bar{v}\bar{z}w)\,dm(v)\right)dm(z)$$

and

$$((f*g)*h)(w) = \int \left(\int f(z)g(\bar{z}v)\,dm(z)\right)h(\bar{v}w)\,dm(v).$$

Since $|f|$ and $|g|$ are in $L_1(\mathbb{T}_1)$ and $|h| \in L_p(\mathbb{T}_1)$; when f, g, and h are replaced by their respective absolute values, the last two integrals converge for almost all w. Fubini's theorem applies and, for almost every w,

$$((f*g)*h)(w) = \int f(z)\left(\int g(\bar{z}v)h(\bar{v}w)\,dm(v)\right)dm(z).$$

Since

$$\int g(\bar{z}v)h(\bar{v}w)\,dm(v) = \int g(v)h(\bar{v}\bar{z}w)\,dm(v),$$

$(f*g)*h = f*(g*h)$.

(v) From (i)–(iv), the L_1-norm and convolution provide $L_1(\mathbb{T}_1)$ with the structure of a commutative Banach algebra (without unit). ∎

3.5.38. With t in $\mathbb{Z}$ and z in $\mathbb{T}_1$, let $\xi_t(z)$ be z^t.

(i) Show that ξ_t is a continuous homomorphism of $\mathbb{T}_1$ into $\mathbb{T}_1$ (a *character* of $\mathbb{T}_1$) and that each such homomorphism has the form ξ_t for some t in $\mathbb{Z}$. [*Hint.* Recall the form of the characters of $\mathbb{R}$.]

(ii) Show that the set $\widehat{\mathbb{T}}_1$ of characters of $\mathbb{T}_1$ provided with pointwise multiplication is a group and that the mapping $t \to \xi_t$ is an isomorphism of $\mathbb{Z}$ onto $\widehat{\mathbb{T}}_1$.

Solution. (i) Since $z \to z^t$ is continuous and

$$\xi_t(zz') = (zz')^t = z^t z'^t = \xi_t(z)\xi_t(z'),$$

ξ_t is a character of $\mathbb{T}_1$. Suppose ξ is a character of $\mathbb{T}_1$. The mapping $t \to \exp 2\pi i t$ of $\mathbb{R}$ onto $\mathbb{T}_1$ is a continuous homomorphism so that the mapping $t \to \xi(\exp 2\pi i t)$ is a character of $\mathbb{R}$. From Lemma 3.2.25, there is an r in $\mathbb{R}$ such that $\xi(\exp 2\pi i t) = \exp 2\pi i t r$. Thus $\exp 2\pi i r = \xi(\exp 2\pi i) = \xi(1) = 1$; so that $r \in \mathbb{Z}$. It follows that $\xi(z) = z^r$ and $\xi = \xi_r$.

(ii) The constant function 1 on $\mathbb{T}_1$ (that is, ξ_0) is a character that serves as the unit element for $\widehat{\mathbb{T}}_1$. If $\xi \in \widehat{\mathbb{T}}_1$, then $\bar{\xi} \in \widehat{\mathbb{T}}_1$ and $\xi\bar{\xi} = \xi_0$, where $\bar{\xi}(z) = \overline{\xi(z)}$. Since multiplication of complex-valued functions (pointwise) is associative, $\widehat{\mathbb{T}}_1$ is a group. Now $\xi_{t+t'}(z) = z^{t+t'} = z^t z^{t'} = \xi_t(z)\xi_{t'}(z)$, so that $\xi_{t+t'} = \xi_t \xi_{t'}$. Hence the mapping $t \to \xi_t$ is a homomorphism from $\mathbb{Z}$ into $\widehat{\mathbb{T}}_1$; and from (i), this mapping is onto. If $\xi_t = \xi_0$, then $z^t = 1$ for each z in $\mathbb{T}_1$ and $t = 0$. Thus the mapping $t \to \xi_t$ is an isomorphism of $\mathbb{Z}$ onto $\widehat{\mathbb{T}}_1$. ∎

3.5.39. Using the notation of Exercises 3.5.37 and 3.5.38, let ρ_t be given by the formula

$$(*) \qquad \rho_t(f) = \int_{\mathbb{T}_1} f(z)\xi_t(z)\, dm(z)\ (= \hat{f}(t)) \qquad (f \in L_1(\mathbb{T}_1, m)).$$

(i) Show that ρ_t is a non-zero multiplicative linear functional on $L_1(\mathbb{T}_1, m)$ for each t in $\mathbb{Z}$.

(ii) Show that each non-zero multiplictive linear functional on $L_1(\mathbb{T}_1, m)$ is ρ_t for some t in $\mathbb{Z}$. [*Hint.* Make use of the fact that $\{\xi_r : r \in \mathbb{Z}\}$ spans a dense linear subspace of $L_1(\mathbb{T}_1, m)$ and $\xi_r * \xi_s = 0$ when $r \neq s$ to give a shorter argument than the one patterned on the proof of Theorem 3.2.26.]

(iii) Show that the set $\mathcal{M}(\mathbb{T}_1)$ of non-zero multiplicative linear functionals on $L_1(\mathbb{T}_1, m)$ is *not* weak* compact. Deduce that $L_1(\mathbb{T}_1, m)$ does not have an identity element. [*Hint.* Compute $\rho_r(\xi_{-t})$ for r and t in $\mathbb{Z}$.]

(iv) Let $\mathcal{A}_0(\mathbb{T}_1)$ be the Banach algebra obtained from $L_1(\mathbb{T}_1, m)$ by adjoining an identity I (as in Remark 3.1.3). Let ρ_∞ be the (non-zero) multiplicative linear functional on $\mathcal{A}_0(\mathbb{T}_1)$ that assigns 1 to I and 0 to each f in $L_1(\mathbb{T}_1, m)$. Denote by ρ'_t the (unique) non-zero multiplicative linear functional on $\mathcal{A}_0(\mathbb{T}_1)$ extending ρ_t. Show that $\{\rho_\infty, \rho'_t : t \in \mathbb{Z}\}$ $(= \mathcal{M}_0(\mathbb{T}_1))$ is the set of all non-zero multiplicative linear functionals on $\mathcal{A}_0(\mathbb{T}_1)$ and that the mapping Λ' of $\mathcal{M}_0(\mathbb{T}_1)$ onto the one-point compactification $\{\mathbb{Z}, \infty\}$ of $\mathbb{Z}$ that assigns ∞ to ρ_∞ and t to ρ'_t is a homeomorphism of $\mathcal{M}_0(\mathbb{T}_1)$ with its weak* topology onto $\{\mathbb{Z}, \infty\}$.

(v)   Show that

$$\lim_{|t| \to \infty} |\hat{f}(t)| = 0 \qquad (f \in L_1(\mathbb{T}_1, m)).$$

Solution. (i)   Since $\rho_t(\xi_{-t}) = 1$, $\rho_t \neq 0$. If f and g are in $L_1(\mathbb{T}_1, m)$, then for each t in $\mathbb{Z}$,

$$\begin{aligned}
(\widehat{f * g})(t) &= \int (f * g)(z) z^t \, dm(z) \\
&= \int \left(\int f(v) \, g(\bar{v}z) \, dm(v) \right) z^t \, dm(z) \\
&= \int f(v) \left(\int g(\bar{v}z) z^t \, dm(z) \right) dm(v) \\
&= \int f(v) v^t \left(\int g(z) z^t \, dm(z) \right) dm(v) \\
&= \int f(v) v^t \, dm(v) \cdot \int g(z) z^t \, dm(z) \\
&= \hat{f}(t) \hat{g}(t).
\end{aligned}$$

Thus $\rho_t(f * g) = \rho_t(f)\rho_t(g)$. Of course ρ_t is linear.

(ii)   As in the proof of Theorem 3.2.26, if ρ is a non-zero multiplicative linear functional on $L_1(\mathbb{T}_1)$, then ρ is in the unit ball of the dual. Since $\rho \neq 0$, there is some f in $L_1(\mathbb{T}_1)$ such that $\rho(f) \neq 0$. From Remark 3.4.15, $\{\xi_r : r \in \mathbb{Z}\}$ spans a dense set in $C(\mathbb{T}_1)$ and $C(\mathbb{T}_1)$ is dense in $L_1(\mathbb{T}_1)$. Thus ρ does not annihilate some ξ_r, say, ξ_{-t}.

With f a function on $\mathbb{T}_1$ and z in $\mathbb{T}_1$, define $f_z(w)$ to be $f(w\bar{z})$ for each w in $\mathbb{T}_1$. Then $(\xi_r)_z = z^{-r}\xi_r = \xi_{-r}(z)\xi_r$ for each z in $\mathbb{T}_1$. If h and g are in $L_1(\mathbb{T}_1)$, then

$$\begin{aligned}
(h_z * g)(w) &= \int h(v\bar{z})g(\bar{v}w) \, dm(v) \\
&= \int h(v)g(\bar{v}\bar{z}w) \, dm(v) \\
&= \int h(v)g_z(\bar{v}w) \, dm(v) \\
&= (h * g_z)(w).
\end{aligned}$$

Thus $h_z * g = h * g_z$ and $\rho(h_z)\rho(g) = \rho(h)\rho(g_z)$. Letting h be ξ_r and g be ξ_{-t}, we have

$$\begin{aligned}
\xi_{-r}(z)\rho(\xi_r)\rho(\xi_{-t}) &= \rho[(\xi_r)_z]\rho(\xi_{-t}) \\
&= \rho(\xi_r)\rho[(\xi_{-t})_z] \\
&= \xi_t(z)\rho(\xi_r)\rho(\xi_{-t});
\end{aligned}$$

and $[\xi_{-r}(z) - \xi_t(z)]\rho(\xi_r) = 0$ for each z in $\mathbb{T}_1$ and r in $\mathbb{Z}$. Thus $\rho(\xi_r) = 0$ unless $\xi_{-r} = \xi_t$; equivalently, unless $r = -t$. Thus ρ takes a non-zero value c on precisely one ξ_r, namely, ξ_{-t}. The same is true of ρ_t (defined in $(*)$), but $\rho_t(\xi_{-t}) = 1$. Hence $\rho - c\rho_t$ annihilates all ξ_r and $\rho = c\rho_t$. It follows that both ρ_t and $c\rho_t$ are multiplicative. Now

$$c = c\rho_t(\xi_{-t})^2 = c\rho_t(\xi_{-t} * \xi_{-t}) = (c\rho_t)(\xi_{-t}) \cdot (c\rho_t)(\xi_{-t}) = c^2;$$

so that $c = 1$ and $\rho = \rho_t$.

   (iii)   We have noted that $\rho_r(\xi_{-t}) = 0$ if $r \neq t$ and $\rho_t(\xi_{-t}) = 1$. Thus V_t is a weak* open subset of $\mathcal{M}(\mathbb{T}_1)$ containing just ρ_t, where

$$V_t = \{\rho :\ \rho \in L_1(\mathbb{T}_1)^{\#},\ |\rho(\xi_{-t}) - 1| < \tfrac{1}{2}\} \cap \mathcal{M}(\mathbb{T}_1).$$

Thus $\{V_t :\ t \in \mathbb{Z}\}$ is an infinite open covering of $\mathcal{M}(\mathbb{T}_1)$ that admits no finite subcovering. If $L_1(\mathbb{T}_1)$ had an identity, Proposition 3.2.20 would imply that $\mathcal{M}(\mathbb{T}_1)$ is weak* compact. Hence $L_1(\mathbb{T}_1)$ has no identity.

   (iv)   With t in $\mathbb{Z}$, if $|(\rho'_r - \rho'_t)(\xi_{-t})| < \tfrac{1}{2}$, then $r = t$ (as we noted in (iii)). Thus Λ' is weak* continuous at each ρ'_t for t in $\mathbb{Z}$. To prove that Λ' is weak* continuous at ρ_∞, we must show that $|t|$ is large when ρ'_t is weak* close to ρ_∞. Let n be a (large) positive integer and f be the characteristic function of $\{\exp is :\ 0 \leq s \leq n^{-1}\}$. From $(*)$,

$$\rho'_t(f) = \int_{\mathbb{T}_1} f(z)\xi_t(z)\, dm(z) = \int_0^{\frac{1}{n}} e^{ist}\, ds.$$

But this last integral is precisely the one we estimated in the last paragraph of the proof of Theorem 3.2.27. From that estimate, we have that if $|\rho'_t(f) - \rho_\infty(f)|\ (= |\rho'_t(f)|) < n^{-3}$ and $|t| \leq n\pi$, then $2n \leq |t|$. Thus $2n \leq |t|$ when $|\rho'_t(f)| < n^{-3}$; and Λ' is weak* continuous at ρ_∞. Hence Λ' is a one-to-one weak* continuous mapping of the weak* compact set $\mathcal{M}_0(\mathbb{T}_1)$ onto the compact space $\{\mathbb{Z}, \infty\}$, and Λ' is a homeomorphism.

(v)   Since Λ'^{-1} is continuous, with $|t|$ large, $\Lambda'^{-1}(t)(= \rho'_t)$ is weak* close to $\Lambda'^{-1}(\infty)(= \rho_\infty)$; and $|\rho'_t(f) - \rho_\infty(f)|$ $(= |\hat{f}(t)|)$ is small. Thus $\lim_{|t|\to\infty} |\hat{f}(t)| = 0$.   ∎

3.5.40.   Let Sf be $\hat{f}$ (as defined in Exercise 3.5.39) for f in $L_2(\mathbb{T}_1, m)$ $(\subseteq L_1(\mathbb{T}_1, m))$, and let T be the unitary transformation of $l_2(\mathbb{Z})$ onto $L_2(\mathbb{T}_1, m)$ described in Exercise 3.5.36(ii).

(i)   Show that S is a unitary transformation of $L_2(\mathbb{T}_1, m)$ onto $l_2(\mathbb{Z})$.

(ii)   Let U be the (self-adjoint) unitary operator on $L_2(\mathbb{T}_1, m)$ that maps ξ_t onto ξ_{-t} for each t in $\mathbb{Z}$. Show that

$$TSU(f) = f \qquad (f \in L_2(\mathbb{T}_1, m))$$

and

$$SUT(g) = g \qquad (g \in l_2(\mathbb{Z})).$$

(iii)   Deduce that

$$T^* = T^{-1} = SU$$

and

$$S^* = S^{-1} = UT.$$

Solution. (i)   From Remark 3.4.15, $\{\xi_t : t \in \mathbb{Z}\}$ is an orthonormal basis for $L_2(\mathbb{T}_1, m)$. With f in $L_2(\mathbb{T}_1, m)$, $f = \sum_{t\in\mathbb{Z}}\langle f, \xi_t\rangle \xi_t$ and

$$\langle f, \xi_t\rangle = \int_{\mathbb{T}_1} f(z)\overline{\xi_t(z)}\, dm(z) = \int_{\mathbb{T}_1} f(z)\xi_{-t}(z)\, dm(z) = \hat{f}(-t).$$

Thus

$$\|f\|_2^2 = \int_{\mathbb{T}_1} |f(z)|^2\, dm(z) = \sum_{t\in\mathbb{Z}} |\langle f, \xi_t\rangle|^2 = \sum_{t\in\mathbb{Z}} |\hat{f}(-t)|^2 = \|\hat{f}\|_2^2$$

and, in particular, $\hat{f} \in l_2(\mathbb{Z})$. It follows that the mapping $f \to \hat{f}$ is a linear isometry of $L_2(\mathbb{T}_1, m)$ into $l_2(\mathbb{Z})$.

We noted in Exercise 3.5.39 that $\hat{\xi}_t(r) = \rho_r(\xi_t) = 0$ if $r \neq -t$ and that $\hat{\xi}_t(-t) = 1$. Thus $\hat{\xi}_t = e_{-t}$, where e_r is the functrion on $\mathbb{Z}$

that takes the value 1 at r and 0 elsewhere. From this we see that the range of the isometry S contains all finite linear combinations of the elements e_r of the orthonormal basis $\{e_r : r \in \mathbb{Z}\}$. Thus S is a unitary transformation of $L_2(\mathbb{T}_1, m)$ onto $l_2(\mathbb{Z})$.

(ii)   It suffices to verify the stated equalities on the orthonormal bases $\{\xi_t : t \in \mathbb{Z}\}$ and $\{e_t : t \in \mathbb{Z}\}$, since T,S, and U, are bounded linear transformations. With the notation of Exercises 3.5.36 and 3.5.39, note that for each t in $\mathbb{Z}$,

$$(Te_t)(z) = \hat{e}_t(z) = \rho_z(e_t) = \sum_{r \in \mathbb{Z}} e_t(r)\xi_z(r) = \xi_z(t) = z^t = \xi_t(z).$$

Hence

$$TSU(\xi_t) = TS(\xi_{-t}) = T(\hat{\xi}_{-t}) = T(e_t) = \xi_t$$

and

$$SUT(e_t) = SU(\xi_t) = S(\xi_{-t}) = e_t.$$

(iii)   Since T and S are unitary transformations, SU is inverse to T, and UT is inverse to S (from (ii)); we have

$$T^* = T^{-1} = SU$$

and

$$S^* = S^{-1} = UT. \qquad \blacksquare$$

3.5.41.   Let $L_1(\mathbb{R})$ denote the Banach algebra formed by providing $L_1(\mathbb{R})$ with convolution multiplication. With f in $L_1(\mathbb{R})$, denote by f, the "translate" of f by the real number r ($f_r(t) = f(t - r)$). Let $\mathcal{I}$ be a norm-closed linear subspace of $L_1(\mathbb{R})$.

(i)   Show that if $\mathcal{I}$ is an ideal in $L_1(\mathbb{R})$, then $\mathcal{I}$ is invariant under translations (that is, $f_r \in \mathcal{I}$ when $f \in \mathcal{I}$). [*Hint.* Use the approximate identity of Lemma 3.2.24 and (5) of Section 3.2.]

(ii)   Show that if $\mathcal{I}$ is invariant under translations, it is a (norm-closed) ideal in $L_1(\mathbb{R})$. [*Hint.* Use the Hahn-Banach theorem to support the "view" of $g * f$ as $\int g(t)f_t \, dt$ and recall the identification of the dual of L_1.]

(iii)   Show that if the span of the translates of a function f in $L_1(\mathbb{R})$ is norm dense in $L_1(\mathbb{R})$, $\hat{f}$ vanishes nowhere on $\mathbb{R}$. (With its converse, this is one of the main theorems in a body of work known as "Wiener's Tauberian theorems.")

Solution. (i)   Let $\{u_n\}$ be an approximate identity for $L_1(\mathbb{R})$ (as in Lemma 3.2.24). As we noted in (5) of Section 3.2, $h_r * g = h * g_r$ when h and g are in $L_1(\mathbb{R})$. Thus $u_n * f_r = (u_n)_r * f \in \mathcal{I}$, if $f \in \mathcal{I}$. But $\|u_n * f_r - f_r\|_1 \to 0$, from Lemma 3.2.24. Since $\mathcal{I}$ is assumed to be normed closed, $f_r \in \mathcal{I}$ for each r in $\mathbb{R}$.

(ii)   Choose f in $\mathcal{I}$ (so that $f_r \in \mathcal{I}$ for each real r, by our present assumption). Let ρ be a bounded linear functional on $L_1(\mathbb{R})$ annihilating $\mathcal{I}$. From Theorem 1.7.8, there is a function h in $L_\infty(\mathbb{R})$ such that $\rho(k) = \int k(s)h(s)\, ds$ for each k in $L_1(\mathbb{R})$. With g in $L_1(\mathbb{R})$, we have for each real t,

$$0 = \rho(g(t)f_t) = \int g(t)f_t(s)h(s)\, ds.$$

Now

$$\int \Big(\int |g(t)||f_t(s)||h(s)|\, ds \Big)\, dt \leq \|h\|_\infty \|f\|_1 \|g\|_1$$

so that $(t,s) \to g(t)f_t(s)h(s)$ is in $L_1(\mathbb{R} \times \mathbb{R})$. From Fubini's theorem and the choice of h,

$$\begin{aligned}
\rho(g * f) &= \int h(s)(g * f)(s)\, ds \\
&= \int h(s)\Big(\int g(t)f(s - t)\, dt \Big)\, ds \\
&= \int \Big(\int g(t)f_t(s)h(s)\, ds \Big)\, dt \\
&= \int g(t)\rho(f_t)\, dt = 0.
\end{aligned}$$

From the Hahn-Banach theorem (see Corollary 1.6.3), $g * f \in \mathcal{I}$. Thus $\mathcal{I}$ is an ideal in $L_1(\mathbb{R})$.

(iii)   If $0 = \hat{f}(t)$ $(= \rho_t(f))$, then f lies in the null space $\mathcal{I}_t$ of ρ_t. Since $\mathcal{I}_t$ is a norm-closed ideal, the span of the translates of f lie in $\mathcal{I}_t$, from (i); so that this span is not dense in $L_1(\mathbb{R})$.   ∎

3.5.42.   Let $L_1(\mathbb{T}_1, m)$ be the Banach algebra described in Exercise 3.5.37. With f in $L_1(\mathbb{T}_1, m)$, denote by f_w the "translate" of f by w $(f_w(z) = f(z\bar{w}))$. Let $\mathcal{I}$ be a norm-closed linear subspace of $L_1(\mathbb{T}_1, m)$.

(i)   Show that the sequence $\{v_n\}$ satisfies

$$\|f * v_n - f\|_p \to 0$$

as $n \to \infty$, where $v_n = 2\pi(u_n \circ \eta^{-1})$, $\{u_n\}$ is the "approximate identity" described in Lemma 3.2.24 and η is defined by $\eta(s) = \exp is$.

(ii) Show that if $\mathcal{I}$ is an ideal in $L_1(\mathbb{T}_1, m)$, then $\mathcal{I}$ is invariant under translation.

(iii) Show that if $\mathcal{I}$ is invariant under translation, it is a (norm-closed) ideal in $L_1(\mathbb{T}_1, m)$.

(iv) Show that if the span of the translates of a function f in $L_1(\mathbb{T}_1, m)$ is norm dense in $L_1(\mathbb{T}_1, m)$, then $\hat{f}(t) \neq 0$ for each t in $\mathbb{Z}$.

Solution. (i) Assume, first, that f is continuous on $\mathbb{T}_1$; and let $g(s)$ be $f(\exp is)$ for s in $[-\pi, \pi)$. Choose a positive ε. By uniform continuity of f, there is a positive integer k with the property that $|f(z) - f(w)| < \varepsilon$ when $|z - w| < k^{-1}$. If n is such that $k < n$, then

$$
\begin{aligned}
&|(f * v_n)(w) - f(w)| \\
&= \left| \int_{\mathbb{T}_1} f(\bar{z}w)v_n(z)\, dm(z) - \int_{\mathbb{T}_1} f(w)v_n(z)\, dm(z) \right| \\
&\leq \int_{\mathbb{T}_1} |f(\bar{z}w) - f(w)|v_n(z)\, dm(z) \\
&\leq \varepsilon.
\end{aligned}
$$

For the last inequality, note that $\int v_n(z)\, dm(z) = 1$ and that $v_n(z) = 0$, unless $z = \exp is$ with $|s| \leq \frac{1}{n}$, whence

$$
|\bar{z}w - w| = |\bar{z} - 1| = |\exp(-is) - 1| \leq \tfrac{1}{n} < \tfrac{1}{k}.
$$

Hence $\|f * v_n - f\|_p \leq \varepsilon$ and $\|f * v_n - f\|_p \to 0$ as $n \to \infty$. For an arbitrary f in $L_p(\mathbb{T}_1, m)$, we can choose $\{f_k\}$ so that $\|f - f_k\|_p \to 0$, where each f_k is continuous on $\mathbb{T}_1$. From Exercise 3.5.37,

$$
\begin{aligned}
\|f * v_n - f\|_p &\leq \|f * v_n - f_k * v_n\|_p + \|f_k * v_n - f_k\|_p + \|f_k - f\|_p \\
&\leq \|v_n\|_1 \|f - f_k\|_p + \|f_k * v_n - f_k\|_p + \|f_k - f\|_p \\
&\leq 2\|f - f_k\|_p + \|f_k * v_n - f_k\|_p.
\end{aligned}
$$

Choosing k large enough, we have, for all n,

$$
\|f * v_n - f\|_p \leq \tfrac{2}{3}\varepsilon + \|f_k * v_n - f_k\|_p.
$$

From what we have proved for continuous functions, with n large enough, $\|f_k * v_n - f_k\|_p < \frac{\varepsilon}{3}$; and $\|f * v_n - f\|_p \le \varepsilon$.

(ii)   We proved in (ii) of Exercise 3.5.39 that $h_z * g = h * g_z$ when h and g are in $L_1(\mathbb{T}_1, m)$ and $z \in \mathbb{T}_1$. Thus $v_n * f_w = (v_n)_w * f \in \mathcal{I}$, if $f \in \mathcal{I}$. But $\|v_n * f_w - f_w\|_1 \to 0$, from (i). Since $\mathcal{I}$ is assumed to be norm closed, $f_w \in \mathcal{I}$ for each w in $\mathbb{T}_1$.

(iii)   Choose f in $\mathcal{I}$ (so that $f_w \in \mathcal{I}$ for w in $\mathbb{T}_1$, by our present assumption). Let ρ be a bounded linear functional on $L_1(\mathbb{T}_1, m)$ annihilating $\mathcal{I}$. From Theorem 1.7.8, there is a function h in $L_\infty(\mathbb{T}_1, m)$ such that $\rho(k) = \int k(z) h(z)\, dm(z)$ for each k in $L_1(\mathbb{T}_1, m)$. With g in $L_1(\mathbb{T}_1, m)$, we have for each w in $\mathbb{T}_1$,

$$0 = \rho(g(w) f_w) = \int_{\mathbb{T}_1} g(w) f_w(z) h(z)\, dm(z).$$

Now

$$\int_{\mathbb{T}_1} \left(\int_{\mathbb{T}_1} |g(w)| |f_w(z)| |h(z)|\, dm(z) \right) dm(w) \le \|h\|_\infty \|f\|_1 \|g\|_1$$

so that $(z, w) \to g(w) f_w(z) h(z)$ is in $L_1(\mathbb{T}_1 \times \mathbb{T}_1, m \times m)$. From Fubini's theorem

$$\begin{aligned}
\rho(g * f) &= \int_{\mathbb{T}_1} h(z)(g * f)(z)\, dm(z) \\
&= \int_{\mathbb{T}_1} h(z) \left(\int_{\mathbb{T}_1} g(w) f(z\bar{w})\, dm(w) \right) dm(z) \\
&= \int_{\mathbb{T}_1} \left(\int_{\mathbb{T}_1} g(w) f_w(z) h(z)\, dm(z) \right) dm(w) \\
&= \int_{\mathbb{T}_1} g(w) \rho(f_w)\, dm(w) \\
&= 0.
\end{aligned}$$

From the Hahn-Banach theorem (see Corollary 1.6.3), $g * f \in \mathcal{I}$. Thus $\mathcal{I}$ is an ideal in $L_1(\mathbb{T}_1, m)$.

(iv)   If $f \in \mathcal{I}$ and $0 = \hat{f}(t)(= \rho_t(f))$, then f lies in the null space $\mathcal{I}_t$ of ρ_t. Since $\mathcal{I}_t$ is a norm-closed ideal, the span of the translates of f lie in $\mathcal{I}_t$, from (ii); so that the span of these translates is not dense in $L_1(\mathbb{T}_1, m)$.   ∎

3.5.43. Let $\mathfrak{A}$ be a Banach algebra and A be an element of $\mathfrak{A}$ such that $\mathrm{sp}_{\mathfrak{A}}(A) \subseteq \mathbb{C} \setminus \mathbb{R}^-$, where $\mathbb{R}^- = \{z : z \in \mathbb{C}, z = -|z|\}$.

(i) Show that there is an element A_0 in $\mathfrak{A}$ such that $(A_0)^2 = A$. (The element A_0 is said to be a *square root* of A in $\mathfrak{A}$.)

(ii) Deduce that if $B \in \mathfrak{A}$ and $\|I - B\| < 1$, then B has a square root in $\mathfrak{A}$.

Solution. (i) Write z in $\mathbb{C} \setminus \mathbb{R}^-$ as $|z| \exp(i\theta)$, where $-\pi < \theta < \pi$, and define $f(z)$ to be $|z|^{\frac{1}{2}} \exp(\frac{1}{2}i\theta)$. Then f is holomorphic on the open set $\mathbb{C} \setminus \mathbb{R}^-$ and $f(z)^2 = z$ for each z in $\mathbb{C} \setminus \mathbb{R}^-$. By assumption $\mathrm{sp}_{\mathfrak{A}}(A) \subseteq \mathbb{C} \setminus \mathbb{R}^-$, so that $f \in \mathcal{H}(A)$. Let A_0 be $f(A)$. From Theorem 3.3.5, $(A_0)^2 = A$.

(ii) From Theorem 3.2.3, $\mathrm{sp}_{\mathfrak{A}}(I - B)$ lies in the interior $\mathbb{D}_0$ of the disk with center 0 and radius 1 in $\mathbb{C}$. But $\mathrm{sp}_{\mathfrak{A}}(I - B) = 1 - \mathrm{sp}_{\mathfrak{A}}(B)$, from Proposition 3.2.10, so that $\mathrm{sp}_{\mathfrak{A}}(B)$ lies in $1 - \mathbb{D}_0$ ($= 1 + \mathbb{D}_0$). In particular, $\mathrm{sp}_{\mathfrak{A}}(B) \subseteq \mathbb{C} \setminus \mathbb{R}^-$. Thus (i) applies and B has a square root. ∎

3.5.44. Let A be an element of the Banach algebra $\mathfrak{A}$.

(i) Show that if $\mathrm{sp}_{\mathfrak{A}}(A) \subseteq \{\lambda : \lambda \in \mathbb{R}, \, 0 < \lambda\}$, then A has a square root in $\mathfrak{A}$ with positive spectrum.

(ii) If the hypothesis of (i) is weakened to

$$\mathrm{sp}_{\mathfrak{A}}(A) \subseteq \{\lambda : \lambda \in \mathbb{R}, \, 0 \le \lambda\} \ \ (= \mathbb{R}^+),$$

does A still have a positive square root in $\mathfrak{A}$? [*Hint.* Consider

$$\begin{bmatrix} 0 & 1 \\ 0 & 0 \end{bmatrix}$$

in the Banach algebra of complex 2×2 matrices.]

(iii) Show that if

$$\mathrm{sp}_{\mathfrak{A}}(A) \subseteq \mathbb{R}^+$$

and $\|B\| = r_{\mathfrak{A}}(B)$ for each B in the Banach subalgebra $\mathfrak{A}_0$ of $\mathfrak{A}$ generated by A and I, then A has a square root in $\mathfrak{A}_0$ with spectrum (relative to $\mathfrak{A}_0$) in $\mathbb{R}^+$. [*Hint.* Study the square roots of $A + n^{-1}I$ as $n \to \infty$ and use Remark 3.2.11 applied to $\mathfrak{A}_0$.]

Solution. (i) The function f described in Exercise 3.5.43(i) maps $\{\lambda : \lambda \in \mathbb{R}, 0 < \lambda\}$ into itself. Since $\mathrm{sp}_{\mathfrak{A}}(A) \subseteq \mathbb{C} \setminus \mathbb{R}^-$,

Exercise 3.5.43(i) applies; and $f(A)$ is a square root of A in $\mathfrak{A}$ From
the spectral mapping theorem (3.3.6),

$$\mathrm{sp}_{\mathfrak{A}}[f(A)] \subseteq \{\lambda : \lambda \in \mathbb{R}, \ 0 < \lambda\}.$$

(ii)   A square root of a (generalized) nilpotent element contained
in a Banach algebra must be a generalized nilpotent element, from
Proposition 3.2.10. With A and $\mathfrak{A}$ as in the hint, A is nilpotent. If
A_0 is a square root of A in $\mathfrak{A}$, then A_0 is a (generalized) nilpotent
matrix in $\mathfrak{A}$. Now A_0 is similar to a matrix in upper triangular
form (with the same spectrum as A), from which we have that A_0^2
$= 0 \neq A$. Thus A has *no* square root in $\mathfrak{A}$.

(iii)   With n a positive integer, by assumption and by using Exercise 3.5.28(v),

$$\begin{aligned}
\mathrm{sp}_{\mathfrak{A}_0}(A + n^{-1}I) &= \mathrm{sp}_{\mathfrak{A}_0}(A) + n^{-1}I \\
&= \mathrm{sp}_{\mathfrak{A}}(A) + n^{-1}I \\
&\subseteq \{\lambda : \lambda \in \mathbb{R}, \ 0 < \lambda\}.
\end{aligned}$$

Thus (i) applies, and $A + n^{-1}I$ has a square root B_n in $\mathfrak{A}_0$ with
positive spectrum (relative to $\mathfrak{A}_0$). From Remark 3.2.11, if $\lambda \in$
$\mathrm{sp}_{\mathfrak{A}_0}(B_n - B_m)$, there is a (non-zero) multiplicative linear functional
ρ on $\mathfrak{A}_0$ such that $\lambda = \rho(B_n - B_m) = \rho(B_n) - \rho(B_m)$. Now

$$\rho(B_n)^2 = \rho(B_n^2) = \rho(A + n^{-1}) = \rho(A) + n^{-1}$$

and $\rho(B_n) \in \mathrm{sp}_{\mathfrak{A}_0}(B_n) \subseteq \mathbb{R}^+$. Thus $\rho(B_n) = [\rho(A) + n^{-1}I]^{\frac{1}{2}}$; and,
similarly, $\rho(B_m) = [\rho(A) + m^{-1}I]^{\frac{1}{2}}$. Hence

$$\begin{aligned}
|\lambda| &= |\rho(B_n - \rho(B_m)| \\
&= |[\rho(A) + n^{-1}]^{\frac{1}{2}} - [\rho(A) + m^{-1}]^{\frac{1}{2}}| \\
&= \frac{|n^{-1} - m^{-1}|}{[\rho(A) + n^{-1}]^{\frac{1}{2}} + [\rho(A) + m^{-1}]^{\frac{1}{2}}} \\
&\leq \frac{|n^{-1} - m^{-1}|}{n^{-\frac{1}{2}} + m^{-\frac{1}{2}}} \\
&= |n^{-\frac{1}{2}} - m^{-\frac{1}{2}}|.
\end{aligned}$$

By assumption and the spectral radius formula (3.3.3),

$$\|B_n - B_m\| = r_{\mathfrak{A}}(B_n - B_m) = r_{\mathfrak{A}_0}(B_n - B_m) \leq |n^{-\frac{1}{2}} - m^{-\frac{1}{2}}|.$$

Thus $\|B_n - B_m\| \to 0$ as $n, m \to \infty$. If A_0 is the limit of the Cauchy sequence $\{B_n\}$ in $\mathfrak{A}_0$, then $B_n^2 = A + n^{-1}I \to A_0^2$ as $n \to \infty$. But $A + n^{-1}I \to A$ as $n \to \infty$. Thus $A = A_0^2$. If $\lambda' \in \mathrm{sp}_{\mathfrak{A}_0}(A_0)$, there is a (non-zero) multiplicative linear functional ρ' on $\mathfrak{A}_0$ such that $\rho'(A_0) = \lambda'$. Since $\rho'(B_n) \in \mathrm{sp}_{\mathfrak{A}_0}(B_n) \subseteq \mathbb{R}^+$ and $\rho'(B_n) \to \rho'(A_0) = \lambda'$, we have that $0 \leq \lambda'$. thus $\mathrm{sp}_{\mathfrak{A}}(A_0) \subseteq \mathrm{sp}_{\mathfrak{A}_0}(A_0) \subseteq \mathbb{R}^+$.   ∎

3.5.45.   Let ρ be a hermitian functional on $C(X)$, where X is a compact Hausdorff space. Suppose $\|\rho\| = \rho(1)$.

(i)   Use Proposition 3.4.11 to show that ρ is a positive linear functional on $C(X)$.

(ii)   Show that ρ is a positive linear functional on $C(X)$ without using Proposition 3.4.11.

Solution. (i)   From Proposition 3.4.11, $\rho = \rho^+ - \rho^-$ and $\|\rho\| = \|\rho^+\| + \|\rho^-\| = \rho^+(1) + \rho^-(1)$. By assumption $\|\rho\| = \rho(1)$; so that $\rho^+(1) - \rho^-(1) = \rho^+(1) + \rho^-(1)$ and $\rho^-(1) = 0$. Since ρ^- is a positive linear functional on $C(X)$, we have (as in the proof of Proposition 3.4.11) that $\rho^- = 0$. Thus $\rho = \rho^+ \geq 0$.

(ii)   If $f \geq 0$ and $\rho(f) < 0$ (we are assuming that ρ is hermitian, so that $\rho(f)$ is real in any case), then

$$\|\rho\| < \rho(1) - \|f\|^{-1}\rho(f) = \rho(1 - \|f\|^{-1}f)$$
$$\leq \|\rho\|\|1 - \|f\|^{-1}f\| \leq \|\rho\|,$$

which is absurd. Thus $0 \leq \rho(f)$, and $0 \leq \rho$.   ∎

3.5.46.   Let X be a compact Hausdorff space and $\mathcal{A}$ be a closed subalglebra of $C(X)$ containing the constant functions and containing $\bar{f}$ when $f \in \mathcal{A}$.

(i)   Show that if f is a real-valued function in $\mathcal{A}$, $\lambda \in \mathrm{sp}_{\mathcal{A}}(f)$, and λ is not real, then there is a real-valued function g in $\mathcal{A}$ such that $i \in \mathrm{sp}_{\mathcal{A}}(g)$.

(ii)   With g as in (i), show that $\sum_{n=0}^{\infty}(-ig)^n/n!$ $(= \exp(-ig))$ is an element of $\mathcal{A}$ of norm 1.

(iii)   Use a multiplicative linear functional on $\mathcal{A}$ to show that $e \in \mathrm{sp}_{\mathcal{A}}[\exp(-ig)]$.

(iv)   Conclude that $\mathrm{sp}_{\mathcal{A}}(f) \subseteq \mathbb{R}$ for each real-valued function f in $\mathcal{A}$.

(v)   Conclude that $\mathrm{sp}_{\mathcal{A}}(f) \subseteq \mathbb{R}$, again, with the aid of Exercise 3.5.28(v).

Solution. (i)   If $\lambda = a + ib$ where $b \neq 0$, then $b^{-1}(f - a1)$ is a real-valued function g in $\mathcal{A}$ and $i \in \mathrm{sp}_{\mathcal{A}}(g)$.

(ii)   Since the sum $\sum_{n=0}^{\infty} \frac{1}{n!}\|g\|^n$ converges and the partial sums of $\sum_{n=0}^{\infty} \frac{1}{n!}(-ig)^n$ lie in $\mathcal{A}$, $\exp(-ig) \in \mathcal{A}$. Now

$$[\exp(-ig)](x) = \sum_{n=0}^{\infty} \frac{1}{n!}(-ig(x))^n = \exp[-ig(x)]$$

for each x in X.   Since $g(x) \in \mathbb{R}$, $|[\exp(-ig)](x)| = 1$. Thus $\|\exp(-ig)\| = 1$.

(iii)   From Remark 3.2.11, there is a (non-zero) multiplicative linear functional ρ on $\mathcal{A}$ such that $\rho(g) = i$. From Proposition 3.2.20, $\|\rho\| = 1$. Thus

$$\rho[\exp(-ig)] = \sum_{n=0}^{\infty} \frac{1}{n!}(-i\rho(g))^n = \sum_{n=0}^{\infty} \frac{1}{n!} = e,$$

and $e \in \mathrm{sp}_{\mathcal{A}}[\exp(-ig)]$.

(iv)   From Theorem 3.2.3 and (iii)

$$e \leq \|\exp(-ig)\| = 1,$$

an absurdity. Thus $\mathrm{sp}_{\mathcal{A}}(f) \subseteq \mathbb{R}$ for each real-valued function f in $\mathcal{A}$.

(v)   Since $\mathrm{sp}_{C(X)}(f) = f(X) \subseteq \mathbb{R}$, we have that $\mathrm{sp}_{\mathcal{A}}(f) = \mathrm{sp}_{C(X)}(f) \subseteq \mathbb{R}$ from Exercise 3.5.28(v).   ∎

3.5.47.   Let $C(X)$ and $\mathcal{A}$ be as in Exercise 3.5.46 and let the partial ordering of the (real) algebra $\mathcal{A}_r$ of real-valued functions in $\mathcal{A}$ be that induced by the partial ordering of $C(X, \mathbb{R})$.

(i)   Show that each (non-zero) multiplicative linear functional ρ on $\mathcal{A}$ has an extension ρ' to $C(X)$ such that ρ' is a state of $C(X)$.

(ii)   Show that ρ is a pure state of $\mathcal{A}$.

(iii)   Show that the set $\mathcal{E}$ of state extensions of ρ to $C(X)$ is convex and weak* compact.

(iv)   Show that each extreme point of $\mathcal{E}$ is a pure state of $C(X)$.

(v)   Conclude that ρ has an extension to $C(X)$ that is a (non-zero) multiplicative linear functional on $C(X)$.

Solution. (i)   With f in $\mathcal{A}_{\mathbf{r}}$, we have $\mathrm{sp}_{\mathcal{A}}(f) \subseteq \mathbb{R}$ from Exercise 3.5.46(iv). Thus $\rho(f) \in \mathbb{R}$ since $\rho(f) \in \mathrm{sp}_{\mathcal{A}}(f)$. From the Hahn-Banach theorem (1.6.1), the restriction ρ_0 of ρ to $\mathcal{A}_{\mathbf{r}}$ has an extension ρ_0' of norm 1 to $C(X,\mathbb{R})$. Since $\|\rho_0'\| = \rho_0'(1)$, from Exercise 3.5.45 we have that ρ_0' is a state of the algebra $C(X,\mathbb{R})$. If $\rho'(g)$ is defined as $\frac{1}{2}[\rho_0'(g + \bar{g}) + i\rho_0'(-i[g - \bar{g}])]$, then ρ' is linear and agrees with ρ_0' on $C(X,\mathbb{R})$. Thus ρ' is an extension of ρ to a state of $C(X)$.

(ii)   Suppose τ is a linear functional on $\mathcal{A}$ such that $0 \leq \tau \leq \rho$. If $\rho(f) = 0$, then $0 = \rho(\bar{f}f) = \rho(|f|^2)$; so that $\tau(|f|^2) = 0$, Thus $\tau(f) = 0$, from the Cauchy-Schwarz inequality (as used in Remark 3.4.9). Since τ annihilates the null space of ρ, τ is a scalar multiple of ρ. From (i), ρ is a state of $\mathcal{A}$. From Lemma 3.4.6, ρ is a pure state of $\mathcal{A}$.

(iii)   The set $\mathcal{E}$ is a subset of the unit ball of $C(X)^{\#}$ since it consists of states of $C(X)$. Thus it suffices to show that $\mathcal{E}$ is weak* closed in order to show that $\mathcal{E}$ is weak* compact. If τ is in the weak* closure of $\mathcal{E}$ and $f \in \mathcal{A}$, then $\tau(f)$ is in the closure of $\{\rho'(f) : \rho' \in \mathcal{E}\}$. But $\rho'(f) = \rho(f)$ for each ρ' in $\mathcal{E}$, so that $\tau(f) = \rho(f)$ for each f in $\mathcal{A}$ and τ is an extension of ρ. Again, with g in $C(X)$, if $g \geq 0$, then $\tau(g)$ is in the closure of $\{\rho'(g) : \rho' \in \mathcal{E}\}$. But $\rho'(g) \geq 0$ for each ρ' in $\mathcal{E}$, so that $\tau(g) \geq 0$. Similarly, $\tau(1) = 1$; and τ is a state extension of ρ. Thus $\tau \in \mathcal{E}$, $\mathcal{E}$ is weak* closed, and $\mathcal{E}$ is weak* compact.

If τ_1 and τ_2 are in $\mathcal{E}$, $0 \leq a \leq 1$, and $f \in \mathcal{A}$, then we have that $[a\tau_1 + (1-a)\tau_2](f) = a\rho(f) + (1-a)\rho(f) = \rho(f)$. Since $a\tau_1 + (1-a)\tau_2$ is a state of $C(X)$, $a\tau_1 + (1-a)\tau_2 \in \mathcal{E}$; and $\mathcal{E}$ is convex. Thus $\mathcal{E}$ is a convex, weak* compact subset of $C(X)^{\#}$.

(iv)   If τ is an extreme point of $\mathcal{E}$ and ρ_1, ρ_2 are states of $C(X)$ such that $\tau = \frac{1}{2}(\rho_1 + \rho_2)$, then $\rho = \tau|\mathcal{A} = \frac{1}{2}(\rho_1|\mathcal{A} + \rho_2|\mathcal{A})$. We proved in (ii) that ρ is a pure state of $\mathcal{A}$. Thus $\rho_1|\mathcal{A} = \rho_2|\mathcal{A} = \rho$. It follows that $\rho_1, \rho_2 \in \mathcal{E}$. Since τ is an extreme point of $\mathcal{E}$, $\tau = \rho_1 = \rho_2$. Hence τ is a pure state of $C(X)$.

(v)   From (i), $\mathcal{E}$ is non-empty. From (iii), $\mathcal{E}$ is weak* compact and convex. From the Krein-Milman theorem (1.4.3), $\mathcal{E}$ has extreme points (and is the weak* closed convex hull of its extreme points). Let τ be such an extreme point of $\mathcal{E}$. From (iv), τ is a pure state of $C(X)$. From Theorem 3.4.7, τ is a (non-zero) multiplicative linear functional on $C(X)$. Thus ρ has an extension to $C(X)$ that is a (non-zero) multiplicative linear functional on $C(X)$.   ∎

3.5.48.   Let $C(X)$ and $\mathcal{A}$ be as in Exercise 3.5.46.

(i)   Show that $\mathrm{sp}_{\mathcal{A}}(f) = \mathrm{sp}_{C(X)}(f)$ for each f in $\mathcal{A}$.

(ii)   Use Exercise 3.5.44 to show that if f is a positive function in $\mathcal{A}$, then the (unique) positive square root of f in $C(X)$ lies in $\mathcal{A}$.

(iii)   Conclude that $|f| \in \mathcal{A}$ when $f \in \mathcal{A}$.

Solution. (i)   Since $\mathcal{A} \subseteq C(X)$, $\mathrm{sp}_{C(X)}(f) \subseteq \mathrm{sp}_{\mathcal{A}}(f)$ for each f in $\mathcal{A}$. Suppose $f \in \mathcal{A}$ and $\lambda \in \mathrm{sp}_{\mathcal{A}}(f)$. From Remark 3.2.11, there is a (non-zero) multiplicative linear functional ρ on $\mathcal{A}$ such that $\rho(f) = \lambda$. From Exercise 3.5.47(v), there is a (non-zero) multiplicative linear functional τ on $C(X)$ that extends ρ. Thus $\tau(f) = \lambda$ and $\lambda \in \mathrm{sp}_{C(X)}(f)$. Hence $\mathrm{sp}_{\mathcal{A}}(f) \subseteq \mathrm{sp}_{C(X)}(f)$. It follows that $\mathrm{sp}_{\mathcal{A}}(f) = \mathrm{sp}_{C(X)}(f)$.

(ii)   If f is a positive function in $\mathcal{A}$ and $\mathcal{A}_0$ is the norm-closed subalgebra of $C(X)$ generated by f and 1, then $\mathrm{sp}_{\mathcal{A}_0}(f) = \mathrm{sp}_{\mathcal{A}}(f) = \mathrm{sp}_{C(X)}(f) \subseteq \mathbb{R}^+$. Moreover, $r_{C(X)}(g) = \|g\|$ for each g in $C(X)$. From Exercise 3.5.44(iii), we see that f has a square root f_0 in $\mathcal{A}_0$ such that $\mathrm{sp}_{\mathcal{A}_0}(f_0) \subseteq \mathbb{R}^+$. But $\mathrm{sp}_{C(X)}(f_0) = \mathrm{sp}_{\mathcal{A}_0}(f_0)$. Thus f_0 is a positive function in $\mathcal{A}$ such that $f_0^2 = f$. Hence f_0 is the unique positive square root of f in $C(X)$ and this square root lies in $\mathcal{A}$.

(iii)   If $f \in \mathcal{A}$, then $\bar{f} \in \mathcal{A}$ by assumption. Thus $|f|^2 \; (= \bar{f}f)$ is in $\mathcal{A}$. From (ii), $|f| \in \mathcal{A}$.   ∎

3.5.49.   Let X be a compact Hausdorff space, and let $\mathcal{L}$ be a subset of $C(X, \mathbb{R})$ such that, with f and g in $\mathcal{L}$, $\mathcal{L}$ contains $f \vee g$ and $f \wedge g$ (in this case, $\mathcal{L}$ is said to be a *sublattice* of $C(X, \mathbb{R})$). Suppose that for each pair r, s of real numbers and each pair p, q of distinct elements of X there is an f in $\mathcal{L}$ such that $f(p) = r$ and $f(q) = s$. Show that $\mathcal{L}$ is norm dense in $C(X, \mathbb{R})$.

Solution.   Let f in $C(X, \mathbb{R})$ and a positive ε be given. With p and q in X, choose f_p in $\mathcal{L}$ such that $f_p(p) = f(p)$ and $f_p(q) = f(q)$. By continuity of $f - f_p$, there is an open set $\mathcal{O}_p$ containing p such that $|f(x) - f_p(x)| < \varepsilon$ for each x in $\mathcal{O}_p$. Select a finite subcovering $\mathcal{O}_{p(1)}, \mathcal{O}_{p(2)}, \ldots, \mathcal{O}_{p(n)}$ from the covering $\{\mathcal{O}_p\}$ of X. Let g_q be $f_{p(1)} \vee \cdots \vee f_{p(n)}$. Then g_q lies in $\mathcal{L}$. With x in X, x lies in some $\mathcal{O}_{p(k)}$ so that $|f(x) - f_{p(k)}(x)| < \varepsilon$ and $f(x) - \varepsilon < f_{p(k)}(x) \le g_q(x)$. Thus $f - \varepsilon \cdot 1 \le g_q$. Moreover, $g_q(q) = f(q)$ since $f_p(q) = f(q)$ for each p in X.

By continuity of $g_q - f$, there is an open set $\mathcal{U}_q$ containing q such that $|g_q(x) - f(x)| < \varepsilon$ for each x in $\mathcal{U}_q$. Select a finite subcovering

$\mathcal{U}_{q(1)}, \mathcal{U}_{q(2)}, \ldots, \mathcal{U}_{q(m)}$ of the covering $\{\mathcal{U}_q\}$ of X. Let h be the function $g_{q(1)} \wedge \cdots \wedge g_{q(m)}$. Then h lies in $\mathcal{L}$. With x in X, x lies in some $\mathcal{U}_{q(j)}$ so that $|g_{q(j)}(x) - f(x)| < \varepsilon$ and $h(x) \leq g_{q(j)}(x) < f(x) + \varepsilon$. Thus $h \leq f + \varepsilon \cdot 1$. Now $f - \varepsilon \cdot 1 \leq h$ since $f - \varepsilon \cdot 1 \leq g_q$ for each q in X. Thus $f - \varepsilon \cdot 1 \leq h \leq f + \varepsilon \cdot 1$ and $-\varepsilon \cdot 1 \leq h - f \leq \varepsilon \cdot 1$. Hence $\|h - f\| \leq \varepsilon$, and $\mathcal{L}$ is norm dense in $C(X, \mathbb{R})$.   ■

    3.5.50.   Combine the results of Exercises 3.5.48 and 3.5.49 to give another proof of the Stone-Weierstrass theorem (3.4.14).

    Solution.   Let $C(X)$ and $\mathcal{A}$ be as in Theorem 3.4.14. With $\mathcal{A}_\mathrm{r}$ the (real, norm-closed) algebra of real-valued functions in $\mathcal{A}$, we have, from Exercise 3.5.48(iii), that $|f| \in \mathcal{A}_\mathrm{r}$ when $f \in \mathcal{A}_\mathrm{r}$. Since $f \vee g = \frac{1}{2}[f + g + |f - g|]$ and $f \wedge g = \frac{1}{2}[f + g - |f - g|]$, both $f \vee g$ and $f \wedge g$ are in $\mathcal{A}_\mathrm{r}$ when f and g are in $\mathcal{A}_\mathrm{r}$. As $\bar{f} \in \mathcal{A}$ when $f \in \mathcal{A}$, $\mathcal{A}_\mathrm{r}$ generates $\mathcal{A}$ linearly. Thus $\mathcal{A}_\mathrm{r}$ separates the points of X. If p and q are distinct points of X, the mapping $f \to (f(p), f(q))$ is a linear mapping of $\mathcal{A}_\mathrm{r}$ into $\mathbb{R}^2$. As noted, there is an f in $\mathcal{A}_\mathrm{r}$ such that $f(p) \neq f(q)$. Since $1 \in \mathcal{A}_\mathrm{r}$, $(1, 1)$ is also in the range of this linear mapping. Thus this mapping is *onto* $\mathbb{R}^2$. It follows that $\mathcal{A}_\mathrm{r}$ is a sublattice of $C(X, \mathbb{R})$ satisfying the conditions of Exercise 3.5.49. Thus $\mathcal{A}_\mathrm{r}$ is norm dense in $C(X, \mathbb{R})$. Since $\mathcal{A}_\mathrm{r}$ is norm closed, $\mathcal{A}_\mathrm{r} = C(X, \mathbb{R})$ and $\mathcal{A} = C(X)$.   ■

CHAPTER 4

ELEMENTARY C*-ALGEBRA THEORY

4.6. Exercises

4.6.1. Suppose that S_1, S_2, T_1, T_2, and A are elements of a C*-algebra $\mathfrak{A}$, and

$$0 \le S_1 \le T_1, \qquad 0 \le S_2 \le T_2.$$

Prove that

$$\|S_1^{1/2} A\| \le \|T_1^{1/2} A\|,$$

and deduce that

$$\|S_1^{1/2} A S_2^{1/2}\| \le \|T_1^{1/2} A T_2^{1/2}\|.$$

Solution. Since

$$0 \le (S_1^{1/2} A)^* (S_1^{1/2} A) = A^* S_1 A \le A^* T_1 A = (T_1^{1/2} A)^* (T_1^{1/2} A),$$

we have

$$\|S_1^{1/2} A\| = \|A^* S_1 A\|^{1/2} \le \|A^* T_1 A\|^{1/2} = \|T_1^{1/2} A\|.$$

A similar argument shows that

$$\|S_2^{1/2} B^*\| \le \|T_2^{1/2} B^*\|,$$

equivalently

$$\|B S_2^{1/2}\| \le \|B T_2^{1/2}\|,$$

for B in $\mathfrak{A}$. Upon replacing A by $A S_2^{1/2}$ and B by $T_1^{1/2} A$, we obtain

$$\|S_1^{1/2} A S_2^{1/2}\| \le \|T_1^{1/2} A S_2^{1/2}\| \le \|T_1^{1/2} A T_2^{1/2}\|. \qquad \blacksquare$$

4.6.2.   Suppose that $\mathfrak{A}$ and $\mathcal{B}$ are C*-algebras and φ is a * homomorphism from $\mathfrak{A}$ onto $\mathcal{B}$. Suppose that $B, K \in \mathcal{B}$, with B self-adjoint and K positive, and let V be the exponential unitary $\exp iB$. Show that there exist A, H, U $(= \exp iA)$ in $\mathfrak{A}$, with A self-adjoint and H positive, such that

$$\varphi(A) = B, \qquad \varphi(H) = K, \qquad \varphi(U) = V.$$

[See also Exercises 4.6.3 and 4.6.59.]

Solution.   Since $\varphi(\mathfrak{A}) = \mathcal{B}$, we have $B = \varphi(A_0)$ for some A_0 in $\mathfrak{A}$. Since

$$\varphi(A_0^*) = \varphi(A_0)^* = B^* = B,$$

we can conclude that $\varphi(A) = B$, where A is the self-adjoint element $\frac{1}{2}(A_0 + A_0^*)$ of $\mathfrak{A}$. By Theorem 4.1.8(ii)

$$V = \exp(iB) = \exp i\varphi(A) = \varphi(\exp iA).$$

With K in place of B, the above argument shows that $K = \varphi(S)$ for some self-adjoint S in $\mathfrak{A}$. With $f : \mathbb{R} \to \mathbb{R}$ the continuous function defined by $f(t) = \max\{t, 0\}$, we have $f(K) = K$, while $f(S)$ is a positive element H $(= S^+)$ of $\mathfrak{A}$. Again by Theorem 4.1.8(ii),

$$K = f(K) = f(\varphi(S)) = \varphi(f(S)) = \varphi(H). \qquad \blacksquare$$

4.6.3.   Suppose that $\mathbb{D}$ is the unit disk $\{z \in \mathbb{C} : |z| \le 1\}$, $\mathbb{T}$ is its boundary $\{z \in \mathbb{C} : |z| = 1\}$, and $\mathbb{S}$ is the union $\{\exp i\theta : \theta \in \mathbb{R}, \pi/4 \le |\theta| \le 3\pi/4\}$ of two closed arcs in $\mathbb{T}$. Consider the C*-algebras $C(\mathbb{D})$, $C(\mathbb{T})$, $C(\mathbb{S})$ and the * homomorphisms φ (from $C(\mathbb{D})$ onto $C(\mathbb{T})$) and ψ (from $C(\mathbb{T})$ onto $C(\mathbb{S})$) defined by restriction; that is,

$$\varphi(f) = f|\mathbb{T}, \qquad \psi(g) = g|\mathbb{S} \qquad (f \in C(\mathbb{D}), \quad g \in C(\mathbb{T})).$$

Find

(i)   a unitary element u of $C(\mathbb{T})$ that is not of the form $\varphi(f)$ for any invertible element f of $C(\mathbb{D})$; [*Hint.* Use the fact (from elementary algebraic topology) that there is no continuous mapping of $\mathbb{D}$ onto $\mathbb{T}$ that leaves each point of $\mathbb{T}$ fixed—that is, $\mathbb{T}$ is not a *retract* of $\mathbb{D}$.]

(ii)   a projection q in $C(\mathbb{S})$ that is not of the form $\psi(p)$ for any projection p in $C(\mathbb{T})$).

[See also Exercise 4.6.59.]

Solution. (i)   Define a unitary element u of $C(\mathbb{T}))$ by $u(z) = z$ ($z \in \mathbb{T}$). If f is an invertible element of $C(\mathbb{D})$ and $\varphi(f) = u$, then f is a continuous complex-valued function on $\mathbb{D}$ that does not take the value 0 and coincides with u on $\mathbb{T}$. Accordingly, the equation $r(z) = f(z)/|f(z)|$ defines a retraction $r : \mathbb{D} \to \mathbb{T}$. This is impossible, however, since $\mathbb{T}$ is not a retract of $\mathbb{D}$.

(ii)   A projection in $C(\mathbb{T})$ is a continuous function on $\mathbb{T}$ with values in $\{0,1\}$. Since $\mathbb{T}$ is connected, there are only two such projections, $0, I$. Hence the projection q in $C(\mathbb{S})$, that is the characteristic function of the arc $\{\exp i\theta : \pi/4 \le \theta \le 3\pi/4\}$, is not of the form $\psi(p)$ with p a projection in $C(\mathbb{T})$.   ∎

4.6.4.   Determine whether the following assertion is true or false: if $\mathfrak{A}$ and $\mathcal{B}$ are abelian C*-algebras, φ is a * homomorphism from $\mathfrak{A}$ onto $\mathcal{B}$, and B is an invertible self-adjoint element of $\mathcal{B}$, there is an invertible self-adjoint element A of $\mathfrak{A}$ such that $\varphi(A) = B$.

Solution.   The following example shows that the assertion is false. Take $\mathbb{T}$, $\mathbb{S}$, and the * homomorphism ψ from $C(\mathbb{T})$ onto $C(\mathbb{S})$, as in Exercise 4.6.3. Let k be the invertible self-adjoint element of $C(\mathbb{S})$ defined by

$$k(e^{i\theta}) = \begin{cases} \ \ 1 & (\pi/4 \le \theta \le 3\pi/4), \\ -1 & (-3\pi/4 \le \theta \le -\pi/4). \end{cases}$$

A self-adjoint element h of $C(\mathbb{T})$ for which $\psi(h) = k$ is a continuous function $h : \mathbb{T} \to \mathbb{R}$ which takes both the values ± 1 and hence (since $\mathbb{T}$ is connected) takes the value 0 at some point of $\mathbb{T}$. Accordingly, such an element h is not invertible in $C(\mathbb{T})$.   ∎

4.6.5.   Use the results of Exercise 4.6.2 and 4.6.3(i) to provide an example of an abelian C*-algebra $\mathcal{B}$ and a unitary element V of $\mathcal{B}$ that is not of the form $\exp iB$ for any self-adjoint B in $\mathcal{B}$.

Solution.   Let φ be the * homomorphism $f \to f|\mathbb{T}$ from $C(\mathbb{D})$ onto $C(\mathbb{T})$. The unitary element u of $C(\mathbb{T})$, defined by $u(z) = z$ for each z in $\mathbb{T}$, is not of the form $\varphi(f)$ for any unitary (indeed, for any invertible) element f of $C(\mathbb{D})$. By the result of Exercise 4.6.2, u is not of the form $\exp ib$, for any self-adjoint b in $C(\mathbb{T})$.   ∎

4.6.6.   Let $\mathcal{U}$ be the (multiplicative) group of all unitary elements in a C*-algebra $\mathfrak{A}$.

(i)   Show that, if $U \in \mathcal{U}$ and $\|I - U\| < 2$, then $U = \exp iH$ for some self-adjoint H in $\mathfrak{A}$.

(ii)   Show that, if $V, W \in \mathcal{U}$ and $\|V - W\| < 2$, then $V = W \exp iH$ for some self-adjoint H in $\mathfrak{A}$.

(iii)   Let $\mathcal{U}_1$ $(\subseteq \mathcal{U})$ be the set of all products of the form

$$(\exp iH_1)(\exp iH_2)\cdots(\exp iH_k),$$

where $\{H_1, H_2, \ldots, H_k\}$ is a finite set of self-adjoint elements of $\mathfrak{A}$. Show that $\mathcal{U}_1$ is open, closed, and arcwise connected, in the (relative) norm topology on $\mathcal{U}$.

Solution. (i)   If $U \in \mathcal{U}$ and $\|I - U\| < 2$, then

$$\mathrm{sp}(U) \subseteq \{z \in \mathbb{C} : |1 - z| < 2\},$$

and thus $-1 \notin \mathrm{sp}(U)$. By Proposition 4.4.10, $U = \exp iH$ for some self-adjoint H in $\mathfrak{A}$.

(ii)   If $V, W \in \mathcal{U}$ and $\|V - W\| < 2$, then $W^{-1}V \in \mathcal{U}$,

$$\|W^{-1}V - I\| \leq \|W^{-1}\|\|V - W\| < 2.$$

By (i), $W^{-1}V = \exp iH$ (and thus $V = W \exp iH$) for some self-adjoint H in $\mathfrak{A}$.

(iii)   From (ii), if $W \in \mathcal{U}_1$, $V \in \mathcal{U}$, and $\|V - W\| < 2$, then $V \in \mathcal{U}_1$. It follows that $\mathcal{U}_1$ is open in $\mathcal{U}$, since it contains the ball $\{V \in \mathcal{U} : \|V - W\| < 2\}$ whenever $W \in \mathcal{U}_1$. Also, $\mathcal{U}_1$ is closed in $\mathcal{U}$; for if a sequence $\{W_n\}$ in $\mathcal{U}_1$ is norm convergent to V (in $\mathcal{U}$), then $\|V - W_n\| < 2$ for some n, and thus $V \in \mathcal{U}_1$. Finally, $\mathcal{U}_1$ is arcwise connected since each element

$$(\exp iH_1)(\exp iH_2)\cdots(\exp iH_k)$$

of $\mathcal{U}_1$ is connected to I by the arc

$$\{(\exp itH_1)(\exp itH_2)\cdots(\exp itH_k) : \ 0 \leq t \leq 1\}$$

in $\mathcal{U}_1$.   ∎

4.6.7. Suppose that $\mathfrak{A}$ is an abelian C*-algebra, $\mathcal{U}$ is its unitary group (considered as a topological group with the norm topology), $\mathcal{U}_I$ is the connected component of $\mathcal{U}$ that contains the identity I, and $U \in \mathcal{U}$. Use the results of Exercise 4.6.6(iii) to show that the following three conditions are equivalent.

(i) $U = \exp iA$ for some self-adjoint A in $\mathfrak{A}$.

(ii) U is connected to I by a continuous arc in $\mathcal{U}$.

(iii) $U \in \mathcal{U}_I$.

Solution. If $U = \exp iA$ for some self-adjoint A in $\mathfrak{A}$, then U is connected to I by the continuous arc $\{\exp itA : 0 \le t \le 1\}$ in $\mathcal{U}$. Hence (i) implies (ii), and it is apparent that (ii) implies (iii).

It remains to prove that (iii) implies (i). The results of Exercise 4.6.6(iii) show that the set $\mathcal{U}_1$ defined there coincides with the connected component $\mathcal{U}_I$. Hence each U in $\mathcal{U}_I$ has the form

$$U = (\exp iH_1)(\exp iH_2) \cdots (\exp iH_n)$$

for some finite collection $\{H_1, H_2, \ldots, H_n\}$ of self-adjoint elements of $\mathfrak{A}$. However, it is easy to prove that

$$(\exp iB_1)(\exp iB_2) = \exp i(B_1 + B_2)$$

for all self-adjoint B_1 and B_2 in the *abelian* C*-algebra $\mathfrak{A}$ (either by multiplication of absolutely convergent power series, as in the scalar case, or by use of the function representation of $\mathfrak{A}$ and the result of Example 4.4.9). Thus $U = \exp iA$, where $A = H_1 + H_2 + \cdots + H_n$; and (iii) implies (i). ∎

4.6.8. Show that, in both of the following cases, each unitary element in the C*-algebra $\mathfrak{A}$ has the form $\exp iA$ for some self-adjoint A in $\mathfrak{A}$.

(i) $\mathfrak{A}$ is the C*-algebra L_∞ associated with a σ-finite measure space.

(ii) $\mathfrak{A}$ is the C*-algebra $C(X)$, where the compact Hausdorff space X is contractible (that is, there is a point x_0 in X and a continuous mapping $f : X \times [0,1] \to X$ such that $f(x,0) = x$, $f(x,1) = x_0$, for each x in X). [*Hint.* Use the result of Exercise 4.6.7.]

Solution. (i)   Suppose that $(S, \mathcal{S}, m)$ is a σ-finite measure space and u is a unitary element of $L_\infty(S, \mathcal{S}, m)$. By changing the values of u on a null set, we may assume that $|u(s)| = 1$ for each s in S. When $s \in \mathcal{S}$, let $g(s)$ be the unique real number such that $0 \le g(s) < 2\pi$ and $\exp ig(s) = u(s)$. Since u is measurable, so is g, and thus g is a self-adjoint element of L_∞. By using the power series expansion for the element $\exp ig$ of L_∞, we have

$$(\exp ig)(s) = \sum_{n=0}^{\infty} \frac{1}{n!}[ig(s)]^n = \exp ig(s) = u(s).$$

Thus $u = \exp ig$.

(ii)   Let u be a unitary element of $C(X)$; that is, u is a continuous mapping from X into $\mathbb{T}$ $(= \{z \in \mathbb{C} : |z| = 1\})$. Since the mapping $(x, t) \to u(f(x, t)) :\ X \times [0, 1] \to \mathbb{T}$ is (uniformly) continuous, we can define a continuous arc $\{u_t :\ 0 \le t \le 1\}$ that connects u to $u(x_0)I$, in the unitary group $\mathcal{U}$ of $C(X)$, by

$$u_t(x) = u(f(x, t)).$$

If we now choose a real number θ such that $\exp i\theta = u(x_0)$, the arc $\{(\exp it\theta)I :\ 0 \le t \le 1\}$ in $\mathcal{U}$ connects I to $u(x_0)I$. Thus u is connected to I by a continuous arc in $\mathcal{U}$; by the result of Exercise 4.6.7, u has the form $\exp ig$, for some self-adjoint g in $C(X)$. ∎

4.6.9.   (i)   Show that, if a, b are complex numbers and P, Q are projections with sum I in a C*-algebra, then the function calculus for the normal element $aP + bQ\ (= N)$ is given by

$$f(N) = f(a)P + f(b)Q.$$

(ii)   Suppose that $\mathbb{T}$ is the unit circle $\{z \in \mathbb{C} :\ |z| = 1\}$ and $\mathcal{M}$ is the algebra of all 2×2 complex matrices, so that $\mathcal{M}$ becomes a C*-algebra when identified in the usual way with the set of all linear operators acting on the two-dimensional Hilbert space $\mathbb{C}^2$. The Banach space $C(\mathbb{T}, \mathcal{M})$ (see Example 1.7.2) becomes a C*-algebra $\mathfrak{A}$ when products and adjoints (as well as the linear structure) are defined pointwise. Let E and F be the projections in $\mathfrak{A}$ given by

$$E(e^{i\theta}) = \begin{bmatrix} 0 & 0 \\ 0 & 1 \end{bmatrix}, \qquad F(e^{i\theta}) = \frac{1}{2}\begin{bmatrix} 1 - \cos\theta & -\sin\theta \\ -\sin\theta & 1 + \cos\theta \end{bmatrix}$$

$(0 \leq \theta \leq 2\pi)$, and let U be the unitary element $(\exp i\pi E)(\exp i\pi F)$ of $\mathfrak{A}$. Show that

$$U(e^{i\theta}) = e^{i\theta}P + e^{-i\theta}Q \qquad (0 \leq \theta \leq 2\pi),$$

where P and Q are the projections in $\mathcal{M}$ given by

$$P = \frac{1}{2}\begin{bmatrix} 1 & -i \\ i & 1 \end{bmatrix}, \qquad Q = \frac{1}{2}\begin{bmatrix} 1 & i \\ -i & 1 \end{bmatrix}.$$

Deduce that U, a product of two exponential unitary elements of $\mathfrak{A}$, is not itself an exponential unitary.

Solution. (i)   Since $\mathrm{sp}(N) = \{a, b\}$, each f in $C(\mathrm{sp}(N))$ is the restriction to $\mathrm{sp}(N)$ of an affine mapping $t \to ct + d : \mathbb{C} \to \mathbb{C}$, and

$$f(N) = cN + dI = c(aP + bQ) + d(P + Q)$$
$$= (ca + d)P + (cb + d)Q = f(a)P + f(b)Q.$$

(ii)   Since E and F are projections in $\mathfrak{A}$, it follows from (i) that

$$\exp i\pi E = \exp i\pi(1 \cdot E + 0 \cdot (I - E))$$
$$= (\exp i\pi)E + (\exp 0)(I - E)$$
$$= I - 2E,$$

and $\exp i\pi F = I - 2F$. Thus $U = (I - 2E)(I - 2F)$, and

$$U(e^{i\theta}) = \left(\begin{bmatrix} 1 & 0 \\ 0 & 1 \end{bmatrix} - 2\begin{bmatrix} 0 & 0 \\ 0 & 1 \end{bmatrix}\right)\left(\begin{bmatrix} 1 & 0 \\ 0 & 1 \end{bmatrix} - \begin{bmatrix} 1 - \cos\theta & -\sin\theta \\ -\sin\theta & 1 + \cos\theta \end{bmatrix}\right)$$
$$= \begin{bmatrix} 1 & 0 \\ 0 & -1 \end{bmatrix}\begin{bmatrix} \cos\theta & \sin\theta \\ \sin\theta & -\cos\theta \end{bmatrix} = \begin{bmatrix} \cos\theta & \sin\theta \\ -\sin\theta & \cos\theta \end{bmatrix}$$
$$= \begin{bmatrix} \frac{1}{2}(e^{i\theta} + e^{-i\theta}) & -\frac{1}{2}i(e^{i\theta} - e^{-i\theta}) \\ \frac{1}{2}i(e^{i\theta} - e^{-i\theta}) & \frac{1}{2}(e^{i\theta} + e^{-i\theta}) \end{bmatrix}$$
$$= e^{i\theta}P + e^{-i\theta}Q.$$

Note that P and Q are minimal projections in $\mathcal{M}$, with sum I, whence $P\mathcal{M}P = \{zP : z \in \mathbb{C}\}$ and $Q\mathcal{M}Q = \{zQ : z \in \mathbb{C}\}$.

Suppose that $U = \exp iH$ for some self-adjoint H in $\mathfrak{A}$. For each z in $\mathbb{T}$, the mapping $A \to A(z) : \mathfrak{A} \to \mathcal{M}$ is a * homomorphism; so, by Theorem 4.1.8(ii),

$$U(z) = (\exp iH)(z) = \exp iH(z).$$

From this, $H(z)$ commutes with $U(z)$. Since

$$U(z) = zP + \bar{z}Q = zP + \bar{z}(I - P) = (z - \bar{z})P + \bar{z}I,$$

it follows that $H(z)$ commutes with P (hence, also, with Q) for all z in $\mathbb{T}$ except perhaps for the two points where $z = \bar{z}$. By continuity, $H(z)$ commutes with P and Q for all z in $\mathbb{T}$. Hence

$$H(z) = (P + Q)H(z) = PH(z)P + QH(z)Q$$
$$= f(z)P + g(z)Q,$$

where f and g are continuous complex-valued functions on $\mathbb{T}$. Thus

$$zP = PU(z) = P \exp iH(z)$$
$$= P[(\exp if(z))P + \exp(ig(z))Q]$$
$$= (\exp if(z))P,$$

and

$$z = \exp if(z) \qquad (z \in \mathbb{T}).$$

Since $|z| = 1$, it follows that f is a continuous *real*-valued function on $\mathbb{T}$, and the mapping

$$(z, t) \to \exp i(1 - t)f(z) : \quad \mathbb{T} \times [0, 1] \to \mathbb{T}$$

contracts $\mathbb{T}$ to the point 1 (a contradiction, since $\mathbb{T}$ is not contractible). $\blacksquare$

4.6.10.   Let $\mathfrak{A}$ be a C*-algebra and $\mathcal{B}$ be a norm-closed (though not necessarily self-adjoint) subalgebra of $\mathfrak{A}$. Let A be a self-adjoint element of $\mathfrak{A}$ in $\mathcal{B}$. Show that $\mathrm{sp}_{\mathcal{B}}(A) = \mathrm{sp}_{\mathfrak{A}}(A)$. [*Hint.* See Exercise 3.5.28.]

Solution.   From Proposition 4.1.1(ii), $\mathrm{sp}_{\mathfrak{A}}(A)$ is a (compact) subset of $\mathbb{R}$ since A is a self-adjoint element of $\mathfrak{A}$. From Exercise 3.5.28(v), $\mathrm{sp}_{\mathfrak{A}}(A) = \mathrm{sp}_{\mathcal{B}}(A)$.

4.6.11.   Suppose that A is a positive element of a C*-algebra $\mathfrak{A}$, E and F are orthogonal projections in $\mathfrak{A}$, and $EAE = 0$. Show that $EAF = 0$.

Solution.   Replacing A by a suitable positive scalar multiple, we may assume that $0 \leq A \leq I$. In this case, $A^2 \leq A$; and

$$0 \leq EA(I - E)AE \leq EAIAE \leq EAE = 0.$$

Thus

$$0 = EA(I - E)AE = [EA(I - E)][EA(I - E)]^*$$

and $0 = EA(I - E)$. It follows that

$$0 = EA(I - E)F = EAF. \qquad \blacksquare$$

4.6.12.   Suppose that a C*-algebra $\mathfrak{A}$ has a maximal abelian * subalgebra $\mathcal{A}$ that is finite dimensional.
   (i)   Show that $\mathcal{A}$ is the linear span of a finite orthogonal family $\{E_1, \ldots, E_n\}$ of projections in $\mathfrak{A}$ with sum I (the identity of $\mathfrak{A}$).
   (ii)   By considering the family $\{E_1, \ldots, E_n, E_j A E_j\}$, where $A = A^* \in \mathfrak{A}$, show that $E_j \mathfrak{A} E_j = \{aE_j : a \in \mathbb{C}\}$ for $j = 1, \ldots, n$.
   (iii)   Suppose that $j, k \in \{1, \ldots, n\}$ and $j \neq k$. Given A and B in $\mathfrak{A}$, let a, b, c, d be the scalars determined by $E_j A^* E_k A E_j = aE_j$, $E_j A^* E_k B E_j = bE_j$, $E_j B^* E_k B E_j = cE_j$, $E_k B E_j A^* E_k = dE_k$. Prove that $a \geq 0$, $c \geq 0$. By considering suitable expressions for bdE_j, bdE_k, and acE_j, show that $b = d$ and $ac = |b|^2$. Deduce that

$$(sE_k A E_j + tE_k B E_j)^*(sE_k A E_j + tE_k B E_j) = 0$$

for suitable scalars s and t (not both 0). Deduce that $E_k \mathfrak{A} E_j$ is at most one dimensional.
   (iv)   Prove that $\mathfrak{A}$ is finite dimensional.

Solution. (i)   By maximality, $I \in \mathcal{A}$ and $\mathcal{A}$ is closed in $\mathfrak{A}$; so $\mathcal{A}$ is an abelian C*-algebra. Thus $\mathfrak{A}$ is * isomorphic to $C(X)$ for some compact Hausdorff space X. The points of X give rise to *evaluation functionals* on $\mathcal{A}$, and these form a linearly independent set in the (finite-dimensional) dual space $\mathcal{A}^{\#}$; so X is finite, and $C(X)$ consists of all complex-valued functions on X. Let X be $\{x_1, \ldots, x_n\}$. For each $j = 1, \ldots, n$, the function that is 1 at x_j and 0 elsewhere on X corresponds to a projection E_j in $\mathcal{A}$. Moreover, $\{E_1, \ldots, E_n\}$ is an orthogonal family of projections with sum I and linear span $\mathcal{A}$.

(ii) Since $E_j A E_j$ is self-adjoint and commutes with $E_1, \ldots, E_n$, the * subalgebra of $\mathfrak{A}$ generated by $\{E_1, \ldots, E_n, E_j A E_j\}$ is abelian and contains $\mathcal{A}$, and so coincides with $\mathcal{A}$. Hence $E_j A E_j \in \mathcal{A}$, and thus $E_j A E_j = \sum_{k=1}^{n} a_k E_k$ for suitable scalars $a_1, \ldots, a_n$. Since $0 = E_j A E_j E_k = a_k E_k$ when $k \neq j$, $E_j A E_j = a_j E_j$. We have now shown that $E_j A E_j$ is a scalar multiple of E_j, for each self-adjoint A in $\mathfrak{A}$. By linearity, this remains true for all A in $\mathfrak{A}$; so $E_j \mathfrak{A} E_j = \{a E_j : a \in \mathbb{C}\}$.

(iii) When $j \neq k$ and $A, B \in \mathfrak{A}$, we can choose scalars a, b, c, d with the stated properties, by (ii). Then,

$$bd E_j = E_j A^*(d E_k) B E_j = E_j A^* E_k B E_j A^* E_k B E_j$$
$$= (b E_j)(b E_j) = b^2 E_j,$$

$$bd E_k = E_k B(b E_j) A^* E_k = E_k B E_j A^* E_k B E_j A^* E_k$$
$$= (d E_k)(d E_k) = d^2 E_k,$$

whence $bd = b^2 = d^2$ and $b = d$. Also,

$$ac E_j = E_j A^* E_k A E_j B^* E_k B E_j$$
$$= E_j A^*(E_k B E_j A^* E_k)^* B E_j$$
$$= \bar{d} E_j A^* E_k B E_j = \bar{d} b E_j = |b|^2 E_j,$$

whence $ac = |b|^2$.

Since $a E_j = (E_k A E_j)^*(E_k A E_j) \geq 0$, it follows that $a \geq 0$; similarly, $c \geq 0$. Note also that, if s and t are scalars,

$$(s E_k A E_j + t E_k B E_j)^*(s E_k A E_j + t E_k B E_j)$$
$$= (\bar{s} E_j A^* E_k + \bar{t} E_j B^* E_k)(s E_k A E_j + t E_k B E_j)$$
$$= \bar{s} s a E_j + \bar{s} t b E_j + s \bar{t} \bar{b} E_j + \bar{t} t c E_j$$
$$= (a|s|^2 + b \bar{s} t + \bar{b} s \bar{t} + c|t|^2) E_j.$$
$$= \langle A z, z \rangle E_j,$$

where A is the operator on $\mathbb{C}^2$ with matrix

$$\begin{bmatrix} a & b \\ \bar{b} & c \end{bmatrix}$$

and z is the vector (s,t). Since $ac = |b|^2$, A is singular, and we can choose complex numbers s, t (not both 0) so that the right-hand-side of the above equations is 0. Then

$$sE_kAE_j + tE_kBE_j = 0.$$

The above argument shows that any two members of $E_k\mathfrak{A}E_j$ are linearly dependent; so $E_k\mathfrak{A}E_j$ is at most one dimensional.

(iv)   By (ii) and (iii), $E_k\mathfrak{A}E_j$ is at most one dimensional, for all $j,k = 1,\ldots,n$. Since $E_1 + \cdots + E_n = I$,

$$A = \sum_{j,k=1}^{n} E_kAE_j \in \sum_{j,k=1}^{n} E_k\mathfrak{A}E_j,$$

for each A in $\mathfrak{A}$. Thus $\mathfrak{A}$ has finite dimension at most n^2.   ■   [83]

4.6.13.   Suppose that $\mathfrak{A}$ is an infinite-dimensional C*-algebra. By using the result of Exercise 4.6.12, show that there is an infinite sequence $\{A_1, A_2, \ldots, \}$ of non-zero elements of $\mathfrak{A}^+$ such that $A_jA_k = 0$ when $j \neq k$.

Solution.   The family of all abelian * subalgebras of $\mathfrak{A}$ is partially ordered by inclusion, and a simple application of Zorn's lemma shows that $\mathfrak{A}$ has a *maximal* abelian * subalgebra $\mathcal{A}$. By maximality, $\mathcal{A}$ is closed in $\mathfrak{A}$ and contains I, and is therefore an abelian C*-algebra. Thus $\mathcal{A}$ is * isomorphic to $C(X)$, for some compact Hausdorff space X. By the result of Exercise 4.6.12, $\mathcal{A}$ is infinite-dimensional; so X is infinite.

It now suffices to prove the existence of an infinite sequence $\{V_1, V_2, \ldots\}$ of disjoint non-empty open sets in X; for then, we can take for A_n the operator in $\mathcal{A}$ corresponding to a non-zero element f_n of $C(X)^+$ that vanishes outside V_n.

If each x in X has a finite neighborhood V_x, there is a finite subcovering of the covering $\{V_x : x \in X\}$ of X, and X is finite, a contradiction. Hence there is at least one point x_0 in X, such that every neighborhood of x_0 is infinite. Choose x_1 ($\neq x_0$) in X, and disjoint neighborhoods V_1 and W_1 of x_1 and x_0, respectively. Choose x_2 ($\neq x_0$) in W_1, and disjoint neighborhoods V_2 and W_2 (both, subsets of W_1) of x_2 and x_0, respectively. Choose $x_3(\neq x_0)$ in W_2, and disjoint neighborhoods V_3 and W_3 (both subsets of W_2) of x_3 and x_0, respectively. By continuing in this way, we obtain a sequence $\{V_n\}$ with the required properties.   ■

4.6.14.   By using the result of Exercise 4.6.12, show that, in an infinite-dimensional C*-algebra, there is a positive element with infinite spectrum.

Solution.   Given an infinite-dimensional C*-algebra $\mathfrak{A}$, construct $\mathcal{A}$, X, $\{V_1, V_2, V_3, \ldots\}$ as in the solution to Exercise 4.6.13. Let $f_n : X \to [0, n^{-2}]$ be a continuous function that takes the value n^{-2} at some point of V_n and vanishes outside V_n. Then $\sum f_n$ is an element of $C(X)^+$ that takes each of the values n^{-2} ($n = 1, 2, 3, \ldots$). There is a corresponding element of $\mathcal{A}^+$ ($\subseteq \mathfrak{A}^+$) whose spectrum (in $\mathcal{A}$; hence, in $\mathfrak{A}$) contains all these values.   ∎

4.6.15.   Let ρ be a state of a C*-algebra $\mathfrak{A}$. We say that ρ is *faithful* if $A = 0$ when $A \in \mathfrak{A}^+$ and $\rho(A) = 0$.

(i)   Show that $\mathcal{L}_\rho$, the left kernel of ρ, is $\{0\}$ when ρ is a faithful state of $\mathfrak{A}$. Deduce that $\mathcal{H}_\rho$ is the completion of $\mathfrak{A}$ relative to the inner product $(A, B) \to \rho(B^*A)$ $(= \langle A, B \rangle)$, and that π_ρ is faithful.

(ii)   Let ρ be a state of $\mathfrak{A}$ such that π_ρ is faithful. Must ρ be faithful?

Solution. (i)   Since ρ is faithful and $A^*A \in \mathfrak{A}^+$ for each A in $\mathfrak{A}$, $A^*A = 0$ if $\rho(A^*A) = 0$; and $A = 0$ in this case. Thus $\mathcal{L}_\rho = \{0\}$ and $\mathcal{H}_\rho$ is the completion of $\mathfrak{A}$ $(= \mathfrak{A}/\mathcal{L}_\rho)$ relative to the inner product: $(A, B) \to \rho(B^*A)$. (See Proposition 4.5.1 and Theorem 4.5.2.) To see that π_ρ is faithful in this case, note that if $\pi_\rho(A) = 0$, then $0 = \|\pi_\rho(A)x_\rho\|^2 = \rho(A^*A)$, whence $A = 0$.

(ii)   No! Let $\mathfrak{A}$ be $\mathcal{B}(\mathcal{H})$ and x be a unit vector in $\mathcal{H}$. Then x is a cyclic vector for $\mathcal{B}(\mathcal{H})$ acting on $\mathcal{H}$. Let ρ be the state ω_x of $\mathcal{B}(\mathcal{H})$. Then $\rho(A) = \langle Ax, x \rangle$ for each A in $\mathcal{B}(\mathcal{H})$ and, from Proposition 4.5.3, the identity representation of $\mathcal{B}(\mathcal{H})$ on $\mathcal{H}$ is unitarily equivalent to π_ρ. Thus π_ρ is faithful. If E is a projection in $\mathcal{B}(\mathcal{H})$ such that x is orthogonal to $E(\mathcal{H})$, then $\rho(E^*E) = \rho(E) = \langle Ex, x \rangle = 0$. If $\mathcal{H}$ has dimension 2 or greater, we can choose E such that $Ex = 0$ and $E \neq 0$. Since $\rho(E^*E) = 0$, ρ is not faithful.   ∎

4.6.16.   Let A be a self-adjoint element in a C*-algebra $\mathfrak{A}$. Let ρ be a state of $\mathfrak{A}$ such that $\rho(A^2) = \rho(A)^2$. (We say that ρ is *definite* on A in this case.) Show that $\rho(AB) = \rho(BA) = \rho(A)\rho(B)$ for each B in $\mathfrak{A}$.

Solution.   Note, first that

$$\rho([A - \rho(A)I]^2) = \rho(A^2) - 2\rho(A)^2 + \rho(A)^2 = 0,$$

so that $A - \rho(A)I$ is in $\mathcal{L}_\rho$, the left kernel of ρ. (See Proposition 4.5.1.) Thus $BA - \rho(A)B \in \mathcal{L}_\rho$, from which we conclude that $0 = \rho(BA - \rho(A)B) = \rho(BA) - \rho(A)\rho(B)$. At the same time,

$$\overline{\rho(AB)} = \rho(B^*A) = \rho(B^*)\rho(A) = \overline{\rho(B)}\rho(A),$$

whence $\rho(AB) = \rho(A)\rho(B)$.   ■   [58]

4.6.17.   Suppose that $\mathfrak{A}$, $\mathcal{A}$, and $\{E_1, \ldots, E_n\}$ are as in Exercise 4.6.12. Let ρ_j be a state of $\mathfrak{A}$ that extends the state of $\mathcal{A}$ assigning 1 to E_j and 0 to E_k when $k \neq j$. Let ρ be $n^{-1} \sum_{j=1}^n \rho_j$.

(i)   Show that ρ is a faithful state of $\mathfrak{A}$.

(ii)   Choose A_{jk} in $\mathfrak{A}$ such that $(E_j A_{jk} E_k)^*(E_j A_{jk} E_k) = nE_k$ for those k and j for which $E_j \mathfrak{A} E_k$ is one dimensional. Show that the set of $E_j A_{jk} E_k$ forms an orthonormal basis for $\mathcal{H}_\rho$.

Solution. (i)   Suppose $A \in \mathfrak{A}^+$ and

$$0 = \rho(A) = \rho\Big(\sum_{j,k=1}^n E_j A E_k \Big)$$
$$= \frac{1}{n} \sum_{j,k=1}^n \sum_{h=1}^n \rho_h(E_j A E_k).$$

Since ρ_h is definite on each E_j for all h and j in $\{1, \ldots, n\}$, $\rho_h(E_j A E_k) = \rho_h(E_j)\rho_h(A)\rho_h(E_k)$. Thus $\rho_h(E_j A E_k) = 0$ unless $h = k = j$. (See Exercise 4.6.16.) It follows that

$$0 = \rho(A) = \frac{1}{n} \sum_{j=1}^n \rho_j(E_j A E_j) = \frac{1}{n} \sum_{j=1}^n a_j,$$

where $E_j A E_j = a_j E_j$. Now $0 \leq a_j$ since $0 \leq E_j A E_j$. Thus $a_j = 0$ and $E_j A E_j = 0$ for all j in $\{1, \ldots, n\}$. From Exercise 4.6.11, $E_j A E_k = 0$ for all j and k in $\{1, \ldots, n\}$, and $0 = \sum_{j,k=1}^n E_j A E_k = A$. Hence ρ is faithful.

(ii)   Note that

$$\rho([E_j A_{jk} E_k]^*[E_{j'} A_{j'k'} E_{k'}]) = \frac{1}{n} \sum_{h=1}^{n} \rho_h(E_k A_{jk}^* E_j E_{j'} A_{j'k'} E_{k'})$$

is 0 unless $k = k'$ and $j = j'$, in which case it is 1 $(= \rho_k(E_k))$.
Thus $\{E_j A_{jk} E_k\}$ is an orthonormal set. For each A in $\mathfrak{A}$ and j, k
in $\{1,\ldots,n\}$, $E_j A E_k$ is 0 or $a_{jk} E_j A_{jk} E_k$ for some scalar a_{jk}, since
$E_j \mathfrak{A} E_k$ is at most one dimensional. Hence $A (= \sum\sum E_j A E_k)$ lies in
the linear span of the set $\{E_j A_{jk} E_k\}$, and this set is an orthonormal
basis of $\mathcal{H}_\rho$.   ∎

4.6.18.   Let $\mathcal{H}$ be a Hilbert space of finite dimension n,
$\{e_1,\ldots,e_n\}$ be an orthonormal basis for $\mathcal{H}$, E_{jk} be the element
of $\mathcal{B}(\mathcal{H})$ that maps e_k to e_j and $e_{k'}$ to 0 when $k' \neq k$, and ρ
be an element of the Banach dual space $\mathcal{B}(\mathcal{H})^\#$. Define A to be
the element of $\mathcal{B}(\mathcal{H})$ with the matrix representation $[a_{jk}]$ relative to
$\{e_1,\ldots,e_n\}$, where $a_{jk} = \rho(E_{kj})$, and τ to be the *normalized trace*
$(\tau(B) = n^{-1} \sum_{j=1}^{n} b_{jj} = n^{-1}\mathrm{tr}(B)$, where B has matrix $[b_{jk}]$ relative
to $\{e_1,\ldots,e_n\})$. Show that:
   (i) τ is a faithful state of $\mathcal{B}(\mathcal{H})$;
   (ii) $\rho(B) = \mathrm{tr}(AB)$ for each B in $\mathcal{B}(\mathcal{H})$, and ρ is a state if and
only if $0 \leq A$ and $\mathrm{tr}(A) = 1$;
   (iii) A is independent of the basis $\{e_1,\ldots,e_n\}$;
   (iv) ρ is a pure state of $\mathcal{B}(\mathcal{H})$ if and only if A is a projection
with one-dimensional range (in which case $\rho = \omega_x$, where x is a unit
vector in the range of A);
   (v) ρ is a faithful state of $\mathcal{B}(\mathcal{H})$ if and only if $\mathrm{tr}(A) = 1$ and
$A \geq aI$ for some positive scalar a.

   Solution.   (i)   By definition, $\tau(I) = n^{-1} \sum_{j=1}^{n} 1 = 1$. If B
is positive, then $b_{jj} = \langle Be_j, e_j \rangle \geq 0$ for j in $\{1,\ldots,n\}$ and $\tau(B) =$
$n^{-1} \sum_{j=1}^{n} b_{jj} \geq 0$. Thus τ is a state of $\mathcal{B}(\mathcal{H})$. If $\tau(B) = 0$, then $b_{jj} =$
0 and $E_{jj} B E_{jj} = 0$ for all j. From Exercise 4.6.11, $E_{jj} B E_{kk} = 0$
for all j and k in $\{1,\ldots,n\}$. Thus $0 = \sum_{j,k=1}^{n} E_{jj} B E_{kk} = B$, and τ
is faithful.
   (ii)   Note that $A E_{kj} e_{j'} = 0$ unless $j = j'$ and that $A E_{kj} e_j = A e_k$.
Thus $A E_{kj}$ is represented by the matrix whose j th column coincides
with the k th column of the matrix representing A and has all other
entries 0. Hence $\mathrm{tr}(A E_{kj}) = a_{jk} = \rho(E_{kj})$ for all j and k. Since

$\{E_{jk}\}$ forms a (linear) basis for $\mathcal{B}(\mathcal{H})$, $\rho(B) = \text{tr}(AB)$ for each B in $\mathcal{B}(\mathcal{H})$.

If ρ is a state, $1 = \rho(I) = \text{tr}(AI) = \text{tr}(A)$ and

$$0 \le \rho(E) = \text{tr}(AE) = \text{tr}(EAE) = \langle Ax, x \rangle \text{tr}(E) = \langle Ax, x \rangle$$

for each (one-dimensional) projection E with range generated by a unit vector x. Thus $0 \le \langle Ay, y \rangle$ for each y in $\mathcal{H}$, and $A \ge 0$.

If $A \ge 0$ and $\text{tr}(A) = 1$, then $\rho(I) = \text{tr}(AI) = \text{tr}(A) = 1$ and from (i), τ is a state so that

$$\rho(B) = \text{tr}(AB)) = n\tau(A^{\frac{1}{2}}BA^{\frac{1}{2}}) \ge 0$$

when $B \ge 0$. Thus ρ is a state.

(iii)   If a different basis is used and A' is constructed with this basis, then $\rho(B) = \text{tr}(A'B) = \text{tr}(AB)$. Thus $\text{tr}(CB) = 0$ for each B in $\mathcal{B}(\mathcal{H})$, where $C = A - A'$. If C has matrix representation $[c_{jk}]$ relative to $\{e_1, \ldots, e_n\}$ and $B = E_{kj}$, then $c_{jk} = \text{tr}(CE_{kj}) = 0$. Thus $0 = C = A - A'$.

(iv)   If $0 \le \eta \le \rho$ and $\eta(B) = \text{tr}(A'B)$ for each B in $\mathcal{B}(\mathcal{H})$, then $0 \le A'$ and $\text{tr}(CH) = (\rho - \eta)(H) \ge 0$ for each H in $\mathcal{B}(\mathcal{H})^+$, where $C = A - A'$. As in (ii), $0 \le C$ so that $0 \le A' \le A$. Conversely, if $0 \le A' \le A$, then $0 \le \eta \le \rho$ (as in (ii)). Now $\eta = t\rho$ for some scalar t if and only if $A' = tA$ (as in (iii)). Since ρ is a pure state if and only if each functional η on $\mathcal{B}(\mathcal{H})$ such that $0 \le \eta \le \rho$ is a scalar multiple of ρ, it follows now that ρ is a pure state if and only if A' is a scalar multiple of A when $0 \le A' \le A$. Considering the diagonal matrix representation of A, we see that A has at most one non-zero entry if and only if this condition is satisfied. Since $\text{tr}(A) = 1$, A is a one-dimensional projection if and only if ρ is a pure state of $\mathcal{B}(\mathcal{H})$. If ρ is pure and x is a unit vector spanning the range of A, then $ABA = \langle Bx, x \rangle A$ for each B in $\mathcal{B}(\mathcal{H})$; so that

$$\rho(B) = \text{tr}(AB) = \text{tr}(ABA) = \langle Bx, x \rangle \text{tr}(A) = \langle Bx, x \rangle = \omega_x(B).$$

Thus $\rho = \omega_x$, when ρ is a pure state of $\mathcal{B}(\mathcal{H})$.

(v)   Suppose $A \ge aI$ for some positive scalar a, and suppose $0 < H$. Then

$$\rho(H) = \text{tr}(AH) = \text{tr}(H^{\frac{1}{2}}AH^{\frac{1}{2}}) \ge \text{tr}(H^{\frac{1}{2}}(aI)H^{\frac{1}{2}}) = a\text{tr}(H) > 0,$$

and ρ is faithful. Conversely, if $aI \not\le A$ for each positive scalar a, then since $\mathcal{H}$ is finite dimensional and $0 \le A$, $Ax = 0$ for some unit vector x. If E is the projection with range $[x]$, then $\rho(E) = \text{tr}(AE) = \langle Ax, x \rangle = 0$; and ρ is not faithful.   ∎

4.6.19.   With the notation of Exercise 4.6.18, suppose ρ is a state and E is the projection in $\mathcal{B}(\mathcal{H})$ with range the range of A. Show that the left kernel $\mathcal{L}$ of ρ is $\mathcal{B}(\mathcal{H})(I - E)$.

Solution.   Since $A = A^*$ and $R(A) = E$, we have $N(A) = I - E$ from Proposition 2.5.13. Thus $A(I - E) = 0$ and $\rho(I - E) = \mathrm{tr}(A(I - E)) = \mathrm{tr}(0) = 0$. Since $(I - E)^*(I - E) = (I - E)$, it follows that $I - E \in \mathcal{L}$. Thus $\mathcal{B}(\mathcal{H})(I - E) \subseteq \mathcal{L}$.

Suppose $B \in \mathcal{L}$. Then

$$0 = \rho(B^*B) = \mathrm{tr}(B^*BA) = \mathrm{tr}(A^{\frac{1}{2}}B^*BA^{\frac{1}{2}}) = n\tau(A^{\frac{1}{2}}B^*BA^{\frac{1}{2}}).$$

From Exercise 4.6.18(i), τ is a faithful state of $\mathcal{B}(\mathcal{H})$. Hence $0 = A^{\frac{1}{2}}B^*BA^{\frac{1}{2}} = (BA^{\frac{1}{2}})^*BA^{\frac{1}{2}}$ and $BE = 0$. It follows from this that $B = B(I - E) \in \mathcal{B}(\mathcal{H})(I - E)$. $\blacksquare$

4.6.20.   Suppose that $\mathfrak{A}$ and $\mathcal{B}$ are C*-algebras, and φ is a * homomorphism from $\mathfrak{A}$ onto $\mathcal{B}$.

(i)   Let B_1 and B_2 be elements of $\mathcal{B}^+$ such that $B_1 B_2 = 0$. By considering $B_1 - B_2$, show that there exist elements A_1 and A_2 of $\mathfrak{A}^+$ such that $A_1 A_2 = 0$, $\varphi(A_1) = B_1$, and $\varphi(A_2) = B_2$.

(ii)   Suppose that $\{B_1, B_2, B_3, \ldots\}$ is a sequence of elements of $\mathcal{B}^+$ such that $B_j B_k = 0$ when $j \neq k$. Prove that there is a sequence $\{A_1, A_2, A_3, \ldots\}$ of elements of $\mathfrak{A}^+$ such that $A_j A_k = 0$ when $j \neq k$, and $\varphi(A_j) = B_j$ for each $j = 1, 2, 3, \ldots$. [*Hint.* Upon replacing B_j by $b_j B_j$ for a suitable positive scalar b_j, we may assume that the series $\sum B_j$ and $\sum B_j^{1/2}$ converge to elements of $\mathcal{B}$. Prove, by induction on n, the following statement: there exist elements $A_1, A_2, \ldots, A_n, X_n$ of $\mathfrak{A}^+$ such that $A_j A_k = A_j X_n = 0$ for all $j, k = 1, \ldots, n$ with $j \neq k$, and

$$\varphi(A_1) = B_1, \ldots, \varphi(A_n) = B_n, \quad \varphi(X_n) = B_{n+1} + B_{n+2} + \cdots.]$$

Solution. (i)   Let $B = B_1 - B_2$. By the uniqueness clause in Proposition 4.2.3(iii), $B_1 = B^+$ and $B_2 = B^-$. We can choose A in $\mathfrak{A}$ such that $\varphi(A) = B$; upon replacing A by $\frac{1}{2}(A + A^*)$, we may suppose that A is self-adjoint. From Theorem 4.1.8(ii),

$$\varphi(A^+) = B^+ = B_1, \qquad \varphi(A^-) = B^- = B_2.$$

It now suffices to take A^+ for A_1 and A^- for A_2.

(ii) We proceed as indicated in the hint. By (i), there exist A_1 and X_1 in $\mathfrak{A}^+$ such that $A_1 X_1 = 0$ and

$$\varphi(A_1) = B_1, \qquad \varphi(X_1) = B_2 + B_3 + \cdots.$$

This establishes the assertion in the hint, in the case $n = 1$.

Now suppose that the assertion has been proved in the case $n = r$, and let $A_1, \ldots, A_r, X_r$ satisfy the stated conditions. Choose a self-adjoint Y in $\mathfrak{A}$ such that

$$\varphi(Y) = B_{r+1}^{1/2} - (B_{r+2}^{1/2} + B_{r+3}^{1/2} + \cdots),$$

and let $Z = X_r^{1/4} Y X_r^{1/4}$. Then

$$\begin{aligned}
\varphi(Z) &= \varphi(X_r)^{1/4} \varphi(Y) \varphi(X_r)^{1/4} \\
&= (B_{r+1} + \cdots)^{1/4} [B_{r+1}^{1/2} - (B_{r+2}^{1/2} + B_{r+3}^{1/2} + \cdots)] (B_{r+1} + \cdots)^{1/4} \\
&= B_{r+1} - (B_{r+2} + B_{r+3} + \cdots).
\end{aligned}$$

The final step in this calculation follows by considering the function representation of the abelian C*-algebra generated by $\{I, B_1, B_2, \ldots\}$, noting that $B_1, B_2, B_3, \ldots$ correspond to non-negative functions, only one of which is non-zero at any given point. We now deduce that

$$\varphi(Z^+) = B_{r+1}, \qquad \varphi(Z^-) = (B_{r+2} + B_{r+3} + \cdots).$$

Since $A_j X_r = 0$ for all $j = 1, \ldots, r$, it follows that $A_j f(X_r) = 0$ when f is a polynomial with zero constant term, and hence also when $f : \mathbb{R} \to \mathbb{R}$ is continuous and $f(0) = 0$. In particular, $A_j X_r^{1/4} = 0$, and thus $A_j Z = 0$. Repeating the above argument, $A_j f(Z) = 0$ ($j = 1, \ldots, r$) for all f of the type just described; so $A_j Z^+ = A_j Z^- = 0$. We now obtain $A_1, \ldots, A_r, A_{r+1}, X_{r+1}$ with the required properties by taking $A_{r+1} = Z^+$, $X_{r+1} = Z^-$. Induction now produces the sequence $\{A_1, A_2, \ldots\}$. ∎

4.6.21.   Suppose that $\mathfrak{A}$ and $\mathcal{B}$ are C*-algebras, φ is a * homomorphism from $\mathfrak{A}$ onto $\mathcal{B}$, and

$$A \in \mathfrak{A}^+, \qquad B \in \mathcal{B}, \qquad 0 \le B \le \varphi(A).$$

(i)   Show that there is a self-adjoint element R of $\mathfrak{A}$ such that $R \leq A$ and $\varphi(R) = B$. Deduce that there exist positive elements S and T of $\mathfrak{A}$ such that

$$\varphi(S) = B, \qquad \varphi(T) = 0, \qquad S \leq T + A.$$

(ii)   For $n = 1, 2, \ldots$, let

$$U_n = S^{1/2}(n^{-1}I + T + A)^{-1}(T + A)^{1/2}A^{1/2}.$$

Show that

$$U_n^* U_n \leq A, \qquad \varphi(U_n) = B^{1/2}[n^{-1}I + \varphi(A)]^{-1}\varphi(A).$$

(iii)   By using the result of Exercise 4.6.1, show that

$$\|U_m - U_n\|$$
$$\leq \|(n^{-1} - m^{-1})(m^{-1}I + T + A)^{-1}(n^{-1}I + T + A)^{-1}(T + A)^{\frac{3}{2}}\|$$
$$\leq |n^{-1} - m^{-1}|^{\frac{1}{2}},$$

$$\|B^{1/2} - \varphi(U_n)\| \leq \|n^{-1}\varphi(A)^{1/2}[n^{-1}I + \varphi(A)]^{-1}\| \leq \tfrac{1}{2}n^{-1/2}.$$

(iv)   Deduce that the sequence $\{U_n^* U_n\}$ converges to an element A_0 of $\mathfrak{A}$ such that

$$0 \leq A_0 \leq A, \qquad \varphi(A_0) = B.$$

Solution. (i)   Let K be the element $\varphi(A) - B$ of $\mathcal{B}^+$. By the result of Exercise 4.6.2, $K = \varphi(H)$ for some H in $\mathfrak{A}^+$. Define $R = A - H$ $(\leq A)$, $S = R^+$, $T = R^-$. Then

$$S = T + R \leq T + A,$$
$$\varphi(R) = \varphi(A) - \varphi(H) = \varphi(A) - K = B,$$
$$\varphi(S) = \varphi(R^+) = B^+ = B,$$
$$\varphi(T) = \varphi(R^-) = B^- = 0.$$

(ii)   Let B_n be $(n^{-1}I + T + A)^{-1}$. Then

$$U_n^* U_n = A^{\frac{1}{2}}(T + A)^{\frac{1}{2}}B_n S B_n(T + A)^{\frac{1}{2}}A^{\frac{1}{2}}$$
$$\leq A^{\frac{1}{2}}(T + A)^{\frac{1}{2}}B_n(T + A)B_n(T + A)^{\frac{1}{2}}A^{\frac{1}{2}}$$
$$= A^{\frac{1}{2}}f_n(T + A)A^{\frac{1}{2}},$$

where $f_n(t) = t^2(n^{-1}+t)^{-2}$. Since $0 \le f_n(t) \le 1$ for all t in $\mathrm{sp}(T+A)$ $(\subseteq \mathbb{R}^+)$, we have $0 \le f_n(T+A) \le I$, and thus $U_n^* U_n \le A$.

From Theorem 4.1.8(ii), and since $\varphi(S) = B$ and $\varphi(T+A) = \varphi(A)$, we have

$$\varphi(U_n) = B^{\frac{1}{2}}[n^{-1}I + \varphi(A)]^{-1}\varphi(A)^{\frac{1}{2}}\varphi(A)^{\frac{1}{2}}$$
$$= B^{\frac{1}{2}}[n^{-1}I + \varphi(A)]^{-1}\varphi(A).$$

(iii) As with B_n, let B_m be $(m^{-1}I + T + A)^{-1}$. Then, since $B_m - B_n = (n^{-1} - m^{-1})B_m B_n$, we have

$$U_m - U_n = (n^{-1} - m^{-1})S^{\frac{1}{2}}B_m B_n(T+A)^{\frac{1}{2}}A^{\frac{1}{2}}.$$

Since $0 \le S \le T + A$ and $0 \le A \le T + A$, it now follows from the result of Exercise 4.6.1 that

$$\|U_m - U_n\| \le \|(n^{-1} - m^{-1})(T+A)^{\frac{1}{2}}B_m B_n(T+A)^{\frac{1}{2}}(T+A)^{\frac{1}{2}}\|;$$

that is,

$$\|U_m - U_n\| \le \|f_{mn}(T+A)\|,$$

where

$$f_{mn}(t) = (n^{-1} - m^{-1})t^{\frac{3}{2}}(m^{-1} + t)^{-1}(n^{-1} + t)^{-1} \qquad (t \in \mathbb{R}^+).$$

When $0 < n \le m$, we have $0 < m^{-1} \le n^{-1}$, and

$$0 \le f_{mn}(t) = (n^{-1} - m^{-1})^{\frac{1}{2}}\left(\frac{n^{-1} - m^{-1}}{n^{-1} + t}\right)^{\frac{1}{2}}\left(\frac{t}{n^{-1} + t}\right)^{\frac{1}{2}}\frac{t}{m^{-1} + t}$$
$$< (n^{-1} - m^{-1})^{\frac{1}{2}} \qquad\qquad\qquad (t \in \mathbb{R}^+).$$

Thus $\|U_m - U_n\| \le \|f_{mn}(T+A)\| \le |n^{-1} - m^{-1}|^{\frac{1}{2}}$.

Since $0 \le B \le \varphi(A)$ and

$$B^{\frac{1}{2}} - \varphi(U_n) = B^{\frac{1}{2}}[I - (n^{-1}I + \varphi(A))^{-1}\varphi(A)]$$
$$= n^{-1}B^{\frac{1}{2}}[n^{-1}I + \varphi(A)]^{-1},$$

it follows from the result of Exercise 4.6.1 that

$$\|B^{\frac{1}{2}} - \varphi(U_n)\| \le \|n^{-1}\varphi(A)^{\frac{1}{2}}[n^{-1}I + \varphi(A)]^{-1}\|$$
$$= \|g_n(\varphi(A))\|,$$

where $g_n(t) = n^{-1}t^{\frac{1}{2}}(n^{-1} + t)^{-1}$. Since $n^{-1} + t \geq 2t^{\frac{1}{2}}n^{-\frac{1}{2}}$ for all t in $\mathrm{sp}(\varphi(A))$ ($\subseteq \mathbb{R}^+$), we have $|g_n(t)| < \frac{1}{2}n^{-\frac{1}{2}}$; so $\|B^{\frac{1}{2}} - \varphi(U_n)\| < \frac{1}{2}n^{-\frac{1}{2}}$.

(iv) From (iii), $\{U_n\}$ is a Cauchy sequence in $\mathfrak{A}$ and $\varphi(U_n) \to B^{\frac{1}{2}}$. With U defined as $\lim U_n$, we have

$$0 \leq U^*U = \lim_n U_n^*U_n \leq A,$$

$$\varphi(U^*U) = \lim \varphi(U_n^*)\varphi(U_n) = B^{\frac{1}{2}}B^{\frac{1}{2}} = B.$$

This proves (iv), with $A_0 = U^*U$. ■ [86]

4.6.22.   Let $\mathfrak{A}$ be the C*-algebra $l_\infty(\mathbf{A})$, where $\mathbf{A}$ is the set $\{1,2,3,4\}$, and define a self-adjoint subspace $\mathcal{M}$ of $\mathfrak{A}$ (that contains the identity of $\mathfrak{A}$) by

$$\mathcal{M} = \{f \in \mathfrak{A}: \quad f(1) + f(2) = f(3) + f(4)\}.$$

Find a hermitian linear functional ρ on $\mathcal{M}$ that has more than one expression in the form $\rho = \rho_1 - \rho_2$, where ρ_1 and ρ_2 are positive linear functionals on $\mathcal{M}$ and $\|\rho\| = \|\rho_1\| + \|\rho_2\|$. [See the discussion following Corollary 4.3.7.]

Solution.   For $j = 1,2,3,4$, let σ_j be the state of $\mathcal{M}$ defined by $\sigma_j(f) = f(j)$. Then $\sigma_1 + \sigma_2 = \sigma_3 + \sigma_4$. Let

$$\rho = \sigma_1 - \sigma_3 = \sigma_4 - \sigma_2.$$

Then $\|\rho\| \leq 2$; and $\|\rho\| = 2$, since $\rho(f) = 2$ when f (in the unit ball of $\mathcal{M}$) is defined by $f(1) = f(4) = 1$, $f(2) = f(3) = -1$. Thus

$$\|\rho\| = \|\sigma_1\| + \|\sigma_3\| = \|\sigma_4\| + \|\sigma_2\|.$$

However, $\sigma_1 \neq \sigma_4$ (and, hence, $\sigma_3 \neq \sigma_2$) since $\sigma_1(f) = 1$ and $\sigma_4(f) = 0$ when f (in $\mathcal{M}$) is defined by $f(1) = f(3) = 1$, $f(2) = f(4) = 0$. ■

4.6.23.   Let φ be a * homomorphism of one C*-algebra $\mathfrak{A}$ onto another C*-algebra $\mathcal{B}$, and let ρ be a linear functional on $\mathcal{B}$.

(i)   Show that $\rho \circ \varphi$ is a state of $\mathfrak{A}$ if ρ is a state of $\mathcal{B}$.

(ii)   Show that ρ is a state of $\mathcal{B}$ if $\rho \circ \varphi$ is a state of $\mathfrak{A}$.

(iii)   Show that $\rho \circ \varphi$ is a pure state of $\mathfrak{A}$ if ρ is a pure state of $\mathcal{B}$.

(iv)   Show that ρ is a pure state of $\mathcal{B}$ if $\rho \circ \varphi$ is a pure state of $\mathfrak{A}$.

Solution. (i)   If $A \in \mathfrak{A}^+$, then $\varphi(A) \geq 0$. (See the discussion preceding Lemma 4.2.1.) If ρ is a state of $\mathcal{B}$, then $(\rho \circ \varphi)(A) = \rho[\varphi(A)] \geq 0$ and $(\rho \circ \varphi)(I) = \rho[\varphi(I)] = \rho(I) = 1$. Thus $\rho \circ \varphi$ is a state of $\mathfrak{A}$.

(ii)   Suppose $\rho \circ \varphi$ is a state of $\mathfrak{A}$. If $B \in \mathcal{B}^+$, then there is an A in $\mathfrak{A}^+$ such that $\varphi(A) = B$, from Exercise 4.6.2. Thus $\rho(B) = \rho[\varphi(A)] = (\rho \circ \varphi)(A) \geq 0$. Since $\varphi(I) = I$, we have that $\rho(I) = \rho[\varphi(I)] = (\rho \circ \varphi)(I) = 1$. Hence ρ is a state of $\mathcal{B}$.

(iii)   If ρ is a pure state of $\mathcal{B}$, then ρ is, in particular, a state of $\mathcal{B}$, and from (i), $\rho \circ \varphi$ is a state of $\mathfrak{A}$. Suppose $0 < \tau \leq \rho \circ \varphi$ for some τ in $\mathfrak{A}^\#$. If $\varphi(A) = 0$, then $\varphi(A^*A) = 0$ and $0 \leq \tau(A^*A) \leq \rho[\varphi(A^*A)] = 0$. From the Schwarz inequality for states, $|\tau(A)|^2 = |\tau(I \cdot A)|^2 \leq \tau(A^*A)\tau(I) = 0$. Thus $\tau(A) = 0$ when $\varphi(A) = 0$ and $\tau = \tau' \circ \varphi$ for some linear functional τ' on $\mathcal{B}$. Since $\tau(I)^{-1}\tau$ $(= (\tau(I)^{-1}\tau') \circ \varphi)$ is a state of $\mathfrak{A}$, $\tau(I)^{-1}\tau'$ is a state of $\mathcal{B}$ (from (ii)); and $0 \leq \tau'$. Now $0 \leq \rho \circ \varphi - \tau = \rho \circ \varphi - \tau' \circ \varphi = (\rho - \tau') \circ \varphi$; and, as before, $0 \leq \rho - \tau'$. Thus $0 \leq \tau' \leq \rho$. By assumption, ρ is pure so that $\tau' = \tau'(I)\rho$; and $\tau = \tau' \circ \varphi = \tau'(I)(\rho \circ \varphi)$. Thus $\rho \circ \varphi$ is a pure state of $\mathfrak{A}$.

(iv)   Suppose $\rho \circ \varphi$ is a pure state of $\mathfrak{A}$. From (ii), ρ is a state of $\mathcal{B}$. If $0 \leq \tau' \leq \rho$ for some τ' in $\mathcal{B}^\#$, then $0 \leq \tau \leq \rho \circ \varphi$, where $\tau = \tau' \circ \varphi$. Since $\rho \circ \varphi$ is pure, $\tau = \tau(I)(\rho \circ \varphi)$. Thus $\tau' = \tau(I)\rho$ (for φ maps $\mathfrak{A}$ *onto* $\mathcal{B}$). Hence ρ is a pure state of $\mathcal{B}$.   ■

4.6.24.   Let $\mathcal{S}$ be the set of states of a C*-algebra $\mathfrak{A}$. With A in $\mathfrak{A}$ let b be $\sup\{|\rho(A)| : \ \rho \in \mathcal{S}\}$.

(i)   Show that there is a pure state ρ of $\mathfrak{A}$ such that $|\rho(A)| = b$.

(ii)   When A is a normal element of $\mathfrak{A}$, $b = \|A\|$, from Theorem 4.3.4. Find an example of a C*-algebra $\mathfrak{A}$ and an element A of $\mathfrak{A}$ for which $b < \|A\|$.

Solution. (i)   From the discusion following Theorem 4.3.2, $\mathcal{S}$ is a convex, weak * compact subset of $\mathfrak{A}^\#$. By definition of the weak * topology, the mapping $\rho \to |\rho(A)| : \ \mathcal{S} \to \mathbb{R}$ is continuous. The set at which this mapping attains its maximum b is a non-null subset of $\mathcal{S}$. Let ρ_0 lie in this set and let F be $\{\rho : \ \rho(A) = \rho_0(A), \rho \in \mathcal{S}\}$. Then F is a weak * closed, convex subset of $\mathcal{S}$. We show that F is a *face* of $\mathcal{S}$, whence F contains a pure state ρ and $|\rho(A)| = |\rho_0(A)| = b$. (See Section 1.4.) Suppose $\rho_1, \rho_2 \in \mathcal{S}$, $0 < a < 1$, and $(a\rho_1 + (1 - a)\rho_2)(A) = \rho_0(A)$. Since $|\rho_1(A)| \leq b$, $|\rho_2(A)| \leq b$, and $|\rho_0(A)| = b$, we have $\rho_1(A) = \rho_2(A) = \rho_0(A)$ (for $\rho_0(A)$ is an extreme point of the closed disk in $\mathbb{C}$ of radius b with 0 as center). Thus $\rho_1, \rho_2 \in F$, and F is a (closed) face of $\mathcal{S}$.

(ii)   Let $\mathfrak{A}$ be $\mathcal{B}(\mathcal{H})$ where $\mathcal{H}$ is a two-dimensional Hilbert space. Let $\{e_1, e_2\}$ be an orthonormal basis for $\mathcal{H}$ and let A be the operator on $\mathcal{H}$ that maps e_1 to e_2 and e_2 to 0. Then A^* maps e_2 to e_1 and e_1 to 0. Thus A^*A is the projection with range $[e_1]$ and $\|A\| = \|A^*A\|^{\frac{1}{2}} = 1$. From (i),

$$\sup\{|\rho(A)| : \rho \in \mathcal{S}\} = \sup\{|\rho(A)| : \rho \in \mathcal{P}\},$$

where $\mathcal{P}$ is the set of pure states of $\mathfrak{A}$. From Exercise 4.6.18(iv), $\rho \in \mathcal{P}$ if and only if $\rho = \omega_x$ for some unit vector x in $\mathcal{H}$. If $x = ae_1 + be_2$, then $|a|^2 + |b|^2 = 1$ and $\omega_x(A) = \langle Ax, x \rangle = a\bar{b}$. Since $2|a||b| \le |a|^2 + |b|^2 = 1$, $|\omega_x(A)| \le \frac{1}{2} < 1 = \|A\|$ for each unit vector x in $\mathcal{H}$, and $b < \|A\|$.   ∎

4.6.25.   Suppose that U is a unitary element of a C*-algebra $\mathfrak{A}$ and ρ is a state of $\mathfrak{A}$. Show that the equation

$$\rho_U(A) = \rho(UAU^*)   (A \in \mathfrak{A})$$

defines a state ρ_U of $\mathfrak{A}$ and that ρ_U is pure if and only if ρ is pure.

Solution.   When $A \in \mathfrak{A}^+$, we have $UAU^* \in \mathfrak{A}^+$ and thus $\rho_U(A) = \rho(UAU^*) \ge 0$. Since, also, $\rho_U(I) = \rho(UIU^*) = \rho(I) = 1$, it follows that the linear functional ρ_U is a state of $\mathfrak{A}$.

With $\mathcal{S}$ the state space of $\mathfrak{A}$, the affine mapping $\rho \to \rho_U$: $\mathcal{S} \to \mathcal{S}$ is bijective (since it has an inverse mapping $\rho \to \rho_{U^*}$), and so maps the set $\mathcal{P}$ of extreme points of $\mathcal{S}$ onto itself. In other words, ρ_U is pure if and only if ρ is pure.   ∎

4.6.26.   Let $\mathfrak{A}$ be a C*-algebra. Show that:

(i)   if $\mathfrak{A}$ is abelian, then $\|\rho_1 - \rho_2\| = 2$ whenever ρ_1 and ρ_2 are distinct pure states of $\mathfrak{A}$;

(ii)   if there is a positive real number δ such that $\|\rho_1 - \rho_2\| \ge \delta$ whenever ρ_1 and ρ_2 are distinct pure states of $\mathfrak{A}$, then $\mathfrak{A}$ is abelian. [*Hint.* Let H be a self-adjoint element of $\mathfrak{A}$, and define $U(t) = \exp itH$ for all real t. With the notation of Exercise 4.6.25, show that $\rho = \rho_{U(t)}$ (and, hence, that $\rho(AU(t)) = \rho(U(t)A)$ for all A in $\mathfrak{A}$) whenever ρ is a pure state of $\mathfrak{A}$ and $|t|$ is sufficiently small. Deduce that $\rho(AH - HA) = 0$.]

Solution. (i) If $\mathfrak{A}$ is abelian, we use the notation and results of Theorem 4.4.3. Given the distinct points ρ_1 and ρ_2 in the compact Hausdorff space $\mathcal{P}(\mathfrak{A})$, there is a continuous mapping f from $\mathcal{P}(\mathfrak{A})$ into the interval $[-1, 1]$ such that $f(\rho_1) = -1$ and $f(\rho_2) = 1$. Moreover, $f = \widehat{A}$ for some self-adjoint A in the unit ball of $\mathfrak{A}$, and

$$\|\rho_1 - \rho_2\| \geq |\rho_1(A) - \rho_2(A)| = |f(\rho_1) - f(\rho_2)| = 2.$$

Thus $\|\rho_1 - \rho_2\| = 2$.

(ii) With the notation introduced in the hint, we have

$$\begin{aligned}
|\rho_{U(t)}(A) - \rho(A)| &= |\rho(U(t)AU(t)^* - A)| \\
&\leq |\rho(U(t)A(U(t)^* - I))| + |\rho((U(t) - I)A| \\
&\leq 2\|A\|\|U(t) - I\|
\end{aligned}$$

for all A in $\mathfrak{A}$ and all pure states ρ of $\mathfrak{A}$. Thus

$$\|\rho_{U(t)} - \rho\| \leq 2\|U(t) - I\|,$$

and the right-hand side is less than δ for all sufficiently small $|t|$. Since $\rho_{U(t)}$ (as well as ρ) is a pure state of $\mathfrak{A}$ (Exercise 4.6.25), it now follows that $\rho_{U(t)} = \rho$ for all pure states ρ and all sufficiently small $|t|$. Accordingly,

$$\begin{aligned}
\rho(AU(t)) &= \rho_{U(t)}(AU(t)) \\
&= \rho(U(t)AU(t)U(t)^*) \\
&= \rho(U(t)A),
\end{aligned}$$

and

$$0 = \rho(AU(t) - U(t)A) = \sum_{n=1}^{\infty} \frac{1}{n!}(it)^n \rho(AH^n - H^n A),$$

for all A in $\mathfrak{A}$ and all sufficiently small $|t|$. By considering the coefficient of t on the right-hand side, we have

$$\rho(AH - HA) = 0.$$

Since this holds for all pure states ρ of $\mathfrak{A}$, $AH = HA$. The last equation has been proved for all A and (self-adjoint) H in $\mathfrak{A}$, so $\mathfrak{A}$ is abelian. ∎

4.6.27.   Suppose that $\mathcal{P}(\mathfrak{A})$ is the set of all pure states of a C*-algebra $\mathfrak{A}$, and

$$\varphi: \quad A \to \widehat{A}: \quad \mathfrak{A} \to C(\mathcal{P}(\mathfrak{A})^-)$$

is the function representation of $\mathfrak{A}$ on its pure state space $\mathcal{P}(\mathfrak{A})^-$ (see the discussion preceding Theorem 4.3.11).

(i)   Show that $\varphi(\mathfrak{A}) = C(\mathcal{P}(\mathfrak{A})^-)$ if and only if $\mathfrak{A}$ is abelian. [*Hint.* Use the result of Exercise 4.6.26(ii).]

(ii)   Show that $\mathfrak{A}$ is abelian if and only if $\varphi(\mathfrak{A})$ is a subalgebra of $C(\mathcal{P}(\mathfrak{A})^-)$.

Solution. (i)   If $\mathfrak{A}$ is abelian, then $\mathcal{P}(\mathfrak{A})^- = \mathcal{P}(\mathfrak{A})$ and $\varphi(\mathfrak{A}) = C(\mathcal{P}(\mathfrak{A}))$, by Theorem 4.4.3.

Conversely, suppose that $\varphi(\mathfrak{A}) = C(\mathcal{P}(\mathfrak{A})^-)$.   Given distinct pure states ρ_1 and ρ_2 of $\mathfrak{A}$, we can find a continuous mapping $f :$ $\mathcal{P}(\mathfrak{A})^- \to [-1,1]$ such that $f(\rho_1) = -1$, $f(\rho_2) = +1$. Then $f = \widehat{A}$ for some self-adjoint A in $\mathfrak{A}$, and $\|A\| = 1$ since φ is isometric on the self-adjoint elements of $\mathfrak{A}$. Thus

$$\|\rho_1 - \rho_2\| \geq |\rho_1(A) - \rho_2(A)| = |f(\rho_1) - f(\rho_2)| = 2.$$

By the result of Exercise 4.6.26(ii), $\mathfrak{A}$ is abelian.

(ii)   Suppose that $\varphi(\mathfrak{A})$ is a subalgebra of $C(\mathcal{P}(\mathfrak{A})^-)$. Since $\varphi(I)$ is the constant function 1, $\varphi(\mathfrak{A})$ contains the constants. Since states of $\mathfrak{A}$ are hermitian, $\varphi(\mathfrak{A})$ is stable under complex conjugation. If ρ_1, $\rho_2 \in \mathcal{P}(\mathfrak{A})^-$ and $\rho_1 \neq \rho_2$, then $\rho_1(A) \neq \rho_2(A)$ for some A in $\mathfrak{A}$. Thus $\varphi(A)(\rho_1) \neq \varphi(A)(\rho_2)$, and $\varphi(\mathfrak{A})$ separates the points of $\mathcal{P}(\mathfrak{A})^-$. Since $\varphi(\mathfrak{A})$ is norm closed in $C(\mathcal{P}(\mathfrak{A})^-)$, $\varphi(\mathfrak{A}) = C(\mathcal{P}(\mathfrak{A})^-)$, from the Stone-Weierstrass theorem. Thus $\mathfrak{A}$ is abelian.

If $\mathfrak{A}$ is abelian, (i) applies and $\varphi(\mathfrak{A}) = C(\mathcal{P}(\mathfrak{A}))$ (and $\mathcal{P}(\mathfrak{A}) = \mathcal{P}(\mathfrak{A})^-$). Thus $\varphi(\mathfrak{A})$ is (a subalgebra of) $C(\mathcal{P}(\mathfrak{A})^-)$.   ■   [54]

4.6.28.   Let $\mathfrak{A}$ be a C*-algebra.

(i)   Show that if $A \in \mathfrak{A}$ and $\lambda \in \mathrm{sp}(A^*A)$, then $A^*A - \lambda I$ has neither a left nor right inverse.

(ii)   Show that the intersection of the maximal left ideals in $\mathfrak{A}$ is $\{0\}$.

Solution. (i)   From Proposition 4.1.1(ii), λ is real; and $A^*A - \lambda I$ is self-adjoint. If $I = (A^*A - \lambda I)B$, then $I = B^*(A^*A - \lambda I)$.

Thus $A^*A - \lambda I$ has a left inverse if it has a right inverse; so that $A^*A - \lambda I$ has a two-sided inverse in this case. But $\lambda \in \mathrm{sp}(A^*A)$ so that $A^*A - \lambda I$ does not have a right inverse. Similarly, $A^*A - \lambda I$ does not have a left inverse.

(ii) Suppose A lies in each maximal left ideal of $\mathfrak{A}$ and $\lambda \in \mathrm{sp}(A^*A)$. From (i), $\mathfrak{A}(A^*A - \lambda I)$ is a proper left ideal in $\mathfrak{A}$. A simple Zorn's lemma argument shows that $\mathfrak{A}(A^*A - \lambda I)$ is contained in a proper maximal left ideal $\mathcal{L}$ of $\mathfrak{A}$. By assumption $A \in \mathcal{L}$. Thus A^*A and, hence, λI are in $\mathcal{L}$. Hence $\lambda = 0$, $r(A^*A) = 0$, $A^*A = 0$, and $A = 0$.   ∎

4.6.29.   Suppose that $\mathfrak{A}$ is a C*-algebra, and each closed left ideal in $\mathfrak{A}$ is a two-sided ideal. Prove that:

(i)   each maximal left ideal in $\mathfrak{A}$ is also a maximal right ideal;

(ii)   if $\mathcal{I}$ is a maximal left ideal in $\mathfrak{A}$ (and, hence, a two-sided ideal), then each non-zero element of the quotient Banach algebra $\mathfrak{A}/\mathcal{I}$ is invertible;

(iii)   each maximal left ideal in $\mathfrak{A}$ is the kernel of a multiplicative linear functional on $\mathfrak{A}$;

(iv)   if $A \in \mathfrak{A}$ and $a \in \mathrm{sp}(A)$, then $a = \rho(A)$ for some multiplicative linear functional ρ on $\mathfrak{A}$.

Deduce that $\mathfrak{A}$ is abelian.

Solution. (i)(ii)   If $\mathcal{K}$ is a closed right ideal in $\mathfrak{A}$, then $\mathcal{K}^*$ is a closed left ideal and is therefore a two-sided ideal; so $\mathcal{K}$ is a two-sided ideal. Hence the three classes of closed ideals in $\mathfrak{A}$ (namely, left, right, two-sided) coincide. Since maximal ideals are closed (Proposition 3.1.8), it now follows that a maximal left ideal $\mathcal{I}$ of $\mathfrak{A}$ is also a maximal right ideal (and a maximal two-sided ideal). Since $\mathfrak{A}$ has no one-sided proper ideals strictly larger than $\mathcal{I}$, the Banach algebra $\mathfrak{A}/\mathcal{I} \, (= \mathcal{B})$ has no non-zero proper one-sided ideals. If $B \in \mathcal{B}$ and $B \neq 0$, the one-sided ideals $B\mathcal{B}$ and $\mathcal{B}B$ both coincide with $\mathcal{B}$ (and so contains I). Thus B is invertible.

(iii)   Given B in $\mathcal{B}$, choose b in $\mathrm{sp}(B)$. Since $B - bI$ is not invertible in $\mathcal{B}$, $B = bI$ by (ii). Thus $\mathcal{B}$ can be identified with the scalar field $\mathbb{C}$, and the quotient mapping from $\mathfrak{A}$ onto $\mathfrak{A}/\mathcal{I} \, (= \mathcal{B})$ is a multiplicative linear functional on $\mathfrak{A}$ with kernel $\mathcal{I}$.

(iv)   If $A \in \mathfrak{A}$ and $a \in \mathrm{sp}(A)$, then $A - aI$ is not invertible in $\mathfrak{A}$. Thus at least one of the one-sided ideals $(A - aI)\mathfrak{A}$, $\mathfrak{A}(A - aI)$ is proper, and has closure a proper (left, or right; in either case, left) ideal. This, in turn, is contained in a maximal left ideal $\mathcal{I}$; and $\mathcal{I}$ is the kernel of a multiplicative linear functional ρ on $\mathfrak{A}$, by (iii). Since $A - aI \in \mathcal{I}$,

$$0 = \rho(A - aI) = \rho(A) - a.$$

Suppose that A and B are self-adjoint elements of $\mathfrak{A}$. Since $\rho(AB - BA) = 0$ for each multiplicative linear functional ρ on $\mathfrak{A}$, it follows from (iv) that the self-adjoint element $i(AB - BA)$ of $\mathfrak{A}$ has spectrum $\{0\}$. Thus $AB = BA$, whenever A and B are self-adjoint in $\mathfrak{A}$; and by linearity, this remains true for all A and B in $\mathfrak{A}$.   ∎

4.6.30.   Suppose that $\mathfrak{A}$ is a C*-algebra with the following property: if $A \in \mathfrak{A}$ and $A^2 = 0$, then $A = 0$.

(i)   For each positive integer n, let $f_n : \mathbb{R} \to [0,1]$ be a continuous function such that $f_n(t) = 0$ when $|t| \leq 1/2n$ and $f_n(t) = 1$ when $|t| \geq 1/n$. Prove that

$$f_n(A^*A) = f_{2n}(A^*A)f_n(A^*A), \quad \|A - Af_n(A^*A)\| \leq n^{-1/2},$$

$$f_n(A^*A)B = f_n(A^*A)Bf_{2n}(A^*A).$$

for all A and B in $\mathfrak{A}$.

(ii)   By using the result of Exercise 4.6.29, show that $\mathfrak{A}$ is abelian.

Solution. (i)   Since $f_n(t) = 0$ when $|t| \leq 1/2n$ and $f_{2n}(t) = 1$ when $|t| \geq 1/2n$, we have $f_{2n}(t)f_n(t) = f_n(t)$ for all real t, and thus $f_{2n}(A^*A)f_n(A^*A) = f_n(A^*A)$. Let

$$C = f_n(A^*A)B(I - f_{2n}(A^*A)).$$

Since $(I - f_{2n}(A^*A))f_n(A^*A) = 0$, we have $C^2 = 0$ and thus $C = 0$; so $f_n(A^*A)B = f_n(A^*A)Bf_{2n}(A^*A)$. Also

$$\begin{aligned}
\|A - Af_n(A^*A)\|^2 &= \|[A - Af_n(A^*A)]^*[A - Af_n(A^*A)]\| \\
&= \|[I - f_n(A^*A)]A^*A[I - f_n(A^*A)]\| \\
&\leq \frac{1}{n},
\end{aligned}$$

since $0 \leq (1 - f_n(t))^2 t \leq 1/n$ for all positive t.

(ii) Suppose that $\mathcal{L}$ is a closed left ideal in $\mathfrak{A}$. Given A in $\mathcal{L}$ and B in $\mathfrak{A}$, we have $A^*A \in \mathcal{L}$, and thus $f_n(A^*A) \in \mathcal{L}$ $(n = 1, 2, \ldots)$, since f_n is continuous and $f_n(0) = 0$. By (i)

$$AB = \lim_{n \to \infty} A f_n(A^*A)B,$$

$$A f_n(A^*A)B = A f_n(A^*A)B f_{2n}(A^*A) \in \mathcal{L};$$

so $AB \in \mathcal{L}$, since $\mathcal{L}$ is closed. Thus $\mathcal{L}$ is a right ideal in $\mathfrak{A}$.

We have now shown that each closed left ideal in $\mathfrak{A}$ is a two-sided ideal. By the result of Exercise 4.6.29, $\mathfrak{A}$ is abelian. ∎

4.6.31.   Suppose that A is a self-adjoint element of a C*-algebra $\mathfrak{A}$ and $\lambda \in \mathrm{sp}(A)$. Show that there is a pure state ρ of $\mathfrak{A}$ such that $\rho(A) = \lambda$ and ρ is definite on A (see Exercise 4.6.16). [*Hint.* Show that $\lambda = \rho_0(A) =$ for some pure state ρ_0 of the C*-subalgebra of $\mathfrak{A}$ generated by A and I, and use Theorem 4.3.13(iv).]

Solution.   Let $\mathfrak{A}(A)$ be the (abelian) C*-subalgebra of $\mathfrak{A}$ generated by A and I. From Theorem 4.1.3, there is a multiplicative linear functional ρ_0 on $\mathfrak{A}(A)$ such that $\rho_0(I) = 1$, $\rho_0(A) = \lambda$. By Proposition 4.4.1 and Theorem 4.3.13(iv), ρ_0 is a pure state of $\mathfrak{A}(A)$ and extends to a pure state of $\mathfrak{A}$. Since

$$\rho(A^2) = \rho_0(A^2) = \rho_0(A)^2 = \rho(A)^2,$$

ρ is definite on A; and $\rho(A) = \rho_0(A) = \lambda$. ∎

4.6.32.   Let A be a self-adjoint element of the C*-algebra $\mathfrak{A}$. Suppose that for each non-zero self-adjoint B in $\mathfrak{A}$ there is a state ρ of $\mathfrak{A}$, definite on A, such that $\rho(B) \neq 0$. Show that A lies in the center of $\mathfrak{A}$.

Solution.   With B a self-adjoint element of $\mathfrak{A}$, if ρ is definite on A, then $\rho[i(AB - BA)] = 0$ from Exercise 4.6.16. As $i(AB - BA)$ is a self-adjoint element of $\mathfrak{A}$, by hypothesis, $i(AB - BA) = 0$. Thus $AB = BA$, and A is in the center of $\mathfrak{A}$. ∎

4.6.33.   Suppose that A is a self-adjoint element of the center of a C*-algebra $\mathfrak{A}$. Show that for each non-zero self-adjoint element B of $\mathfrak{A}$ there is a state ρ of $\mathfrak{A}$, definite on A, such that $\rho(B) \neq 0$.

Solution.   From Proposition 4.3.14, each pure state of $\mathfrak{A}$ is multiplicative on the center. From Theorem 4.3.8(i), there is a pure state ρ of $\mathfrak{A}$ such that $\rho(B) \neq 0$; and ρ is definite on A.   ∎

4.6.34.   Let A, B, and C, be elements of a C*-algebra $\mathfrak{A}$. Suppose that A is self-adjoint, C is in the center of $\mathfrak{A}$, and $AB - BA = C$. Show that $C = 0$.

Solution.   If ρ is a state of $\mathfrak{A}$ that is definite on A, then from Exercise 4.6.16

$$0 = \rho(A)\rho(B) - \rho(B)\rho(A) = \rho(AB - BA) = \rho(C).$$

Let $\mathfrak{A}_0$ be the (abelian) C*-subalgebra of $\mathfrak{A}$ generated by A, C, and I. Each pure state of $\mathfrak{A}_0$ is multiplicative so that each state extension to $\mathfrak{A}$ of a pure state of $\mathfrak{A}_0$ is definite on A. Thus each such extension annihilates C. If C were not zero, there would be a pure state of $\mathfrak{A}_0$ and a state extension of it to $\mathfrak{A}$ that does not annihilate C. Hence $C = 0$.   ∎

4.6.35.   Suppose that $\mathcal{L}_0$ is a (not necessarily closed) left ideal in a C*-algebra $\mathfrak{A}$. Given a finite subset $F = \{A_1, \ldots, A_n\}$ of $\mathcal{L}_0$, define H_F and V_F in $\mathcal{L}_0$ by

$$H_F = A_1^* A_1 + \cdots + A_n^* A_n, \qquad V_F = H_F(H_F + n^{-1}I)^{-1}.$$

Show that, with the family $\mathcal{F}$ of all finite subsets of $\mathcal{L}_0$ directed by the inclusion relation $\supseteq$, the net $(V_F, F \in \mathcal{F}, \supseteq)$ is an increasing right approximate identity for $\mathcal{L}_0$.

Solution.   Since $H_F \geq 0$ and $0 \leq t(t + n^{-1})^{-1} \leq 1$ for all t in $\mathbb{R}^+$, it follows that $0 \leq V_F \leq I$. Suppose that $F, G \in \mathcal{F}$ and $F \supseteq G$. We may suppose that $F = \{A_1, \ldots, A_m\}$ and $G = \{A_1, \ldots, A_n\}$, where $n \leq m$. Then $H_F \geq H_G$, $H_F + n^{-1}I \geq H_G + n^{-1}I$, and therefore (Proposition 4.2.8(iii))

$$(H_F + n^{-1}I)^{-1} \leq (H_G + n^{-1}I)^{-1}$$

From this, and since $m^{-1}(t + m^{-1})^{-1} \leq n^{-1}(t + n^{-1})^{-1}$ for all t in $\mathbb{R}^+$, we now have

$$m^{-1}(H_F + m^{-1}I)^{-1} \leq n^{-1}(H_F + n^{-1}I)^{-1} \leq n^{-1}(H_G + n^{-1}I)^{-1};$$

that is, $I - V_F \leq I - V_G$. Hence $V_F \geq V_G$ when $F \supseteq G$.

Suppose that $A \in \mathcal{L}_0$. Given any positive integer n, choose F_0 in $\mathcal{F}$ such that F_0 has n members including A. If $F \in \mathcal{F}$ and $F \supseteq F_0$, then F has $m\,(\geq n)$ members including A. Thus $A^*A \leq H_F$, and

$$(A - AV_F)^*(A - AV_F) = (I - V_F)A^*A(I - V_F)$$
$$\leq (I - V_F)H_F(I - V_F)$$
$$= m^{-2}(H_F + m^{-1}I)^{-2}H_F.$$

From this, and since $m^{-2}(t + m^{-1})^{-2}t \leq (4m)^{-1}$ for all t in $\mathbb{R}^+$, we now have

$$\|A - AV_F\|^2 = \|(A - AV_F)^*(A - AV_F)\|$$
$$\leq \|m^{-2}(H_F + m^{-1}I)^{-2}H_F\|$$
$$\leq \frac{1}{4m} \leq \frac{1}{4n}.$$

Thus $\|A - AV_F\| \leq \frac{1}{2\sqrt{n}}$ when $F \supseteq F_0$; so $\|A - AV_F\| \to 0$ over the net F. ∎ [1,17,28(Lemme 2, pp.222-223),100]

4.6.36.   With the notation of Exercise 4.6.35, let $\mathcal{L}$ be the closure of $\mathcal{L}_0$, so that $\mathcal{L}$ is a closed left ideal in $\mathfrak{A}$. Show that $\{V_F\}$ is an increasing right approximate identity for $\mathcal{L}$. Prove also that, if $\mathcal{L}$ is a two-sided ideal in $\mathfrak{A}$, then $\{V_F\}$ is a two-sided approximate identity for $\mathcal{L}$.

Solution.   We have already shown that $0 \leq V_G \leq V_F \leq I$ when $F, G \in \mathcal{F}$ and $F \supseteq G$, and that $\|A_0 - A_0V_F\| \to 0$ when $A_0 \in \mathcal{L}_0$.

Given A in $\mathcal{L}$ and $\varepsilon\,(> 0)$, choose A_0 in $\mathcal{L}_0$ such that $\|A - A_0\| < \frac{1}{2}\varepsilon$. There exists F_0 in $\mathcal{F}$ such that $\|A_0 - A_0V_F\| < \frac{1}{2}\varepsilon$ whenever $F \supseteq F_0$. Thus

$$\|A - AV_F\| \leq \|(A - A_0)(I - V_F)\| + \|A_0 - A_0V_F\|$$
$$< \tfrac{1}{2}\varepsilon\|I - V_F\| + \tfrac{1}{2}\varepsilon \leq \varepsilon,$$

whenever $F \supseteq F_0$. This shows that $\{V_F\}$ is an increasing right approximate identity for $\mathcal{L}$.

If $\mathcal{L}$ is a two-sided ideal in $\mathfrak{A}$, it is self-adjoint by Corollary 4.2.10. Thus, given A in $\mathcal{L}$, we have $A^* \in \mathcal{L}$, and it results from the preceding paragraph that

$$\|A - V_FA\| = \|A^* - A^*V_F\| \to 0.$$

Hence $\{V_F\}$ is a two-sided approximate identity for the ideal $\mathcal{L}$, in this case. ∎ [1,17,28,100]

4.6.37.   Suppose that $\mathcal{H}$ is a Hilbert space with dimension at least 2, y is a unit vector in $\mathcal{H}$, and $\mathcal{L}$ is the closed left ideal in the C*-algebra $\mathcal{B}(\mathcal{H})$ defined by

$$\mathcal{L} = \{A \in \mathcal{B}(\mathcal{H}): \quad Ay = 0\}.$$

Show that $\mathcal{L}$ has no left approximate identity.

Solution.   Let z be a unit vector, in $\mathcal{H}$ and orthogonal to y. Define A_0 in $\mathcal{B}(\mathcal{H})$ by

$$A_0 x = \langle x, z \rangle y \qquad (x \in \mathcal{H}).$$

Then $A_0 \in \mathcal{L}$, since $A_0 y = 0$. Note, also, that $A_0 z = y$.

Given any net $\{V_\lambda\}$ of elements of $\mathcal{L}$, $V_\lambda A_0 z = V_\lambda y = 0$ for all λ, and

$$\|V_\lambda A_0 - A_0\| \geq \|V_\lambda A_0 z - A_0 z\| = \|y\| = 1.$$

Thus $\{V_\lambda\}$ is not a left approximate identity for $\mathcal{L}$.   ∎

4.6.38.   Provide examples, as indicated below, to show that the statements in Corollary 4.2.10 concerning *closed* two sided ideals in a C*-algebra $\mathfrak{A}$ are not in general valid for two-sided ideals that are not closed, even when $\mathfrak{A}$ is abelian.

(i)   Find an ideal that is not self-adjoint in the abelian C*-algebra $C(\mathbb{D})$, where $\mathbb{D}$ is the unit disk $\{z \in \mathbb{C}: |z| \leq 1\}$.

(ii)   Suppose that X is the unit interval $[0, 1]$, $\mathfrak{A}$ is the abelian C*-algebra $C(X)$, and $\mathcal{K}$ is the ideal $u\mathfrak{A}$ in $\mathfrak{A}$, where $u(t) = t$ $(0 \leq t \leq 1)$. Find an ideal in $\mathcal{K}$ that is not an ideal in $\mathfrak{A}$.

Solution. (i)   Let $\mathcal{I}$ be the ideal $fC(\mathbb{D})$ in $C(\mathbb{D})$, where f in $C(\mathbb{D})$ is defined by $f(z) = z$. Then $f^*(z) = \bar{z}$, and there is no element g of $C(\mathbb{D})$ such that $fg = f^*$ since this would entail

$$\frac{\bar{z}}{z} = \frac{f^*(z)}{f(z)} = g(z) \qquad (0 < |z| \leq 1),$$

$$\lim_{z \to 0} \frac{\bar{z}}{z} = g(0)$$

(and $\bar{z}/z$ has no limit as $z \to 0$). Hence $f \in \mathcal{I}$, $f^* \notin \mathcal{I}$, and $\mathcal{I}$ is not self-adjoint.

(ii)   Let $\mathcal{I} = \{cu + fu^2 : c \in \mathbb{C}, f \in \mathfrak{A}\}$. Then $\mathcal{I}$ is an ideal in $\mathcal{K} (= u\mathfrak{A})$, and contains u. Note that, if $g \in \mathcal{I}$ and $\lim_{t \to 0} g(t)/t = 0$, then $g \in u^2\mathfrak{A}$. Hence, while $u \in \mathcal{I}$, $u \cdot u^{\frac{1}{2}} \notin \mathcal{I}$, and $\mathcal{I}$ is not an ideal in $\mathfrak{A}$.   ∎

4.6.39.   Let $\mathcal{I}$ be a closed left ideal in a C*-algebra $\mathfrak{A}$. Suppose that $A \in \mathcal{I}$, $B \in \mathcal{I}^+$, $\|B\| \le 1$, and $AA^* \le B^4$. By considering the sequence $\{C_n\}$, where $C_n = (B + n^{-1}I)^{-1}A$, show that $A = BC$ for some C in $\mathcal{I}$ with $\|C\| \le 1$.

Solution.   Let B_n and B_m be $(B + n^{-1}I)^{-1}$ and $(B + m^{-1}I)^{-1}$, respectively. We have

$$(1) \qquad\qquad C_n \in \mathcal{I}, \qquad A = (B + n^{-1}I)C_n.$$

Since

$$0 \le C_n C_n^* = B_n AA^* B_n \le B_n B^4 B_n \le I$$

(the last inequality following from the fact that, for all t in $\mathrm{sp}(B)\,(\subseteq [0,1])$, we have $(t + n^{-1})^{-2}t^4 \le 1$), it follows that

$$(2) \qquad\qquad\qquad\qquad \|C_n\| \le 1.$$

Also, $C_m - C_n = (n^{-1} - m^{-1})B_m B_n A$, and

$$\begin{aligned}
0 &\le (C_m - C_n)(C_m - C_n)^* \\
&= (n^{-1} - m^{-1})^2 B_m B_n AA^* B_n B_m \\
&\le (n^{-1} - m^{-1})^2 B_m B_n B^4 B_n B_m \\
&= (n^{-1} - m^{-1})^2 B_m^2 B_n^2 B^4 \\
&\le (n^{-1} - m^{-1})^2 I.
\end{aligned}$$

(For the last inequality, note that $(t + m^{-1})^{-2}(t + n^{-1})^{-2}t^4 \le 1$ for all t in $\mathrm{sp}(B)$). It follows that $\|C_m - C_n\| \le |n^{-1} - m^{-1}|$, so $\{C_n\}$ is a Cauchy sequence in $\mathfrak{A}$. With C defined as $\lim C_n$, we deduce from (1) and (2) that

$$C \in \mathcal{I}, \qquad A = BC, \qquad \|C\| \le 1. \qquad \blacksquare$$

4.6.40.   Let $\mathcal{I}$ be a closed two-sided ideal in a C*-algebra $\mathfrak{A}$. Suppose that $A_1, A_2, \ldots \in \mathcal{I}$, and $\sum_{n=1}^{\infty} \|A_n\|^2 \le 1$. Show that there exist elements $B, C_1, C_2, \ldots$ of $\mathcal{I}$ such that $B \ge 0$, $\|C_n\| \le 1$, and $A_n = BC_n$. [*Hint.* Use the result of Exercise 4.6.39.]

Solution.   Since $\mathcal{I}$ is a closed ideal, $A_n \in \mathcal{I}$, and $\sum \|A_n A_n^*\| \le 1$, it follows that the series $\sum A_n A_n^*$ is norm convergent to a positive element K of the unit ball of $\mathcal{I}$. If $B = K^{\frac{1}{4}}$, then

$$B \in \mathcal{I}^+, \quad \|B\| \le 1, \quad A_n A_n^* \le B^4 \quad (n = 1, 2, \ldots).$$

It now follows from Exercise 4.6.39 that $A_n = BC_n$ for some C_n in the unit ball of $\mathcal{I}$. $\blacksquare$

4.6.41.   Suppose that $\mathcal{L}$ is a closed left ideal in a C*-algebra $\mathfrak{A}$, and $\mathcal{S} = \mathcal{L}^+$. Prove that

(i) $\mathcal{S}$ is a closed subset of $\mathfrak{A}^+$,

(ii) $A + B \in \mathcal{S}$ whenever $A, B \in \mathcal{S}$,

(iii) $aA \in \mathcal{S}$ whenever $A \in \mathcal{S}$ and $a \geq 0$,

(iv) $A \in \mathcal{S}$ whenever $A \in \mathfrak{A}$ and $0 \leq A \leq B$ for some B in $\mathcal{S}$.

[*Hint.* Use the result of Exercise 4.6.39.]

Show also that the closed left ideal $\mathcal{L}$ is a two-sided ideal if and only if

(v) $UAU^* \in \mathcal{S}$ whenever $A \in \mathcal{S}$ and U is a unitary element of $\mathfrak{A}$.

Solution.   Properties (i)–(iii) of $\mathcal{S}$ are evident.

(iv)   Suppose that $A \in \mathfrak{A}$, $B \in \mathcal{S}$ and $0 \leq A \leq B$. In proving that $A \in \mathcal{S}$, we may assume that $\|B\| \leq 1$; then, $A^{\frac{1}{2}} = B^{\frac{1}{4}}C$ for some C in $\mathfrak{A}$, by Exercise 4.6.39 (with $\mathcal{I} = \mathfrak{A}$). Thus $A^{\frac{1}{2}} = C^*B^{\frac{1}{4}} \in \mathcal{L}$, since $B^{\frac{1}{4}} \in \mathcal{L}$ and $\mathcal{L}$ is a left ideal. Finally, $A = A^{\frac{1}{2}}A^{\frac{1}{2}} \in \mathcal{L} \cap \mathfrak{A}^+ = \mathcal{S}$, and (iv) is proved.

If $\mathcal{L}$ is a two-sided ideal, (v) is evident. Conversely, suppose that (v) is satisfied. Given A in $\mathcal{L}$ and a unitary element U of $\mathfrak{A}$, $A = VK$ for some V in $\mathfrak{A}$ and K in $\mathcal{L}^+ (= \mathcal{S})$ by Proposition 4.2.9. Since $U^*KU \in \mathcal{S}\,(\subseteq \mathcal{L})$, it follows that

$$AU = VKU = VU(U^*KU) \in \mathcal{L}.$$

Since each element B of $\mathfrak{A}$ is a finite linear combination of unitary elements, we now deduce that $AB \in \mathcal{L}$ whenever $A \in \mathcal{L}$ and $B \in \mathfrak{A}$. Thus $\mathcal{L}$ is a right (as well as left) ideal of $\mathfrak{A}$. ■ [31]

4.6.42.   Suppose that a subset $\mathcal{S}$ of a C*-algebra $\mathfrak{A}$ satisfies conditions (i)–(iv) of Exercise 4.6.41. Show that there is a unique closed left ideal $\mathcal{L}$ in $\mathfrak{A}$ such that $\mathcal{L}^+ = \mathcal{S}$, and that $\mathcal{L} = \{A \in \mathfrak{A} : A^*A \in \mathcal{S}\}$.

Solution.   If $\mathcal{L}_0$ is a closed left ideal in $\mathfrak{A}$ such that $\mathcal{L}_0^+ = \mathcal{S}$, then $\mathcal{L}_0 = \{AS : A \in \mathfrak{A}, S \in \mathcal{S}\}$ by Proposition 4.2.9. Hence there is at most one such ideal $\mathcal{L}_0$.

Since $\mathcal{S}$ is closed in $\mathfrak{A}$, the same is true of $\mathcal{L}$ (as defined at the end of the exercise). Suppose that $A \in \mathfrak{A}$, $B, C \in \mathcal{L}$, and $a \in \mathbb{C}$. Since $B^*B, C^*C \in \mathcal{S}$, and

$$(B + C)^*(B + C) \leq (B + C)^*(B + C) + (B - C)^*(B - C)$$
$$= 2B^*B + 2C^*C \in \mathcal{S},$$

$$(AB)^*(AB) = B^*A^*AB \le \|A\|^2 B^*B \in \mathcal{S}.$$

and $(aB)^*(aB) = |a|^2 B^*B \in \mathcal{S}$ (by properties (ii) and (iii) of $\mathcal{S}$), it follows from property (iv) that

$$(B+C)^*(B+C),\ (AB)^*(AB),\ (aB)^*(aB) \in \mathcal{S}.$$

Thus $B+C$, AB, $aB \in \mathcal{L}$, and $\mathcal{L}$ is a left ideal of $\mathfrak{A}$.

If $A \in \mathcal{L}^+$, then $A^2 \in \mathcal{S}$. Since

$$0 \le A(A + n^{-1}I)^{-1}A \le nA^2,$$

it follows that $A(A + n^{-1}I)^{-1}A \in \mathcal{S}$ for each $n = 1, 2, \ldots$. Now $t^2(t + n^{-1})^{-1} \to t$ uniformly for t in $\mathbb{R}^+$, and $\mathrm{sp}(A) \subseteq \mathbb{R}^+$. Thus $A = \lim_{n \to \infty} A(A + n^{-1}I)^{-1}A$, and $A \in \mathcal{S}$ since $\mathcal{S}$ is closed. This shows that $\mathcal{L}^+ \subseteq \mathcal{S}$. Conversely, if $A \in \mathcal{S}$, then $A^2 \in \mathcal{S}$ since $0 \le A^2 \le \|A\|A$, and so $A \in \mathcal{L} \cap \mathfrak{A}^+ = \mathcal{L}^+$. Hence $\mathcal{L}^+ = \mathcal{S}$. ∎ [31]

4.6.43.   Suppose A is an element in a C*-algebra $\mathfrak{A}$, $|A| = (A^*A)^{1/2}$, and $V_n = A(A^*A + n^{-1}I)^{-1/2}$ for each positive integer n.

(i)   Establish the following two inequalities:

$$\|A^*A[I - (A^*A)^{1/2}(A^*A + n^{-1}I)^{-1/2}]^2\| \le n^{-1},$$

$$\| |A|[I - (A^*A)^{1/2}(A^*A + n^{-1}I)^{-1/2}]\| \le n^{-1/2}.$$

Use these inequalities to show that:

(ii) $\|A - V_n|A|\| \to 0$ as $n \to \infty$;

(iii) $\| |A| - V_n^*A\| \to 0$ as $n \to \infty$.

Solution. (i)   Pass to the function representation of the (commutative) C*-subalgebra of $\mathfrak{A}$ generated by A^*A and I. From this representation it suffices to show that for each non-negative, real t,

$$|t[1 - t^{\frac{1}{2}}(t + n^{-1})^{-\frac{1}{2}}]^2| \le n^{-1}.$$

Note for this last

$$
\begin{aligned}
|t[1 - t^{\frac{1}{2}}(t + n^{-1})^{-\frac{1}{2}}]^2| &= \left[t^{\frac{1}{2}} - \frac{t}{(t + n^{-1})^{\frac{1}{2}}} \right]^2 \\
&= \frac{2t^2 + tn^{-1} - 2t(t^2 + tn^{-1})^{\frac{1}{2}}}{t + n^{-1}} \\
&\le \frac{2t^2 + tn^{-1} - 2t^2}{t + n^{-1}} \le n^{-1}.
\end{aligned}
$$

The second inequality follows from the first by taking the square root of both sides of the first inequality.

(ii) Since

$$\|A - V_n|A|\,\|^2 = \|A - A(A^*A)^{\frac{1}{2}}(A^*A + n^{-1}I)^{-\frac{1}{2}}\|^2$$
$$= \|A^*A[I - (A^*A)^{\frac{1}{2}}(A^*A + n^{-1}I)^{-\frac{1}{2}}]^2\|$$
$$\leq n^{-1},$$

from (i), $\|A - V_n|A|\,\| \to 0$ as $n \to \infty$.

(iii) Since

$$\|\,|A| - V_n^*A\| = \|\,|A| - (A^*A + n^{-1}I)^{-\frac{1}{2}}A^*A\|$$
$$= \|\,|A|[I - |A|(A^*A + n^{-1}I)^{-\frac{1}{2}}]\|$$
$$\leq n^{-\frac{1}{2}},$$

from (i), $\|\,|A| - V_n^*A\| \to 0$ as $n \to \infty$. ■

4.6.44. Suppose $\mathcal{L}$ is a closed left ideal in the C*-algebra $\mathfrak{A}$. Show that $A \in \mathcal{L}$ if and only if $(A^*A)^{1/2}\ (= |A|) \in \mathcal{L}$.

Solution. From Exercise 4.6.43, $A = \lim V_n|A|$ and $|A| = \lim V_n^*A$. Since $\mathcal{L}$ is a norm-closed left ideal, $A \in \mathcal{L}$ if and only if $|A| \in \mathcal{L}$. Alternatively, from Exercises 4.6.41 and 4.6.42, $\mathcal{L} = \{A : A \in \mathfrak{A}, A^*A \in \mathcal{L}^+\}$. Since $|A|$ is the norm limit of polynomials in A^*A without constant term, $|A| \in \mathcal{L}^+$ if and only if $A^*A \in \mathcal{L}^+$; and, again, $A \in \mathcal{L}$ if and only if $|A| \in \mathcal{L}$. ■

4.6.45. Suppose A is an invertible element of a C*-algebra $\mathfrak{A}$.

(i) Show that $A = UH$ for some unitary element U in $\mathfrak{A}$ and some positive element H in $\mathfrak{A}$.

(ii) Show that the elements U and H of $\mathfrak{A}$ occurring in the ("polar") decomposition of A described in (i) are unique.

Solution. (i) From Theorem 4.2.6, $A^*A \in \mathfrak{A}^+$ and $A^*A = H^2$ for some H in $\mathfrak{A}^+$. If H were not invertible (in $\mathfrak{A}$), then 0 would lie in $\mathrm{sp}(H^2)\,(= \mathrm{sp}(A^*A))$ but A^*A has $A^{-1}(A^{-1})^*$ as its inverse in $\mathfrak{A}$. Thus H has an inverse in $\mathfrak{A}$. Let U be AH^{-1}. Then $U^*U =$

$H^{-1}A^*AH^{-1} = H^{-1}H^2H^{-1} = I$. In addition, $U\,(= AH^{-1})$ is invertible, so that U is unitary.

(ii)   If $A = VK$ with V a unitary element in $\mathfrak{A}$ and K a positive element in $\mathfrak{A}$, then $A^*A = K^2$. From Theorem 4.2.6, $K = H$. Hence $V = AH^{-1} = U$.   ∎

4.6.46.   For each positive real number a, let $f_a : \mathbb{R}^+ \to \mathbb{R}^+$ be the continuous function defined by $f_a(t) = t^a$. Prove that f_a is operator-monotonic increasing if $0 < a \leq 1$ but not if $a > 1$. [*Hint.* Let X be the set of all positive real numbers for which f_a is operator-monotonic increasing. Then $1 \in X$ and (from the discussion following Proposition 4.2.8), $\frac{1}{2} \in X$ and $2 \notin X$. Show that X is closed in $\mathbb{R}^+ \setminus \{0\}$. Given a, b in X, prove that $ab \in X$ and (by an argument similar to the proof of Proposition 4.2.8) that $\frac{1}{2}(a + b) \in X$.]

Solution.   The function $f_a : \mathbb{R}^+ \to \mathbb{R}^+$ is operator-monotonic increasing if and only if $A^a \leq B^a$ whenever A and B lie in a C*-algebra $\mathfrak{A}$ and $0 \leq A \leq B$. Here, we write A^a for $f_a(A)$; the laws of exponents, $A^{a+b} = A^a A^b$, $(A^a)^b = A^{ab}$ follow from the relations $f_{a+b} = f_{ab}$, $f_b \circ f_a = f_{ab}$. Since $(t + \varepsilon)^a \downarrow t^a$ as $\varepsilon \downarrow 0$, uniformly on each compact subset of $\mathbb{R}^+$ (by Dini's theorem), it follows that $(A+\varepsilon I)^a \to A^a$ and $(B+\varepsilon I)^a \to B^a$ as $\varepsilon \downarrow 0$. If $(A+\varepsilon I)^a \leq (B+\varepsilon I)^a$ for all positive ε, then $A^a \leq B^a$. Note also that $0 \leq A+\varepsilon I \leq B+\varepsilon I$, *and $B + \varepsilon I$ is invertible in $\mathfrak{A}$*

From the preceding paragraph, f_a is operator-monotonic increasing if and only if $A^a \leq B^a$ whenever

(1) A, B lie in a $\mathrm{C}^* -$ algebra $\mathfrak{A}$, $0 \leq A \leq B$, B is invertible.

In this case, we can form negative powers $B^c\,(= f_c(B))$, with $c < 0$.

Suppose that $a = \lim a_n$, where $a_1, a_2, \ldots \in X$ and $a > 0$. If (1) is satisfied, then

(2) $f_{a_n}(A) \leq f_{a_n}(B)$ $(n = 1, 2, \ldots)$.

Since $f_{a_n}(t) \to f_a(t)$ uniformly on the compact subset $\mathrm{sp}(A) \cup \mathrm{sp}(B)$ of $\mathbb{R}^+$, we have $f_a(A) = \lim f_{a_n}(A)$ and $f_a(B) = \lim f_{a_n}(B)$. When $n \to \infty$, (2) gives $f_a(A) \leq f_a(B)$. Thus f_a is operator-monotonic increasing, $a \in X$, and X is closed in $\mathbb{R}^+ \setminus \{0\}$.

If $a, b \in X$ and (1) is satisfied, we have

$$0 \leq f_a(A) \leq f_a(B), \qquad f_b(f_a(A)) \leq f_b(f_a(B));$$

that is (by Theorem 4.4.8, and since $f_b \circ f_a = f_{ab}$) $f_{ab}(A) \leq f_{ab}(B)$. Thus f_{ab} is operator-monotonic increasing, and $ab \in X$.

Suppose again that (1) is satisfied and $a, b \in X$. Then

$$0 \leq A^a \leq B^a, \qquad 0 \leq A^b \leq B^b.$$

From the first set of inequalities, we have that $0 \leq B^{-\frac{a}{2}} A^a B^{-\frac{a}{2}} \leq I$, and $\|A^{\frac{a}{2}} B^{-\frac{a}{2}}\| = \|B^{-\frac{a}{2}} A^a B^{-\frac{a}{2}}\|^{\frac{1}{2}} \leq 1$. Similarly, $\|A^{\frac{b}{2}} B^{-\frac{b}{2}}\| \leq 1$. Since $r(ST) = r(TS)$ for all S, T in $\mathfrak{A}$, we have

$$
\begin{aligned}
r(B^{-\frac{a+b}{4}} A^{\frac{a+b}{2}} B^{-\frac{a+b}{4}}) &= r(B^{-\frac{a-b}{4}} B^{-\frac{b}{2}} A^{\frac{a+b}{2}} B^{-\frac{a+b}{4}}) \\
&= r(B^{-\frac{b}{2}} A^{\frac{a+b}{2}} B^{-\frac{a+b}{4}} B^{-\frac{a-b}{4}}) \\
&= r(B^{-\frac{b}{2}} A^{\frac{a+b}{2}} B^{-\frac{a}{2}}) \\
&\leq \|(A^{\frac{b}{2}} B^{-\frac{b}{2}})^* (A^{\frac{a}{2}} B^{-\frac{a}{2}})\| \leq 1.
\end{aligned}
$$

Thus $0 \leq B^{-\frac{a+b}{4}} A^{\frac{a+b}{2}} B^{-\frac{a+b}{4}} \leq I$, and hence $0 \leq A^{\frac{a+b}{2}} \leq B^{\frac{a+b}{2}}$. It follows that $\frac{1}{2}(a + b) \in X$.

We have now proved all the properties of X set out in the hint. Since X is closed in $\mathbb{R}^+ \setminus \{0\}$ and $\frac{1}{2}(a + b) \in X$ whenever $a, b \in X$, it follows that X is an interval. Since $\frac{1}{2}, 1 \in X$ and $ab \in X$ whenever $a, b \in X$, the interval X contains $1, \frac{1}{2}, (\frac{1}{2})^2, (\frac{1}{2})^3, \ldots$ and so contains $(0, 1]$. If X contains any element c for which $c > 1$, we can choose a positive integer n such that $c^n > 2$; since $1, c^n \in X$ and $1 < 2 < c^n$, it would then follow that $2 \in X$, a contradiction. Hence X contains no such c, and $X = (0, 1]$. ∎   [84,87]

4.6.47. Let $\{u_1, u_2\}$ be an orthonormal basis in a two-dimensional Hilbert space $\mathcal{H}$, and let $\{v_1, v_2\}$ be the orthonormal basis given by

$$v_1 = 2^{-1/2}(u_1 + u_2), \qquad v_2 = 2^{-1/2}(u_1 - u_2).$$

Define A and B in $\mathcal{B}(\mathcal{H})^+$ by

$$Ax = \langle x, u_1 \rangle u_1, \qquad Bx = \lambda \langle x, v_1 \rangle v_1 + \mu \langle x, v_2 \rangle v_2 \qquad (x \in \mathcal{H}),$$

where λ and μ are positive real numbers. Show that, for each continuous function $f : \mathbb{R}^+ \to \mathbb{R}$ with $f(0) = 0$,

$$f(A)x = f(1)\langle x, u_1 \rangle u_1, \qquad f(B)x = f(\lambda)\langle x, v_1 \rangle v_1 + f(\mu)\langle x, v_2 \rangle v_2.$$

By considering the matrix of $f(B) - f(A)$, relative to the basis $\{u_1, u_2\}$, show that $f(A) \leq f(B)$ if and only if

$$f(\lambda) + f(\mu) \geq \max\{2f(1), 0\}, \qquad f(1)[f(\lambda) + f(\mu)] \leq 2f(\lambda)f(\mu).$$

By considering the case in which $\lambda = 1 + \varepsilon$ and $\mu = 1 - \varepsilon + 2\varepsilon^2$ for a sufficiently small positive real number ε, show that the function

$$f_a: \quad t \to t^a: \quad \mathbb{R}^+ \to \mathbb{R}$$

is not operator-monotonic increasing when $a > 1$ (thus re-proving a part of the result of Exercise 4.6.46).

Solution.   For all a_1 and a_2,

$$A(a_1 u_1 + a_2 u_2) = a_1 u_1, \quad B(a_1 v_1 + a_2 v_2) = \lambda a_1 v_1 + \mu a_2 v_2.$$

From this, it is apparent that

$$f(A)(a_1 u_1 + a_2 u_2) = f(1)a_1 u_1,$$
$$f(B)(a_1 v_1 + a_2 v_2) = f(\lambda)a_1 v_1 + f(\mu)a_2 v_2,$$

for every polynomial f with zero constant term. By continuity, this remains true for each continuous $f : \mathbb{R}^+ \to \mathbb{R}$ such that $f(0) = 0$ (since, on any compact subset of $\mathbb{R}^+$, such a function is the uniform limit of polynomials with zero constant term). Thus

$$f(A)x = f(1)\langle x, u_1 \rangle u_1$$
$$f(B)x = f(\lambda)\langle x, v_1 \rangle v_1 + f(\mu)\langle x, v_2 \rangle v_2,$$

for all x in $\mathcal{H}$.

With respect to the basis $\{u_1, u_2\}$, $f(A)$ has matrix

$$\begin{bmatrix} f(1) & 0 \\ 0 & 0 \end{bmatrix}.$$

Calculation shows that

$$\langle f(B)u_1, u_1 \rangle = \tfrac{1}{2}[f(\lambda) + f(\mu)] = \langle f(B)u_2, u_2 \rangle,$$
$$\langle f(B)u_1, u_2 \rangle = \tfrac{1}{2}[f(\lambda) - f(\mu)] = \langle f(B)u_2, u_1 \rangle,$$

so $f(B)$ has matrix (relative to $\{u_1, u_2\}$)

$$\frac{1}{2}\begin{bmatrix} f(\lambda) + f(\mu) & f(\lambda) - f(\mu) \\ f(\lambda) - f(\mu) & f(\lambda) + f(\mu) \end{bmatrix}$$

and $f(B) - f(A)$ has matrix

$$\frac{1}{2}\begin{bmatrix} f(\lambda) + f(\mu) - 2f(1) & f(\lambda) - f(\mu) \\ f(\lambda) - f(\mu) & f(\lambda) + f(\mu) \end{bmatrix}.$$

Hence $f(B) - f(A) \geq 0$ if and only if

$$f(\lambda) + f(\mu) - 2f(1) \geq 0, \qquad f(\lambda) + f(\mu) \geq 0,$$

$$[f(\lambda) + f(\mu) - 2f(1)][f(\lambda) + f(\mu)] \geq [f(\lambda) - f(\mu)]^2.$$

Equivalently, $f(A) \leq f(B)$ if and only if

$$f(\lambda) + f(\mu) \geq \max\{2f(1), 0\},$$
$$f(1)[f(\lambda) + f(\mu)] \leq 2f(\lambda)f(\mu).$$

By taking $f(t) = t$, it follows that $A \leq B$ if and only if

$$\lambda + \mu \geq 2, \qquad \lambda + \mu \leq 2\lambda\mu.$$

If $\lambda = 1 + \varepsilon$ and $\mu = 1 - \varepsilon + 2\varepsilon^2$ (where $\varepsilon > 0$), we have $\lambda + \mu = 2(1 + \varepsilon^2)$ and $2\lambda\mu = 2(1 + \varepsilon^2 + 2\varepsilon^3)$; so $0 \leq A \leq B$ in this case. If $a > 1$, by taking f_a for f, it follows that $A^a \leq B^a$ if and only if

$$(*) \qquad \lambda^a + \mu^a \geq 2, \qquad \lambda^a + \mu^a \leq 2\lambda^a\mu^a.$$

Now

$$\begin{aligned}
\lambda^a + \mu^a &= (1 + \varepsilon)^a + (1 - \varepsilon + 2\varepsilon^2)^a \\
&= 1 + a\varepsilon + \tfrac{1}{2}a(a - 1)\varepsilon^2 + \cdots \\
&\quad + 1 - a\varepsilon + 2a\varepsilon^2 + \tfrac{1}{2}a(a - 1)\varepsilon^2 + \cdots \\
&= 2 + a(a + 1)\varepsilon^2 + O(\varepsilon^3), \\
2\lambda^a\mu^a &= 2(1 + \varepsilon)^a(1 - \varepsilon + 2\varepsilon^2)^a \\
&= 2(1 + \varepsilon^2 + 2\varepsilon^3)^a \\
&= 2 + 2a\varepsilon^2 + O(\varepsilon^3).
\end{aligned}$$

Since $a(a + 1) > 2a$, it follows that $\lambda^a + \mu^a > 2\lambda^a\mu^a$ when ε is small enough; and then, $A^a \not\leq B^a$. Hence f_a is not operator-monotonic increasing.   ■

4.6.48.   Suppose that V is a partially ordered vector space with positive cone V^+ and with an order unit I, and let $\mathcal{S}$ be the set of all states of V. (The relevant definitions are given in the discussion following Remark 3.4.4.) Prove that:

(i)   each element of V can be expressed as the difference of two positive elements;

(ii)   each positive linear functional on V is a non-negative multiple of a state of V;

(iii)   the set of all positive linear functionals on V is a cone in the algebraic dual space V' of V;

(iv)   for each H in V, the subset $\{\rho(H) : \rho \in \mathcal{S}\}$ of $\mathbb{R}$ is bounded;

(v) $\mathcal{S}$ is a convex subset of V' and is compact in the weak topology $\sigma(V', V)$ obtained by considering V as a separating family of linear functionals on V'.

Solution. (i)(iv)   Given H in V, there is a positive scalar a such that $-aI \leq H \leq aI$. Hence H can be expressed as the difference $\frac{1}{2}(aI + H) - \frac{1}{2}(aI - H)$ of two positive elements of V, and

$$-a = \rho(-aI) \leq \rho(H) \leq \rho(aI) = a \qquad (\rho \in \mathcal{S}).$$

This proves (i) and (iv). Note also that, for any positive linear functional ρ (not necessarily a state),

$$-a\rho(I) \leq \rho(H) \leq a\rho(I);$$

so a positive linear functional that vanishes at I is 0.

(ii)   It suffices to consider a non-zero positive linear functional ρ, and $\rho(I) > 0$ from the last sentence. Thus $\rho = \rho(I)\tau$, where $\tau\,(= \rho(I)^{-1}\rho)$ is a state.

(iii)   It is apparent that finite sums and positive scalar multiples of positive linear functionals yield positive linear functionals. If both ρ and $-\rho$ are positive linear functionals on V, then $\rho(K) = 0$ for each positive K in V, and hence $\rho = 0$ by (i).

(v)   For each H in V, there is a compact real interval Q_H that contains $\{\rho(H) : \rho \in \mathcal{S}\}$, by (iv). The product topological space $Q = \prod_{H \in V} Q_H$ is compact, by Tychonoff's theorem. It consists of all real-valued functions $q : V \to \mathbb{R}$ such that $q(H) \in Q_H$ for each H in V, and so contains $\mathcal{S}$. In fact, $\mathcal{S}$ consists of all q in Q such that

$$q(aA + bB) = aq(A) + bq(B), \quad q(H) \geq 0, \quad q(I) = 1$$

whenever $A, B \in \mathcal{V}$, $H \in \mathcal{V}^+$ and $a, b \in \mathbb{R}$. From this, $\mathcal{S}$ is a closed subset of Q (since the coordinate mappings $q \to q(A) : Q \to \mathbb{R}$ are continuous), and is therefore compact in its relative topology as a subset of Q. However, this topology on $\mathcal{S}$ coincides with the relative $\sigma(\mathcal{V}', \mathcal{V})$ topology on $\mathcal{S}$ as a subset of $\mathcal{V}'$.   ∎

4.6.49.   With the notation of Exercise 4.6.48, suppose that $\mathcal{M}$ is a subspace of $\mathcal{V}$ that contains I, so that $\mathcal{M}$ is a partially ordered vector space with positive cone $\mathcal{M} \cap \mathcal{V}^+$ and I is an order unit for $\mathcal{M}$. Let ρ_0 be a positive linear functional on $\mathcal{M}$, and for each H in $\mathcal{V}$ define

$$l_H = \sup\{\rho_0(B): \ B \in \mathcal{M}, B \le H\},$$
$$u_H = \inf\{\rho_0(B): \ B \in \mathcal{M}, B \ge H\}.$$

  (i)   Prove that l_H and u_H are real numbers satisfying $l_H \le u_H$.

 (ii)   Show that, if $H \in \mathcal{V}$, $c \in \mathbb{R}$, and $l_H \le c \le u_H$, the equation

$$\rho_1(aH + B) = ac + \rho_0(B) \qquad (a \in \mathbb{R}, \quad B \in \mathcal{M})$$

defines a positive linear functional ρ_1 on the subspace

$$\mathcal{M}_1 = \{aH + B : \ a \in \mathbb{R}, \ B \in \mathcal{M}\}$$

of $\mathcal{V}$.

 (iii)   Let $\mathcal{E}$ be the set of all pairs $(\mathcal{N}, \tau)$ in which $\mathcal{N}$ is a subspace of $\mathcal{V}$ that contains $\mathcal{M}$ and τ is a positive linear functional on $\mathcal{N}$ that extends ρ_0. By use of Zorn's lemma, applied to $\mathcal{E}$ with the partial ordering in which "$(\mathcal{N}_1, \tau_1) \le (\mathcal{N}_2, \tau_2)$" means "$\mathcal{N}_1 \subseteq \mathcal{N}_2$ and $\tau_1 = \tau_2|\mathcal{N}_1$", prove that ρ_0 extends to a positive linear functional ρ on $\mathcal{V}$. Prove also that, if $H \in \mathcal{V}$ and $c \in \mathbb{R}$, the extension ρ can be chosen so that $\rho(H) = c$ if and only if $l_H \le c \le u_H$.

 (iv)   Show that each state of $\mathcal{M}$ extends to a state of $\mathcal{V}$, and each pure state of $\mathcal{M}$ extends to a pure state of $\mathcal{V}$. Prove also that, if a pure state of $\mathcal{M}$ has only one extension as a pure state of $\mathcal{V}$, then it has only one extension as a state of $\mathcal{V}$. [*Hint*. Note the analogy with Theorem 4.3.13.]

Solution. (i)   There exist elements B_1, B_2 of $\mathcal{M}$ (for example, suitable scalar multiples of I) such that $B_1 \le H \le B_2$. By varying B_1 while keeping B_2 fixed, and noting that $\rho_0(B_1) \le \rho_0(B_2)$, it follows that l_H is finite and $l_H \leq \rho_0(B_2)$. If we now vary B_2, it follows that u_H is finite and $l_H \leq u_H$.

(ii)   If $H \in \mathcal{M}$, then $u_H = l_H = \rho_0(H)$; so $c = \rho_0(H)$, $\mathcal{M}_1 = \mathcal{M}$, and $\rho_1 = \rho_0$, in this case.

If $H \notin \mathcal{M}$, each element of $\mathcal{M}_1$ has a unique expression in the form $aH + B$, with a in $\mathbb{R}$ and B in $\mathcal{M}$, and ρ_1 (as defined) is a linear functional on $\mathcal{M}_1$. Suppose that $aH + B$ is a positive element of $\mathcal{M}_1$; by considering three cases, we show that $\rho_1(aH + B) \geq 0$ (whence, ρ_1 is a positive linear functional on $\mathcal{M}_1$). First, if $a = 0$, then $B \geq 0$ and $\rho_1(aH + B) = \rho_0(B) \geq 0$. Secondly, consider the case in which $a > 0$. Since $aH + B \geq 0$, we have $-a^{-1}B \in \mathcal{M}$, $-a^{-1}B \leq H$, and therefore $\rho_0(-a^{-1}B) \leq l_H \leq c$. Thus

$$\rho(aH + B) = ac + \rho_0(B) \geq 0.$$

Finally, consider the case in which $a < 0$. Since $aH + B \geq 0$, we have $-a^{-1}B \in \mathcal{M}$, $-a^{-1}B \geq H$, and therefore $\rho_0(-a^{-1}B) \geq u_H \geq c$. Thus

$$\rho(aH + B) = ac + \rho_0(B) \geq 0.$$

(iii)   If $\{(\mathcal{N}_\alpha, \tau_\alpha)\}$ is a totally ordered subset of $\mathcal{E}$, then $\cup \mathcal{N}_\alpha$ is a subspace $\mathcal{N}$ of $\mathcal{V}$ that includes each $\mathcal{N}_\alpha$ (and, hence, $\mathcal{M}$). Moreover there is a positive linear functional τ on $\mathcal{N}$ such that $\tau | \mathcal{N}_\alpha = \tau_\alpha$ for each α (and hence $\tau | \mathcal{M} = \rho_0$). Thus $(\mathcal{N}, \tau)$ is an upper bound of $\{(\mathcal{N}_\alpha, \tau_\alpha)\}$ in $\mathcal{E}$. It now follows from Zorn's lemma that $\mathcal{E}$ has a maximal element, $(\mathcal{N}, \tau)$. If $\mathcal{N} \neq \mathcal{V}$, the result of (ii) would permit the construction of a further extension of τ to a positive linear functional τ_1 on a subspace $\mathcal{N}_1$ strictly larger than $\mathcal{N}$, contradicting the maximality of $(\mathcal{N}, \tau)$. Thus $\mathcal{N} = \mathcal{V}$, and τ is a positive linear functional on $\mathcal{V}$ and extends ρ_0.

If $l_H \leq c \leq u_H$, the positive linear functional ρ_1 on $\mathcal{M}_1$ (occurring in (ii)) extends to a positive linear functional ρ on $\mathcal{V}$: and $\rho | \mathcal{M} = \rho_1 | \mathcal{M} = \rho_0$, $\rho(H) = \rho_1(H) = c$.

If ρ is any positive linear functional on $\mathcal{V}$ that extends ρ_0, and $\rho(H) = c$, then

$$\rho(B_1) \leq \rho(H) \leq \rho(B_2)$$

(that is, $\rho_0(B) \leq c \leq \rho_0(B_2)$) whenever $B_1, B_2 \in \mathcal{M}$ and $B_1 \leq H \leq B_2$; so $l_H \leq c \leq u_H$.

(iv)   By imposing the additional condition that $\rho_0(I) = 1$, it results from (iii) that each state ρ_0 of $\mathcal{M}$ extends to a state ρ of $\mathcal{V}$.

Now suppose that ρ_0 is a pure state of $\mathcal{M}$, and let $\mathcal{S}_0$ be the set of all states of $\mathcal{V}$ that extend ρ_0. Then $\mathcal{S}_0$ is a convex $\sigma(\mathcal{V}', \mathcal{V})$-closed (and, therefore, $\sigma(\mathcal{V}', \mathcal{V})$ -compact) subset of the state space $\mathcal{S}$ of $\mathcal{V}$. By the Krein-Milman theorem, $\mathcal{S}_0$ is the closed convex hull of its extreme points (and so reduces to a single element if and only if it has just one extreme point). It now suffices to show that the extreme points in $\mathcal{S}_0$ are precisely the pure states (that is, extreme points of $\mathcal{S}$) that lie in $\mathcal{S}_0$. Clearly if τ is an extreme point of $\mathcal{S}$ and lies in $\mathcal{S}_0$, it is an extreme point of $\mathcal{S}_0$. If an extreme point τ of $\mathcal{S}_0$ can be expressed as $\tau = a\tau_1 + (1-a)\tau_2$, where $0 < a < 1$ and $\tau_1, \tau_2 \in \mathcal{S}$, then the pure state $\rho_0 \, (= \tau|\mathcal{M})$ of $\mathcal{M}$ has the form $a(\tau_1|\mathcal{M}) + (1-a)(\tau_2|\mathcal{M})$. Thus $\tau_1|\mathcal{M} = \tau_2|\mathcal{M} = \rho_0$, and $\tau_1, \tau_2 \in \mathcal{S}_0$. Since τ is extreme in $\mathcal{S}_0$ and $\tau = a\tau_1 + (1-a)\tau_2$, it now follows that $\tau_1 = \tau_2 = \tau$; so τ is extreme in $\mathcal{S}$. ∎   [52]

4.6.50.   With the notation of Exercise 4.6.48, define

$$\|H\|_I = \inf\{a \in \mathbb{R}^+ : \; -aI \leq H \leq aI\}$$

for each H in $\mathcal{V}$.

(i)   Prove that $\|H\|_I = \sup\{|\rho(H)| : \rho \in \mathcal{S}\}$. [*Hint.* Use the result of Exercise 4.6.49(iii), with $\mathcal{M}$ the subspace $\{aI : a \in \mathbb{R}\}$ of $\mathcal{V}$.]

(ii)   Prove that $\| \; \|_I$ is a semi-norm on $\mathcal{V}$.

(iii)   Show that, if ρ is a positive linear functional on $\mathcal{V}$, then

$$|\rho(H)| \leq \rho(I)\|H\|_I \qquad (H \in \mathcal{V}).$$

(iv)   Show that the subset

$$\mathcal{B} = \{a\rho_1 - b\rho_2 : \; \rho_1, \rho_2 \in \mathcal{S}, \, a, b \in \mathbb{R}^+, \, a + b = 1\}$$

of the algebraic dual space $\mathcal{V}'$ is convex, and is compact in the topology $\sigma(\mathcal{V}', \mathcal{V})$. By means of a Hahn-Banach separation theorem, show that

$$\mathcal{B} = \{\tau \in \mathcal{V}' : \; |\tau(H)| \leq \|H\|_I \; (H \in \mathcal{V})\}.$$

(v)   Show that a linear functional τ on $\mathcal{V}$ can be expressed as the difference of two positive linear functionals on $\mathcal{V}$ if and only if there is a real number k such that

$$|\tau(H)| \leq k\|H\|_I \qquad (H \in \mathcal{V}).$$

Solution. (i)   Let $\mathcal{M}$ be the subspace $\{aI : a \in \mathbb{R}\}$ of $\mathcal{V}$, and let ρ_0 be the positive linear functional $aI \to a$ of $\mathcal{M}$ (so that *all* states of $\mathcal{V}$ extend ρ_0). With the notation of Exercise 4.6.49,

$$u_H = \inf\{a \in \mathbb{R} : \ aI \geq H\},$$
$$l_H = \sup\{a \in \mathbb{R} : \ aI \leq H\},$$
$$-l_H = \inf\{a \in \mathbb{R} : \ -aI \leq H\},$$

Thus

$$\max\{u_H, -l_H\} = \inf\{a \in \mathbb{R} : \ -aI \leq H \leq aI\} = \|H\|_I.$$

From the result of Exercise 4.6.49(iii),

$$\{\rho(H) : \ \rho \in \mathcal{S}\} = [l_H, u_H],$$

so

$$\sup\{|\rho(H)| : \ \rho \in \mathcal{S}\} = \sup\{|c| : \ c \in [l_H, u_H]\}$$
$$= \max\{u_H, -l_H\} = \|H\|_I.$$

(ii)   This is an immediate consequence of (i), given that $\| \ \|_I$ takes finite values because I is an order unit.

(iii)   This is apparent from (i), since $\rho = \rho(I)\tau$ for some state τ of $\mathcal{V}$.

(iv) $\mathcal{B}$ is $\sigma(\mathcal{V}', \mathcal{V})$-compact since it is the image in $\mathcal{V}'$ of the compact set $[0,1] \times \mathcal{S} \times \mathcal{S}$ under the continuous mapping $(a, \rho_1, \rho_2) \to a\rho_1 - (1-a)\rho_2$. It is a routine matter to verify that $\mathcal{B}$ is convex. It is apparent from (i) that $|\tau(H)| \leq \|H\|_I$ whenever $\tau \in \mathcal{B}$ and $H \in \mathcal{V}$.

Now suppose that $\tau_0 \in \mathcal{V}'$ and $|\tau_0(H)| \leq \|H\|_I$ for all H in $\mathcal{V}$. We have to show that $\tau_0 \in \mathcal{B}$. Suppose the contrary. Then there is a $\sigma(\mathcal{V}', \mathcal{V})$-continuous linear functional Ω on $\mathcal{V}'$, and a real number c, such that

$$\Omega(\tau_0) > c \geq \Omega(\tau) \qquad (\tau \in \mathcal{B}).$$

In other words, there is an element H_0 of $\mathcal{V}$ such that

$$\tau_0(H_0) > c \geq \tau(H_0) \qquad (\tau \in \mathcal{B}).$$

Since $\mathcal{S} \cup -\mathcal{S} \subseteq \mathcal{B}$, it follows that $c \geq |\rho(H_0)|$ for each ρ in $\mathcal{S}$. From (i), we now have

$$\tau_0(H_0) > c \geq \sup\{|\rho(H_0)| : \ \rho \in \mathcal{S}\} = \|H_0\|_I,$$

contradicting our assumption concerning τ_0. Thus $\tau_0 \in \mathcal{B}$.

(v)   If τ can be expressed as the difference of two positive linear functionals on $\mathcal{V}$, it is apparent from (iii) that there is a real number k with the stated property.

Conversely, suppose that τ is a linear functional on $\mathcal{V}$, $k > 0$, and $|\tau(H)| \leq k\|H\|_I$ for each H in $\mathcal{V}$. From (iv), $k^{-1}\tau \in \mathcal{B}$. Thus τ has the form $ka\rho_1 - kb\rho_2$, where $\rho_1, \rho_2 \in \mathcal{S}$ and a, b are non-negative scalars with sum 1; so τ is the difference of two positive linear functionals on $\mathcal{V}$. ∎ [39,52]

4.6.51.   In the partially ordered vector space $C([0,1]; \mathbb{R})$, let u be the order unit defined by $u(t) = 1 \ (0 \leq t \leq 1)$, and let $\mathcal{M}$ be the subspace (containing u) that consists of all polynomials with real coefficients. When $f \in \mathcal{M}$, let $\tilde{f}$ be the unique extension of f as a real polynomial defined throughout $\mathbb{R}$. Show that the equation

$$\rho(f) = \tilde{f}(2) \qquad (f \in \mathcal{M})$$

defines a linear functional ρ on $\mathcal{M}$ that cannot be expressed as the difference of two positive linear functionals on $\mathcal{M}$.

Solution.   It is apparent that ρ, as defined, is a linear functional on $\mathcal{M}$. With the notation of Exercise 4.6.50,

$$\begin{aligned}
\|f\|_u &= \inf\{a \in \mathbb{R}^+ : -au \leq f \leq au\} \\
&= \inf\{a \in \mathbb{R} : -a \leq f(t) \leq a \ \ (0 \leq t \leq 1)\} \\
&= \sup\{|f(t)| : 0 \leq t \leq 1\} \qquad\qquad (f \in \mathcal{M}).
\end{aligned}$$

Define $x_0, x_1, \ldots$ in $\mathcal{M}$ by $x_n(t) = t^n \ (0 \leq t \leq 1)$. Since

$$\rho(x_n) = 2^n = 2^n\|x_n\|_u \qquad (n = 0, 1, 2, \ldots),$$

it follows from the result of Exercise 4.6.50(v) that ρ cannot be expressed as the difference of two positive linear functionals on $\mathcal{M}$. ∎

4.6.52.   Suppose that $\mathcal{V}$ is a partially ordered vector space with positive cone $\mathcal{V}^+$ and with an order unit I, and let $\| \ \|_I$ be the semi-norm defined in Exercise 4.6.50. We say that $\mathcal{V}$ is *archimedian* if the following condition is satisfied: if $H \in \mathcal{V}$ and $H \leq \varepsilon I$ for every positive real number ε, then $H \leq 0$.

(i)   Prove that, if $\mathcal{V}$ is archimedian, then $\|\ \|_I$ is a norm on $\mathcal{V}$, $\mathcal{V}^+$ is closed in the associated norm topology on $\mathcal{V}$, and

$$\mathcal{V}^+ = \{H \in \mathcal{V} : \quad \rho(H) \geq 0 \text{ for each state } \rho \text{ of } \mathcal{V}\}.$$

(ii)   Show that the real vector space $\mathbb{R}^2$ $(= \mathcal{V})$ becomes a partially ordered vector space, and has an order unit $(1,0)$ $(= I)$, when the positive cone $\mathcal{V}^+$ is defined by

$$\mathcal{V}^+ = \{(x,y) \in \mathbb{R}^2 : \quad x > 0 \text{ or } x = 0 \text{ and } y \geq 0\}.$$

Show also that $\mathcal{V}$ is not archimedian, $\|\ \|_I$ is not a norm on $\mathcal{V}$, and $\mathcal{V} \setminus \mathcal{V}^+$ contains elements H such that $\rho(H) \geq 0$ for each state ρ of $\mathcal{V}$.

Solution. (i)   If $H \in \mathcal{V}$ and $\|H\|_I = 0$, then $-\varepsilon I \leq H \leq \varepsilon I$ (equivalently, $-H \leq \varepsilon I$ and $H \leq \varepsilon I$) for every positive ε. Since $\mathcal{V}$ is archimedian, $-H \leq 0$ and $H \leq 0$, whence $H, -H \in \mathcal{V}^+$ and therefore $H = 0$. Accordingly, the semi-norm $\|\ \|_I$ is a norm.

Suppose that $A_1, A_2, \ldots \in \mathcal{V}^+$, $H \in \mathcal{V}$, and $\|H - A_n\|_I \to 0$ as $n \to \infty$. Given any positive ε, we can choose n so that $\|H - A_n\|_I < \varepsilon$, and then $-\varepsilon I \leq H - A_n \leq \varepsilon I$. Hence $-H \leq \varepsilon I - A_n \leq \varepsilon I$. Since $\mathcal{V}$ is archimedian, $-H \leq 0$, and $H \in \mathcal{V}^+$. This shows that $\mathcal{V}^+$ is closed in the topology associated with $\|\ \|_I$.

If $H_0 \in \mathcal{V} \setminus \mathcal{V}^+$, it follows from the Hahn-Banach theorem that there is a linear functional ρ on $\mathcal{V}$ and a real number c such that

$$\rho(H_0) < c \leq \rho(H) \qquad (H \in \mathcal{V}^+).$$

Accordingly, $c \leq \rho(tH) = t\rho(H)$ whenever $H \in \mathcal{V}^+$ and $t \geq 0$. Thus

$$\rho(H_0) < c \leq 0 \leq \rho(H) \qquad (H \in \mathcal{V}^+).$$

A suitable positive scalar multiple of ρ is a state ρ_0 of $\mathcal{V}$ such that $\rho_0(H_0) < 0$.

From the preceding paragraph, $\mathcal{V}^+$ contains (and so coincides with) the set

$$\{H \in \mathcal{V} : \quad \rho(H) \geq 0 \text{ for each state } \rho \text{ of } \mathcal{V}\}.$$

(ii)   With $\mathcal{V}$ and $\mathcal{V}^+$ defined as in (ii), suppose that $(x,y) \in \mathcal{V}^+$, $a \in \mathbb{R}$, and $a \geq 0$. Then

(1)   *either* $x > 0$ *or* $x = 0 \leq y$;

so *either* $ax > 0$ *or* $ax = 0 \leq ay$, and $a(x,y) \in \mathcal{V}^+$. If (x,y) and $(-x,-y)$ are in $\mathcal{V}^+$, we have (1), and also *either* $-x > 0$ *or* $-x = 0 < -y$; so $(x,y) = (0,0)$. Finally, if (x_1,y_1), $(x_2,y_2) \in \mathcal{V}^+$, then both (x_1,y_1) and (x_2,y_2) satisfy (1); from this, it follows easily that $(x_1 + x_2, y_1 + y_2)$ satisfies (1) and so lies in $\mathcal{V}^+$. Thus $\mathcal{V}^+$ is a cone.

If $(x,y) \in \mathcal{V}$, $k \in \mathbb{R}$ and $k > |x|$, then $k \pm x > 0$; so $(k \pm x, \pm y) \in \mathcal{V}^+$, and

$$(2) \qquad\qquad -k(1,0) \leq (x,y) \leq k(1,0).$$

Hence $(1,0)$ is an order unit I in $\mathcal{V}$. Note also that if, conversely, (2) is satisfied, then $k \pm x \geq 0$, and $k \geq |x|$. It follows that

$$\|(x,y)\|_I = \inf\{k \in \mathbb{R}^+ : -kI \leq (x,y) \leq kI\} = |x|.$$

Thus $\| \ \|_I$ is not a norm on $\mathcal{V}$, and $\mathcal{V}$ is not archimedian, by (i). The last conclusion can be seen directly; for $(0,1) \leq \varepsilon(1,0)$ for every positive ε, but $(0,1) \not\leq (0,0)$.

A state ρ of $\mathcal{V}$ has the form $(x,y) \rightarrow ax + by$, for some a and b in $\mathbb{R}$. Also, $a = \rho((1,0)) = \rho(I) = 1$, and $\varepsilon \pm b = \rho((\varepsilon, \pm 1)) \geq 0$ whenever $\varepsilon > 0$, since $(\varepsilon, \pm 1) \in \mathcal{V}^+$. Thus $b = 0$, and $\rho_0 : (x,y) \rightarrow x$ is the only state of $\mathcal{V}$. If H is the element $(0,-1)$ of $\mathcal{V}$, then $H \in \mathcal{V} \setminus \mathcal{V}^+$ and $\rho(H) \geq 0$ (in fact, we have that $\rho(H) = 0$) when ρ is a (the only) state of $\mathcal{V}$. ∎

4.6.53.　　　By a *Banach lattice*, we mean a partially ordered vector space $\mathcal{V}$ (with positive cone $\mathcal{V}^+$ and an order unit I) that is archimedian, is a lattice with the partial ordering induced by $\mathcal{V}^+$, and is a Banach space with the norm $\| \ \|_I$ defined in Exercise 4.6.50. When X is a compact Hausdorff space, $C(X, \mathbb{R})$ is a Banach lattice; the present exercise shows that every Banach lattice is isomorphic to one of the form $C(X, \mathbb{R})$.

Suppose that $\mathcal{S}$ is the state space of a Banach lattice $\mathcal{V}$ and $\mathcal{P}^-$ is the closure in $\mathcal{S}$ of the set $\mathcal{P}$ of all pure states of $\mathcal{V}$. When $A \in \mathcal{V}$, define a real-valued function $\widehat{A}$ on $\mathcal{P}^-$ by $\widehat{A}(\rho) = \rho(A)$.

(i)　Prove that $\widehat{A} \in C(\mathcal{P}^-, \mathbb{R})$ for each A in $\mathcal{V}$.

(ii)　Show that the mapping $A \rightarrow \widehat{A} : \mathcal{V} \rightarrow C(\mathcal{P}^-, \mathbb{R})$ is a linear isometry, with range a closed subspace $\mathcal{M}$ of $C(\mathcal{P}^-, \mathbb{R})$ that contains the constant functions and separates the points of $\mathcal{P}^-$. Show also that $A \geq 0$ (in $\mathcal{V}$) if and only if $\widehat{A} \geq 0$ (in $C(\mathcal{P}^-, \mathbb{R})$).

(iii) Show that each pure state ρ_0 of $\mathcal{M}$ extends uniquely to a pure state of $C(\mathcal{P}^-,\mathbb{R})$. By using the results of Exercise 4.6.49, deduce that for all f in $C(\mathcal{P}^-,\mathbb{R})$

$$f(\rho) = \inf\{\widehat{A}(\rho) : \ A \in \mathcal{V}, \ \widehat{A} \geq f\}$$
$$= \sup\{\widehat{A}(\rho) : \ A \in \mathcal{V}, \ \widehat{A} \leq f\} \qquad (\rho \in \mathcal{P}).$$

(iv) Let $A, B \in \mathcal{V}$, and define $C = A \vee B$, $D = A \wedge B$ (where $\vee$, $\wedge$ denote the lattice operations in $\mathcal{V}$). Show that

$$\widehat{C}(\rho) = \max\{\widehat{A}(\rho), \widehat{B}(\rho)\}, \qquad \widehat{D}(\rho) = \min\{\widehat{A}(\rho), \widehat{B}(\rho)\}$$

for all ρ in $\mathcal{P}^-$. [*Hint.* Since $C \geq A$ and $C \geq B$, we have $\widehat{C} \geq \widehat{A}$ and $\widehat{C} \geq \widehat{B}$; so $\widehat{C} \geq f$, where f (in $C(\mathcal{P}^-,\mathbb{R})$) is defined by $f(\rho) = \max\{\widehat{A}(\rho), \widehat{B}(\rho)\}$. Use (iii) to show that we cannot have $\widehat{C}(\rho) > f(\rho)$ when $\rho \in \mathcal{P}$.]

(v) Use the result of Exercise 3.5.49 to show that $\mathcal{M} = C(\mathcal{P}^-,\mathbb{R})$ (and deduce that $\mathcal{P} = \mathcal{P}^-$).

Solution. (i) This is immediate from the fact that $\mathcal{P}^-$ $(\subseteq \mathcal{S} \subseteq \mathcal{V}')$ has the topology $\sigma(\mathcal{V}',\mathcal{V})$.

(ii) It is apparent that the mapping $A \to \widehat{A} : \mathcal{V} \to C(\mathcal{P}^-,\mathbb{R})$ is linear, and that $\widehat{A} \geq 0$ when $A \geq 0$. From Exercises 4.6.50(i) and 4.6.52(i), and the fact that $\mathcal{S}$ is the closed convex hull of $\mathcal{P}^-$, it follows that

$$\|A\|_I = \sup\{|\rho(A)| : \ \rho \in \mathcal{S}\}$$
$$= \sup\{|\rho(A)| : \ \rho \in \mathcal{P}^-\}$$
$$= \sup\{|\widehat{A}(\rho)| : \ \rho \in \mathcal{P}^-\}$$
$$= \|\widehat{A}\|,$$

and that $A \geq 0$ if $\widehat{A} \geq 0$. Thus $A \geq 0$ if and only if $\widehat{A} \geq 0$, the mapping $A \to \widehat{A}$ is a linear isometry, and (since $\mathcal{V}$ is complete) the subspace $\mathcal{M} (= \{\widehat{A} : \ A \in \mathcal{V}\})$ of $C(\mathcal{P}^-,\mathbb{R})$ is closed. Since $\widehat{I}(\rho) = \rho(I) = 1$ for each ρ in $\mathcal{P}^-$, $\mathcal{M}$ contains the contant functions. If ρ_1 and ρ_2 are distinct points of $\mathcal{P}^-$, then $\rho_1(A) \neq \rho_2(A)$ (that is, $\widehat{A}(\rho_1) \neq \widehat{A}(\rho_2)$) for some A in $\mathcal{V}$; so $\mathcal{M}$ separates the points of $\mathcal{P}^-$.

(iii) Each pure state of $C(\mathcal{P}^-,\mathbb{R})$ is an evaluation mapping $f \to f(\rho)$, where ρ can be any point of $\mathcal{P}^-$. Since the mapping $A \to \widehat{A}$ is an order-preserving linear isometry from $\mathcal{V}$ onto $\mathcal{M}$, each pure state ρ_0 of $\mathcal{M}$ has the form $\widehat{A} \to \rho(A)$ for some pure state ρ of $\mathcal{V}$; so $\rho \in \mathcal{P}$ and $\rho_0(\widehat{A}) = \widehat{A}(\rho)$. Since $\mathcal{M}$ separates the points of $\mathcal{P}^-$, it now follows that ρ_0 extends to exactly one pure state of $C(\mathcal{P}^-, \mathbb{R})$, the evaluation mapping $f \to f(\rho)$ at the point ρ of $\mathcal{P}$. Once we have proved that $\mathcal{M} = C(\mathcal{P}^-, \mathbb{R})$, the last sentence will imply that every pure state of $C(\mathcal{P}^-, \mathbb{R})$ is evaluation at a point of $\mathcal{P}$, whence $\mathcal{P} = \mathcal{P}^-$.

Suppose $\rho \in \mathcal{P}$. Since the pure state $\widehat{A} \to \widehat{A}(\rho)$ of $\mathcal{M}$ extends to a unique pure state, $f \to f(\rho)$, of $C(\mathcal{P}^-, \mathbb{R})$, this is its only extension to a state of $C(\mathcal{P}^-, \mathbb{R})$. Accordingly, with the notation of Exercise 4.6.49, it follows from part (iii) of that exercise that $l_f = u_f = f(\rho)$ for each f in $C(\mathcal{P}^-, \mathbb{R})$. In other words,

$$f(\rho) = \inf\{\widehat{A}(\rho) : \ A \in \mathcal{V},\ \widehat{A} \geq f\}$$
$$= \sup\{\widehat{A}(\rho) : \ A \in \mathcal{V},\ \widehat{A} \leq f\} \qquad (f \in C(\mathcal{P}^-, \mathbb{R})).$$

(iv)   Following the argument suggested in the hint, we know that $\widehat{C} \geq f$, and it suffices to prove that $\widehat{C}(\rho) = f(\rho)$ when $\rho \in \mathcal{P}$, since $\mathcal{P}$ is dense in $\mathcal{P}^-$.

Suppose that $\rho_0 \in \mathcal{P}$ and $\widehat{C}(\rho_0) > f(\rho_0)$. From (iii), we can choose A_0 in $\mathcal{V}$ such that $\widehat{A}_0(\rho_0) < \widehat{C}(\rho_0)$ and $\widehat{A}_0 \geq f$. Then $\widehat{A}_0(\rho) \geq f(\rho) = \max\{\widehat{A}(\rho), \widehat{B}(\rho)\}$ for all ρ in $\mathcal{P}^-$; so $\widehat{A}_0 \geq \widehat{A}$, $\widehat{A}_0 \geq \widehat{B}$, $A_0 \geq A$, $A_0 \geq B$, and therefore $A_0 \geq A \vee B = C$. But this contradicts the earlier assertion that $\widehat{A}_0(\rho_0) < \widehat{C}(\rho_0)$; so $\widehat{C}(\rho) = f(\rho) = \max\{\widehat{A}(\rho), \widehat{B}(\rho)\}$. A similar argument shows that $\widehat{D}(\rho) = \min\{\widehat{A}(\rho), \widehat{B}(\rho)\}$.

(v)   We now know that $\mathcal{M}$ is a closed subspace and also a sublattice of $C(\mathcal{P}^-, \mathbb{R})$. Since $\mathcal{M}$ separates the points of $\mathcal{P}^-$ and contains the contant functions, it is easily verified that, given distinct points ρ_1, ρ_2 in $\mathcal{P}^-$ and real numbers r, s, there is an element g of $\mathcal{M}$ satisfying $g(\rho_1) = r$, $g(\rho_2) = s$. From the result of Exercise 3.5.49, $\mathcal{M} = C(\mathcal{P}^-, \mathbb{R})$. As already noted, this implies that $\mathcal{P} = \mathcal{P}^-$.   ∎
[52,118]

4.6.54.   Suppose that $\mathfrak{A}$ is a C*-algebra and, with the usual partial ordering, the set $\mathfrak{A}_h$ of all self-adjoint elements of $\mathfrak{A}$ is a lattice. Prove that $\mathfrak{A}$ is abelian. [*Hint.* Use the results of Exercises 4.6.53 and 4.6.27.]

Solution. $\mathfrak{A}_h$ is a partially ordered vector space, with positive cone $\mathfrak{A}^+$ and with an order unit I. By hypothesis, it is a lattice relative to this partial ordering. It is archimedian, and the norm $\| \ \|_I$ coincides with the usual norm relative to which $\mathfrak{A}_h$ is a Banach space. Thus $\mathfrak{A}_h$ is a Banach lattice.

Since each bounded linear functional on $\mathfrak{A}_h$ extends (uniquely, and without change of norm) to a bounded hermitian functional on $\mathfrak{A}$, we can identify the Banach dual space of $\mathfrak{A}_h$ with the set of bounded hermitian functionals on $\mathfrak{A}$. In this way, the state space S of $\mathfrak{A}_h$, the set $\mathcal{P}$ of all pure states of $\mathfrak{A}_h$, and the pure state space $\mathcal{P}^-$ of $\mathfrak{A}_h$, can each be identified with the corresponding mathematical construct for $\mathfrak{A}$.

From Exercise 4.6.53, the function representation $A \to \widehat{A} : \mathfrak{A}_h \to C(\mathcal{P}^-, \mathbb{R})$ has range $C(\mathcal{P}^-, \mathbb{R})$. Hence the function representation $A \to \widehat{A} : \mathfrak{A} \to C(\mathcal{P}^-)$ has range $C(\mathcal{P}^-)$, and $\mathfrak{A}$ is a abelian by the result of Exercise 4.6.27(i). ∎ [103]

4.6.55. Show that, if x and y are unit vectors in a Hilbert space $\mathcal{H}$ and the corresponding vector states of $\mathcal{B}(\mathcal{H})$ satisfy $\|\omega_x - \omega_y\| < \varepsilon$, then $\|x - cy\| < \varepsilon^{1/2}$ for some complex number c for which $|c| = 1$.

Deduce that if a state ω of $\mathcal{B}(\mathcal{H})$ is the norm limit of a sequence of vector states of $\mathcal{B}(\mathcal{H})$, then ω is a vector state. [This is a special case of a result proved in Theorem 7.3.11.]

Solution. If $c \in \mathbb{C}$ and $|c| = 1$, then

$$\|x - cy\|^2 = 2[1 - \mathbf{Re}\, c\langle y, x \rangle].$$

For a suitably chosen c (with $|c| = 1$),

$$(1) \qquad \|x - cy\|^2 = 2[1 - |\langle x, y \rangle|].$$

If P is the projection from $\mathcal{H}$ onto the one-dimensional subspace that contains x, then $2P - I$ is unitary, and

$$\begin{aligned}
\varepsilon > \|\omega_x - \omega_y\| &\geq |\omega_x(2P - I) - \omega_y(2P - I)| \\
&= 2|\omega_x(P) - \omega_y(P)| = 2[1 - |\langle x, y \rangle|^2] \\
&\geq 2[1 - |\langle x, y \rangle|] = \|x - cy\|^2,
\end{aligned}$$

when c is chosen as in (1).

Now suppose that $\{y(1), y(2), \ldots\}$ is a sequence of unit vectors in $\mathcal{H}$ and $\|\omega - \omega_{y(n)}\| \to 0$. Upon replacing $\{y(n)\}$ by a subsequence, we may suppose that

$$\|\omega_{y(n)} - \omega_{y(n+1)}\| < 2^{-2n} \quad (n = 1, 2, \ldots).$$

From the preceding paragraph, we can choose a complex number c_n such that

$$|c_n| = 1, \quad \|y(n) - c_n y(n+1)\| < 2^{-n}.$$

If $\{x(n)\}$ is the sequence of unit vectors defined by

$$x(1) = y(1), \quad x(2) = c_1 y(2), \quad x(3) = c_1 c_2 y(3), \ldots,$$

then $\omega_{x(n)} = \omega_{y(n)}$ and

$$\|x(n) - x(n+1)\| < 2^{-n}.$$

It follows that $\{x(n)\}$ is a Cauchy sequence in $\mathcal{H}$, and so converges to a unit vector x in $\mathcal{H}$.

For each A in $\mathcal{B}(\mathcal{H})$,

$$\begin{aligned}
|\omega_x(A) - \omega_{x(n)}(A)| &= |\langle Ax, x \rangle - \langle Ax(n), x(n) \rangle| \\
&\leq |\langle A(x - x(n)), x \rangle| + |\langle Ax(n), x - x(n) \rangle| \\
&\leq 2\|A\|\|x - x(n)\| \quad (n = 1, 2, \ldots).
\end{aligned}$$

Thus $\|\omega_x - \omega_{x(n)}\| \leq 2\|x - x(n)\| \to 0$, and

$$\omega_x = \lim_{n \to \infty} \omega_{x(n)} = \lim_{n \to \infty} \omega_{y(n)} = \omega. \quad \blacksquare$$

4.6.56.   With the notation of Exercises 1.9.19 and 3.5.4, observe that l_∞ becomes an abelian C*-algebra in which c is a C*-subalgebra and c_0 ($\subseteq c$) is a closed ideal when the involution in l_∞ is pointwise complex conjugation (that is, $\{x_n\}^*$ is $\{\bar{x}_n\}$ for each bounded complex sequence $\{x_n\}$). Note also that the equation

$$\rho_0(\{x_n\}) = \lim_{n \to \infty} x_n \quad (\{x_n\} \in c)$$

defines a pure state ρ_0 of c.

(i)   Let ρ be any pure state of l_∞ that extends ρ_0 (Theorem 4.3.13(iv)). Show that ρ is a multiplicative linear functional on l_∞, and that

$$\liminf x_n \leq \rho(\{x_n\}) \leq \limsup x_n$$

for every bounded real sequence $\{x_n\}$.

(ii)   Show that the multiplicative linear functionals ρ satisfying the conditions set out in (i) are precisely the elements of $\beta(\mathbf{N}) \setminus \mathbf{N}$ (see Exercises 3.5.5 and 3.5.6).

Solution. (i) By Proposition 4.4.1, the multiplicative linear functional ρ_0 on c is a pure state of c, and any extension of ρ_0 to a pure state of l_∞ is a multiplicative linear functional on l_∞. Given a bounded real sequence $\{x_n\}$, let $m = \liminf x_n$ and $M = \limsup x_n$. Let $z_n = x_n - y_n$ where, for $n = 1, 2, \ldots$,

$$y_n = \begin{cases} m & \text{if } x_n < m, \\ x_n & \text{if } m \le x_n \le M, \\ M & \text{if } M < x_n. \end{cases}$$

Then $\{x_n\} = \{y_n\} + \{z_n\}$, $\rho(\{z_n\}) = 0$ since $\lim z_n = 0$ and ρ extends ρ_0, and $m \le \rho(\{y_n\}) \le M$ since $\rho(\{1,1,1,\ldots\}) = 1$ and $m \le y_n \le M$ for all n. Hence $\rho(\{x_n\}) = \rho(\{y_n\})$ and $m \le \rho(\{x_n\}) \le M$.

(ii) If $\rho_1 \in \beta(\mathbf{N}) \setminus \mathbf{N}$, then ρ_1 is a multiplicative linear functional on (that is, a pure state of) l_∞, and vanishes on c_0 (see Exercise 3.5.5, 3.5.6(iv)). Hence ρ_1 coincides with ρ_0 at all points of c_0 and also at $\{1,1,1,\ldots\}$; so ρ_1 coincides with ρ_0 through the linear span c of c_0 and $\{1,1,1,\ldots\}$, and thus ρ_1 extends ρ_0.

Conversely, if ρ satisfies the conditions set out in (i), then ρ is a multiplicative linear functional on l_∞ (that is, $\rho \in \beta(\mathbf{N})$). Since ρ extends ρ_0,

$$0 = \rho(\{0, 0, \ldots, 0, 1, 0, 0, \ldots\})$$
$$\ne n^{\#}(\{0, 0, \ldots, 0, 1, 0, 0, \ldots\}) = 1 \qquad (n = 1, 2, \ldots)$$

(where the 1 appears in the nth term of the sequence). Thus ρ is not of the form $n^{\#}$, and thus $\rho \in \beta(\mathbf{N}) \setminus \mathbf{N}$. ∎

4.6.57. Let $\mathcal{H}$ be a Hilbert space and let $\mathcal{L}_0$ be the set of all sequences $\{u_1, u_2, \ldots\}$ of elements of $\mathcal{H}$ that are weakly convergent to 0. Let ρ be a pure state of the C*-algebra l_∞, with the properties set out in Exercise 4.6.56(i).

(i) Show that $\mathcal{L}_0$ becomes a complex vector space when $a\{u_n\} + b\{v_n\}$ is defined to be $\{au_n + bv_n\}$ for all $\{u_n\}$ and $\{v_n\}$ in $\mathcal{L}_0$.

(ii) Show that, if $\{u_n\}, \{v_n\} \in \mathcal{L}_0$, then the complex sequences $\{\langle u_n, v_n \rangle\}$, $\{\|u_n\|\}$ are in l_∞. Show also that the equation

$$\langle \{u_n\}, \{v_n\} \rangle_0 = \rho(\{\langle u_n, v_n \rangle\})$$

defines an inner product $\langle \ , \ \rangle_0$ on $\mathcal{L}_0$, and the corresponding semi-norm $\| \ \|_0$ on $\mathcal{L}_0$ is given by

$$\|\{u_n\}\|_0 = \rho(\{\|u_n\|\}).$$

(iii) Let $\mathcal{N}_0$ be the subspace $\{\{u_n\} \in \mathcal{L}_0 : \ \|\{u_n\}\|_0 = 0\}$ of $\mathcal{L}_0$. With $\mathcal{L}_1$ the quotient space $\mathcal{L}_0/\mathcal{N}_0$, let $\langle \ , \ \rangle_1$ be the definite inner product on $\mathcal{L}_1$, and $\| \ \|_1$ the corresponding norm, derived (as in Proposition 2.1.1) from $\langle \ , \ \rangle_0$. Let $\mathcal{L}$ be the completion of the pre-Hilbert space $\mathcal{L}_0/\mathcal{N}_0$ so obtained, and use the same symbols, $\langle \ , \ \rangle_1$ and $\| \ \|_1$, for its inner product and norm. Show that, for each T in $\mathcal{B}(\mathcal{H})$, the equation

$$\pi_0(T)\{u_n\} = \{Tu_n\} \qquad (\{u_n\} \in \mathcal{L}_0)$$

defines a linear operator $\pi_0(T)$ acting on $\mathcal{L}_0$, and

$$\|\pi_0(T)\{u_n\}\|_0 \leq \|T\|\|\{u_n\}\|_0.$$

Deduce that the mapping

$$\{u_n\} + \mathcal{N}_0 \to \{Tu_n\} + \mathcal{N}_0 : \quad \mathcal{L}_1 \to \mathcal{L}_1$$

is well defined and extends uniquely to a bounded linear operator $\pi(T)$ acting on $\mathcal{L}$ with $\|\pi(T)\| \leq \|T\|$.

(iv) Show that the mapping

$$\pi : \ T \to \pi(T) : \quad \mathcal{B}(\mathcal{H}) \to \mathcal{B}(\mathcal{L})$$

is a representation of $\mathcal{B}(\mathcal{H})$.

(v) Show that the kernel of π is the ideal $\mathcal{K}$ consisting of all compact linear operators acting on $\mathcal{H}$. [*Hint.* Use condition (ii) in Exercise 2.8.20 as the defining property of a compact linear operator.]

Solution. (i) When $\{u_n\}, \{v_n\} \in \mathcal{L}_0$, these sequences converge weakly to 0, and hence so does $\{au_n + bv_n\}$, for all scalars a, b. Thus $\{au_n + bv_n\} \in \mathcal{L}_0$, and $\mathcal{L}_0$ is a complex vector space.

(ii) The principle of uniform boundedness shows that the sequences $\{\|u_n\|\}, \{\|v_n\|\}$, and (since $|\langle u_n, v_n \rangle| \leq \|u_n\|\|v_n\|$) $\{\langle u_n, v_n \rangle\}$ are bounded, when $\{u_n\}, \{v_n\} \in \mathcal{L}_0$. Since ρ is a positive (hence, hermitian) linear funtional, the complex number $\rho(\{\langle u_n, v_n \rangle\})$ depends linearly on $\{u_n\}$, is replaced by its conjugate complex number if $\{u_n\}$ and $\{v_n\}$ are exchanged, and is non-negative when $\{u_n\} = \{v_n\}$. Hence $\langle \ , \ \rangle_0$ is an inner product on $\mathcal{L}_0$. Since ρ is multiplicative,

$$\begin{aligned}
\|\{u_n\}\|_0^2 &= \langle \{u_n\}, \{u_n\} \rangle_0 \\
&= \rho(\{\langle u_n, u_n \rangle\}) \\
&= \rho(\{\|u_n\|^2\}) \\
&= [\rho(\{\|u_n\|\})]^2,
\end{aligned}$$

so $\|\{u_n\}\|_0 = \rho(\{\|u_n\|\})$.

(iii) Suppose that $T \in \mathcal{B}(\mathcal{H})$. If $\{u_n\} \in \mathcal{L}_0$ then $\{u_n\}$ converges weakly to 0, and hence the same is true of $\{Tu_n\}$; so $\{Tu_n\} \in \mathcal{L}_0$. Moreover $\|Tu_n\| \leq \|T\|\|u_n\|$, so that (in l_∞)$\{\|Tu_n\|\} \leq \|T\|\{\|u_n\|\}$; and by applying the positive linear functional ρ, we obtain

$$\|\{Tu_n\}\|_0 \leq \|T\|\|\{u_n\}\|_0.$$

Hence $\pi_0(T)$, as defined, is a mapping from $\mathcal{L}_0$ into $\mathcal{L}_0$, satisfies

$$\|\pi_0(T)\{u_n\}\|_0 \leq \|T\|\|\{u_n\}\|_0 \qquad (\{u_n\} \in \mathcal{L}_0),$$

and hence maps $\mathcal{N}_0$ into $\mathcal{N}_0$. Clearly $\pi_0(T) : \mathcal{L}_0 \to \mathcal{L}_0$ is linear.

From the preceding paragraph, the null-space of the linear mapping $\{u_n\} \to \pi_0(T)\{u_n\} + \mathcal{N}_0 : \mathcal{L}_0 \to \mathcal{L}_0/\mathcal{N}_0$ contains $\mathcal{N}_0$; so this mapping factors through $\mathcal{L}_0/\mathcal{N}_0$. In other words, the mapping $\{u_n\} + \mathcal{N}_0 \to \{Tu_n\} + \mathcal{N}_0 : \mathcal{L}_1 \to \mathcal{L}_1$ is well-defined. Clearly, it is linear, moreover

$$\begin{aligned}
\|\{Tu_n\} + \mathcal{N}_0\|_1 &= \|\{Tu_n\}\|_0 \\
&\leq \|T\| \; \|\{u_n\}\|_0 \\
&= \|T\| \; \|\{u_n\} + \mathcal{N}_0\|_1.
\end{aligned}$$

Accordingly, there is a bounded linear operator $\pi(T) : \mathcal{L} \to \mathcal{L}$, with $\|\pi(T)\| \leq \|T\|$, that is determined by the condition

$$\pi(T)(\{u_n\} + \mathcal{N}_0) = \{Tu_n\} + \mathcal{N}_0 \qquad (\{u_n\} \in \mathcal{L}_0).$$

(iv) When $S, T \in \mathcal{B}(\mathcal{H})$ and $a, b \in \mathbb{C}$, we have

$$\begin{aligned}
\pi(aS + bT)(\{u_n\} + \mathcal{N}_0) &= \{(aS + bT)u_n\} + \mathcal{N}_0 \\
&= a[\{Su_n\} + \mathcal{N}_0] + b[\{Tu_n\} + \mathcal{N}_0] \\
&= (a\pi(S) + b\pi(T))(\{u_n\} + \mathcal{N}_0), \\
\pi(ST)(\{u_n\} + \mathcal{N}_0) &= \{STu_n\} + \mathcal{N}_0 \\
&= \pi(S)(\{Tu_n\} + \mathcal{N}_0) \\
&= \pi(S)\pi(T)(\{u_n\} + \mathcal{N}_0),
\end{aligned}$$

and

$$\langle \pi(S)(\{u_n\} + \mathcal{N}_0), \{v_n\} + \mathcal{N}_0\rangle_1 = \langle \{Su_n\} + \mathcal{N}_0, \{v_n\} + \mathcal{N}_0\rangle_1$$
$$= \langle \{Su_n\}, \{v_n\}\rangle_0$$
$$= \rho(\{\langle Su_n, v_n\rangle\})$$
$$= \rho(\{\langle u_n, S^*v_n\rangle\})$$
$$= \langle \{u_n\}, \{S^*v_n\}\rangle_0$$
$$= \langle \{u_n\} + \mathcal{N}_0, \{S^*v_n\} + \mathcal{N}_0\rangle_1$$
$$= \langle \{u_n\} + \mathcal{N}_0, \pi(S^*)(\{v_n\} + \mathcal{N}_0)\rangle_1,$$

whenever $\{u_n\}, \{v_n\} \in \mathcal{L}_0$. Since $\mathcal{L}_0/\mathcal{N}_0$ is dense in $\mathcal{L}$, it follows that

$$\pi(aS + bT) = a\pi(S) + b\pi(T), \ \ \pi(ST) = \pi(S)\pi(T), \ \ \pi(S)^* = \pi(S^*).$$

It is apparent that $\pi(I)$ is the identity operator on $\mathcal{L}$, so π is a representation of $\mathcal{B}(\mathcal{H})$ on $\mathcal{L}$.

(v)   If S is a compact linear operator acting on $\mathcal{H}$, then $\{\|Su_n\|\}$ converges to 0 whenever $\{u_n\} \in \mathcal{L}_0$, by condition (ii) in Exercise 2.8.20. Since ρ extends ρ_0 (in the notation of Exercise 4.6.56), we have

$$\|\pi(S)(\{u_n\} + \mathcal{N}_0)\|_1 = \|\{Su_n\} + \mathcal{N}_0\|_1 = \|\{Su_n\}\|_0$$
$$= \rho(\{\|Su_n\|\}) = 0.$$

Then $\pi(S)$ vanishes on the dense subset $\mathcal{L}_0/\mathcal{N}_0$ of $\mathcal{L}$, and hence $\pi(S) = 0$ in this case.

If $S \in \mathcal{B}(\mathcal{H})$ and S is not compact, there is a weakly convergent sequence $\{x_n\}$ in $\mathcal{H}$, with limit x, such that $\{Sx_n\}$ is not norm convergent to Sx. Upon passing to a subsequence, we may assume that (for some positive ε), $\|Sx_n - Sx\| \geq \varepsilon$ for all n. With u_n defined as $x_n - x$, $\{u_n\} \in \mathcal{L}_0$. Since ρ is positive and (in l_∞) $\{\|Su_n\|\} \geq \varepsilon\{1, 1, \ldots\}$,

$$\|\pi(S)(\{u_n\} + \mathcal{N}_0)\|_1 = \|\{Su_n\} + \mathcal{N}_0\|_1 = \|\{Su_n\}\|_0$$
$$= \rho(\{\|Su_n\|\}) \geq \varepsilon > 0;$$

so $\pi(S) \neq 0$ in this case. Thus π has kernel $\mathcal{K}$. ∎   [15]

4.6.58.   Suppose that $\mathcal{H}$ is a separable Hilbert space, $\mathcal{K}$ ($\subseteq$ $\mathcal{B}(\mathcal{H})$) is the ideal consisting of all compact linear operators acting on $\mathcal{H}$, and $\{e_q : q \in \mathbb{Q}\}$ is an orthonormal basis of $\mathcal{H}$ indexed by the (countable) set $\mathbb{Q}$ of all rational numbers. Let φ be a representation of $\mathcal{B}(\mathcal{H})$ that has kernel $\mathcal{K}$ (see Exercise 4.6.57). For each real number t, choose a sequence $\{q(1), q(2), \dots\}$ of rational numbers (with no repetitions) that converges to t, and let E_t be the projection from $\mathcal{H}$ onto the subspace generated by $\{e_{q(1)}, e_{q(2)}, \dots\}$. Show that:

(i) $\{E_t : t \in \mathbb{R}\}$ is a commuting family of projections such that $E_t \notin \mathcal{K}, E_s E_t \in \mathcal{K}$, whenever $s, t \in \mathbb{R}$ and $s \neq t$;

(ii) the Hilbert space on which $\varphi(\mathcal{B}(\mathcal{H}))$ acts is not separable;

(iii) the result of Exercise 4.6.20(ii) cannot be extended to the case in which $\{B_1, B_2, \dots\}$ is replaced by an uncountable family, even when $\mathfrak{A}$ is abelian.

Solution. (i) Assume $s, t \in \mathbb{R}$ and $s \neq t$. Then the sequence $\{e_{q(1)}, e_{q(2)}, \dots\}$ converges weakly to 0, but $\{E_t e_{q(1)}, E_t e_{q(2)}, \dots\}$ ($= \{e_{q(1)}, e_{q(2)}, \dots\}$) does not converge to 0 in norm. Thus E_t does not have property (ii) of Exercise 2.8.20, and E_t is not compact. However, since $s \neq t$, the rational sequences converging to s and t, respectively, can have only a finite number of elements in common. The corresponding finite subset of $\{e_q\}$ generates the range of $E_s E_t$. Accordingly, $E_s E_t$ has finite-dimensional range, and is therefore compact.

(ii) From (i), $\{\varphi(E_t) : t \in \mathbb{R}\}$ is an orthogonal family of non-zero projections in the Hilbert space $\mathcal{H}_\varphi$ on which $\varphi(\mathcal{B}(\mathcal{H}))$ acts. With x_t a unit vector in the range of $\varphi(E_t)$, $\{x_t : t \in \mathbb{R}\}$ is an uncountable orthonormal system in $\mathcal{H}_\varphi$, whence $\mathcal{H}_\varphi$ is not separable.

(iii) Let $\mathcal{A}$ be the abelian C*-subalgebra of $\mathcal{B}(\mathcal{H})$ generated by $\{I\} \cup \{E_t : t \in \mathbb{R}\}$, and let $\psi = \varphi | \mathcal{A}$. Since $\psi : \mathcal{A} \to \mathcal{B}(\mathcal{H}_\varphi)$ is a * homomorphism, $\psi(\mathcal{A})$ is a C*-subalgebra of $\mathcal{B}(\mathcal{H}_\varphi)$ by Theorem 4.1.9. In $\psi(\mathcal{A})$, we have the family $\{\psi(E_t) : t \in \mathbb{R}\}$ of non-zero, pairwise-orthogonal, projections. If the result of Exercise 4.6.20(ii) extends so as to apply to uncountable families, we can find positive operators A_t ($t \in \mathbb{R}$) in $\mathcal{A}$ ($\subseteq \mathcal{B}(\mathcal{H})$) such that $A_s A_t = 0$ when $s \neq t$ and $\psi(A_t) = \psi(E_t) \neq 0$. With y_t a unit vector in the range of A_t, $\{y_t : t \in \mathbb{R}\}$ is an uncountable orthonormal system in the separable Hilbert space $\mathcal{H}$ — a contradiction. Hence the result of Exercise 4.6.20(ii) cannot be so extended, even in the case of abelian C*-algebras. ∎ [15(Lemma 4.2, p.859)]

4.6.59. Suppose that $\{e_0, e_1, e_2, \ldots\}$ is an orthonormal basis in a separable Hilbert space $\mathcal{H}$, and W is the isometric linear operator on $\mathcal{H}$ defined (as in Example 3.2.18) by $We_j = e_{j+1}$ $(j = 0, 1, 2, \ldots)$. Let $\mathcal{K}$ be the ideal in $\mathcal{B}(\mathcal{H})$ that consists of all compact linear operators acting on $\mathcal{H}$, and let π be a representation of $\mathcal{B}(\mathcal{H})$ that has kernel $\mathcal{K}$ (see Exercise 4.6.57).

(i) Show that $W^*W = I$ and $WW^* = I - E_0$, where E_0 is the projection from $\mathcal{H}$ onto the one-dimensional subspace containing e_0.

(ii) Show that $\pi(W)$ is a unitary operator U.

(iii) Show that there is no invertible operator T in $\mathcal{B}(\mathcal{H})$ such that $\pi(T) = U$. [*Hint.* If $K \in \mathcal{K}$ and $T = W + K$, use the relation $T^*W = I + K^*W$ and the result of Exercise 3.5.18(iv) to show that T is not invertible.]

(iv) Show that there is no normal operator N in $\mathcal{B}(\mathcal{H})$ such that $\pi(N) = U$. [*Hint.* Suppose that $K \in \mathcal{K}$ and $W + K$ is a normal operator N. Let $\mathcal{M}$ be the null space of N, so that $\mathcal{M}$ is also the null space of N^* (Proposition 2.4.6(iii)). Use the relation $W^*N = I + W^*K$ and the properties of compact linear operators to show that $\mathcal{M}$ is finite dimensional. Prove also that $N + E\,(= W + K + E)$ is normal and one-to-one, where E is the projection from $\mathcal{H}$ onto $\mathcal{M}$. Hence reduce to the case in which $\mathcal{M} = (0)$. In this case, use the relation $N^*W = I + K^*W$ and the properties of compact linear operators to show that N is invertible in $\mathcal{B}(\mathcal{H})$, contradicting the conclusion of (iii).]

Solution. (i) For all x and y in $\mathcal{H}$, we have

$$x = \sum_{j=0}^{\infty} \langle x, e_j \rangle e_j, \quad y = \sum_{j=0}^{\infty} \langle y, e_j \rangle e_j, \quad Wx = \sum_{j=0}^{\infty} \langle x, e_j \rangle e_{j+1}$$

(the last, because $We_j = e_{j+1}$). Thus

$$\langle x, W^*y \rangle = \langle Wx, y \rangle = \sum_{j=0}^{\infty} \langle x, e_j \rangle \langle e_{j+1}, y \rangle = \langle x, \sum_{j=0}^{\infty} \langle y, e_{j+1} \rangle e_j \rangle.$$

Hence

$$W^*y = \sum_{j=0}^{\infty} \langle y, e_{j+1} \rangle e_j \qquad (y \in \mathcal{H}).$$

In particular, $W^*e_0 = 0$ and $W^*e_{j+1} = e_j$. Accordingly, for $j = 0, 1, 2, \ldots$,

$$W^*We_j = e_j, \quad WW^*e_{j+1} = e_{j+1}, \quad WW^*e_0 = 0.$$

Hence $W^*W = I$ and $WW^* = I - E_0$.

(ii)   Since $E_0 \in \mathcal{K}$, $\pi(E_0) = 0$. Hence, with U defined as $\pi(W)$,

$$U^*U = \pi(W^*W) = \pi(I) = I, \quad UU^* = \pi(WW^*) = \pi(I - E_0) = I.$$

(iii)   If $T \in \mathcal{B}(\mathcal{H})$ and $\pi(T) = U \, (= \pi(W))$, then $\pi(T - W) = 0$ and hence $T - W \in \mathcal{K}$. Thus $T = W + K$, for some K in $\mathcal{K}$, and

$$T^*W = (W^* + K^*)W = I + K^*W.$$

If T is invertible, then so is T^*; hence T^*W is one-to-one (because T^* and W are one-to-one), and -1 is not an eigenvalue of the compact linear operator K^*W. However $I + K^*W \, (= T^*W)$ does not have range $\mathcal{H}$ (because T^* is one-to-one and $W(\mathcal{H}) \neq \mathcal{H}$). This contradicts the conclusion of Exercise 3.5.18(iv); so T is not invertible.

(iv)   *If N is a normal element of $\mathcal{B}(\mathcal{H})$ and $\pi(N) = U$, then* $N = W + K$ for some K in $\mathcal{K}$. Since

$$W^*N = I + W^*K$$

and W^*K is compact, the null space of W^*N is finite-dimensional (Execise 3.5.18(vi)) and contains the null space $\mathcal{M}$ of N. Hence $\mathcal{M}$ is finite-dimensional, and $E \in \mathcal{K}$ (with the notation of the hint).

Since $NE = N^*E = 0 = (NE)^* = (N^*E)^*$, it follows that E commutes with both N and N^*; so $N + E$ is normal. If $x \in \mathcal{H}$ and $(N + E)x = 0$, we have $Ex = E(N + E)x = 0$, whence $0 = (N + E)x = Nx$, $x \in \mathcal{M}$, and $x = Ex = 0$. Thus $N + E$ is one-to-one.

Upon replacing K by $K + E$, we may now assume that $K \in \mathcal{K}$, and $W + K$ is a *one-to-one* normal operator N; and N^* is one-to-one by Proposition 2.4.6(iii). Thus $N^*W \, (= I + K^*W)$ is one-to-one, and so has range $\mathcal{H}$ (Exercise 3.5.18(iv)) since K^*W is compact. Hence the one-to-one operator N^* has range $\mathcal{H}$, and is therefore invertible in $\mathcal{B}(\mathcal{H})$ (and so is N) by the Banach inversion theorem. However, this contradicts the conclusion of (iii); so there is no normal element N of $\mathcal{B}(\mathcal{H})$ for which $\pi(N) = U$.   ∎

4.6.60. Suppose that $\mathcal{I}$ is a proper closed two-sided ideal in a C*-algebra $\mathfrak{A}$, $\{V_\lambda\}$ is an increasing two-sided approximate identity for $\mathcal{I}$, and $\varphi : \mathfrak{A} \to \mathfrak{A}/\mathcal{I}$ is the quotient mapping from $\mathfrak{A}$ onto the Banach algebra $\mathfrak{A}/\mathcal{I}$ (Proposition 3.1.8). Prove that:

 (i) $\mathfrak{A}/\mathcal{I}$ has an involution defined by $\varphi(A)^* = \varphi(A^*)$ $(A \in \mathfrak{A})$;

 (ii) the usual quotient norm on $\mathfrak{A}/\mathcal{I}$ satisfies

$$\|\varphi(A)\| = \lim_\lambda \|A - AV_\lambda\| = \lim_\lambda \|A - V_\lambda A\| \qquad (A \in \mathfrak{A});$$

 (iii) with the quotient norm and the involution defined in (i), $\mathfrak{A}/\mathcal{I}$ is a C*-algebra. [This important result will be proved by another method in Theorem 10.1.7.]

Solution. (i) By Corollary 4.2.10, $\mathcal{I}$ is self-adjoint. If $A_1, A_2 \in \mathfrak{A}$ and $\varphi(A_1) = \varphi(A_2)$, then $A_1 - A_2 \in \mathcal{I}$, hence $A_1^* - A_2^* \in \mathcal{I}$, and thus $\varphi(A_1^*) = \varphi(A_2^*)$. Accordingly, the mapping $\varphi(A) \to \varphi(A^*) : \mathfrak{A}/\mathcal{I} \to \mathfrak{A}/\mathcal{I}$ is well-defined. Routine algebraic manipulation shows that it is an involution.

 (ii) For A in $\mathfrak{A}$, $\|\varphi(A)\| = \inf\{\|A + K\| : K \in \mathcal{I}\}$. Thus

$$(1) \qquad \|\varphi(A)\| \le \inf_\lambda \|A - AV_\lambda\|.$$

Given any positive real number ε, we can choose K in $\mathcal{I}$ so that $\|A + K\| < \|\varphi(A)\| + \frac{1}{2}\varepsilon$; and then, we can choose an index λ_0 so that $\|K - KV_\lambda\| < \frac{1}{2}\varepsilon$ whenever $\lambda \ge \lambda_0$. Since $0 \le V_\lambda \le I$,

$$\begin{aligned}
\|A - AV_\lambda\| &\le \|(A + K)(I - V_\lambda)\| + \|K - KV_\lambda\| \\
&\le \|A + K\| + \|K - KV_\lambda\| \\
&< \|\varphi(A)\| + \frac{1}{2}\varepsilon + \frac{1}{2}\varepsilon \\
&= \|\varphi(A)\| + \varepsilon,
\end{aligned}$$

whenever $\lambda \ge \lambda_0$. This, with (1), shows that

$$\|\varphi(A)\| = \lim_\lambda \|A - AV_\lambda\|;$$

a similar argument gives

$$\|\varphi(A)\| = \lim_\lambda \|A - V_\lambda A\|.$$

(iii)   Since

$$\|A^*A(I - V_\lambda)\| \geq \|(I - V_\lambda)A^*A(I - V_\lambda)\| = \|A(I - V_\lambda)\|^2,$$

by taking limits over $\{\lambda\}$ we obtain

$$\|\varphi(A)^*\varphi(A)\| \geq \|\varphi(A)\|^2 \qquad (A \in \mathfrak{A}).$$

From this, it follows easily that $\|\varphi(A)^*\| = \|\varphi(A)\|$ and also that $\|\varphi(A)^*\varphi(A)\| = \|\varphi(A)\|^2$. ■ [101]

4.6.61.   Show that, if $\mathcal{I}$ is a proper closed two-sided ideal in a C*-algebra $\mathfrak{A}$, then there is a representation of $\mathfrak{A}$ that has kernel $\mathcal{I}$.

Solution.   Let $\psi : \mathfrak{A} \to \mathfrak{A}/\mathcal{I}$ be the quotient mapping, let ψ_1 be a faithful representation of the C^*-algebra $\mathfrak{A}/\mathcal{I}$ (Exercise 4.6.60), and let $\varphi = \psi_1 \circ \psi$.   ■

4.6.62.   Suppose that $\mathfrak{A}$ and $\mathcal{B}$ are C*-algebras, φ is a * homomorphism from $\mathfrak{A}$ into $\mathcal{B}$, and $\mathcal{I}$ is a closed two-sided ideal in $\mathfrak{A}$.
(i)   By adapting the proof of Theorem 4.1.9, show that $\varphi(\mathcal{I})$ is closed in $\mathcal{B}$.
(ii)   Let $\mathcal{C}$ be the C*-subalgebra $\{cI + S : c \in \mathbb{C}, S \in \mathcal{I}\}$ of $\mathfrak{A}$, and let $\mathcal{K}$ be the kernel of the * homomorphism $\varphi|\mathcal{C} : \mathcal{C} \to \mathcal{B}$. By considering the induced * isomorphism ψ from the C*-algebra $\mathcal{C}/\mathcal{K}$ into $\mathcal{B}$, give a second proof that $\varphi(\mathcal{I})$ is closed in $\mathcal{B}$.

Solution.   (i)   Suppose that $B \in \mathcal{B}$ and $B = \lim \varphi(A_n)$ for some sequence $\{A_n\}$ of elements of $\mathcal{I}$; we have to show that $B \in \varphi(\mathcal{I})$. Since $\mathcal{I}$ is self-adjoint, it suffices to consider the case in which $B, A_1, A_2, \ldots$ are all self-adjoint. The argument used to prove Theorem 4.1.9 now applies as it stands, subject to two further observations. First, since $A_{n+1} - A_n \in \mathcal{I}$ and the continuous function $f_n : \mathbb{R} \to [-2^{-n}, 2^{-n}]$ vanishes at 0, we have that the element $f_n(A_{n+1} - A_n)$ lies in $\mathcal{I}$; and therefore, second, the sum of the convergent series $A_1 + \sum f_n(A_{n+1} - A_n)$ lies in $\mathcal{I}$.
(ii)   The * isomorphism

$$\psi : C + \mathcal{K} \to \varphi(C) : \mathcal{C}/\mathcal{K} \to \mathcal{B}$$

is isometric by Theorem 4.1.8(iii). If $\mathcal{K} \subseteq \mathcal{I}$, the ideal $\mathcal{I}/\mathcal{K}$ in $\mathcal{C}/\mathcal{K}$ is complete relative to the quotient norm (Theorem 1.5.3), and

$\psi(\mathcal{I}/\mathcal{K}) = \varphi(\mathcal{I})$. If $\mathcal{K} \nsubseteq \mathcal{I}$, then $I - S_0 \in \mathcal{K}$ for some S_0 in $\mathcal{I}$; in this case,

$$\varphi(cI + S) = \varphi(cS_0 + S) \in \varphi(\mathcal{I}) \quad (c \in \mathbb{C}, \ S \in \mathcal{I}),$$

and $\varphi(\mathcal{I}) = \varphi(\mathcal{C}) = \psi(\mathcal{C}/\mathcal{K})$. In both cases, $\varphi(\mathcal{I})$ is the isometric image (under ψ) of a complete space, and is therefore closed in the C*-algebra $\mathcal{B}$.   ∎

4.6.63.   Suppose that $\mathcal{I}$ and $\mathcal{J}$ are closed two-sided ideals in a C*-algebra $\mathfrak{A}$. By using the results of Exercises 4.6.60(iii) and 4.6.62, show that the ideal $\mathcal{I} + \mathcal{J}$ is closed in $\mathfrak{A}$.

Solution.   The quotient mapping $\varphi : \mathfrak{A} \to \mathfrak{A}/\mathcal{J}$ is a * homomorphism between C*-algebras, so $\varphi(\mathcal{I})$ is closed in $\mathfrak{A}/\mathcal{J}$. Since φ is continuous, $\mathcal{I} + \mathcal{J} \ (= \varphi^{-1}(\varphi(\mathcal{I})))$ is closed in $\mathfrak{A}$.   ∎

4.6.64.   Suppose that $\mathcal{I}$ and $\mathcal{J}$ are closed two-sided ideals in a C*-algebra $\mathfrak{A}$, and $A = B + C \in \mathfrak{A}^+$, where $B \in \mathcal{I}$ and $C \in \mathcal{J}$.

(i)   Show that $A = S + T$ for suitably chosen self-adjoint elements S of $\mathcal{I}$ and T of $\mathcal{J}$.

(ii)   Suppose that $\varepsilon > 0$, and define

$$H = |S| + |T| + \varepsilon I, \qquad D = A^{1/2}H^{-1/2},$$
$$S_1 = D|S|D^*, \qquad\qquad T_1 = D|T|D^*,$$

where $|S|$ and $|T|$ denote the positive square roots of S^2 and T^2, respectively. Prove that $D^*D \leq I$, and deduce that

$$A - \varepsilon I \leq S_1 + T_1 \leq A.$$

Prove also that

$$S_1 \in \mathcal{I}^+, \qquad T_1 \in \mathcal{J}^+, \qquad \|S_1\| \leq \|A\|, \qquad \|T_1\| \leq \|A\|,$$

and $0 \leq A_1 \leq \varepsilon I$, where

$$A_1 = A - S_1 - T_1 = (S - S_1) + (T - T_1) \ \in (\mathcal{I} + \mathcal{J})^+.$$

(iii)   By repeated application of the result of (ii), show that A can be expressed in the form $X + Y$, with X in $\mathcal{I}^+$ and Y in $\mathcal{J}^+$. [This exercise shows that $(\mathcal{I} + \mathcal{J})^+ = \mathcal{I}^+ + \mathcal{J}^+$ when $\mathcal{I}$ and $\mathcal{J}$ are closed two-sided ideals in a C*-algebra.]

Solution. (i) Since $\mathcal{I}$ and $\mathcal{J}$ are self-adjoint and $A = A^*$, it suffices to take $\frac{1}{2}(B + B^*)$ for S and $\frac{1}{2}(C + C^*)$ for T.

(ii) Since $t \le |t| \,(= (t^2)^{1/2})$ for all real t, we have $S \le |S|$, $T \le |T|$, and

$$A = S + T < |S| + |T| + \varepsilon I = H.$$

Thus

$$D^*D = H^{-\frac{1}{2}}AH^{-\frac{1}{2}} \le H^{-\frac{1}{2}}HH^{-\frac{1}{2}} = I,$$

and $0 \le DD^* \le I$ since $\|DD^*\| = \|D\|^2 = \|D^*D\| \le 1$.

Since $A^{1/2} = DH^{1/2}$ and $0 \le DD^* \le I$, we have

$$A = DHD^* = D[|S| + |T| + \varepsilon I]D^*$$
$$= S_1 + T_1 + \varepsilon DD^*,$$

and

$$A - \varepsilon I \le S_1 + T_1 \le A.$$

Thus $0 \le A - S_1 - T_1 \le \varepsilon I$; that is, $0 \le A_1 \le \varepsilon I$.

Since $S \in \mathcal{I}$, we have $S^2 \in \mathcal{I}$ and thus $|S| = (S^2)^{1/2} \in \mathcal{I}$; so $|S| \in \mathcal{I}^+$, and $S_1 = D|S|D^* \in \mathcal{I}^+$. Also,

$$0 \le S_1 \le S_1 + T_1 \le A,$$

and thus $\|S_1\| \le \|A\|$. A similar argument shows that $T_1 \in \mathcal{J}^+$ and $\|T_1\| \le \|A\|$.

(iii) The result of (ii) can be expressed as follows: if $A \in (\mathcal{I}+\mathcal{J})^+$ and $\varepsilon > 0$, there exist S_1 in $\mathcal{I}^+$ and T_1 in $\mathcal{J}^+$ such that $\|S_1\| \le \|A\|$, $\|T_1\| \le \|A\|$ and

$$A_1 \in (\mathcal{I} + \mathcal{J})^+, \quad \|A_1\| \le \varepsilon,$$

where $A_1 = A - S_1 - T_1$.

Given A in $(\mathcal{I}+\mathcal{J})^+$, repeated use of this result yields sequences $\{A_0, A_1, A_2, \ldots\}$ in $(\mathcal{I} + \mathcal{J})^+$, $\{S_1, S_2, \ldots\}$ in $\mathcal{I}^+$, and $\{T_1, T_2, \ldots\}$ in $\mathcal{J}^+$, such that $A_0 = A$ and

$$A_n = A_{n-1} - S_n - T_n, \quad \|A_n\| \le 2^{-n},$$
$$\|S_n\| \le \|A_{n-1}\|, \qquad\qquad \|T_n\| \le \|A_{n-1}\|,$$

for $n = 1, 2, \ldots$. Since $\|S_n\| \le 2^{1-n}$ and $\|T_n\| \le 2^{1-n}$ when $n > 1$, we can define X in $\mathcal{I}^+$ and Y in $\mathcal{J}^+$ by

$$X = \sum_{n=1}^{\infty} S_n, \qquad Y = \sum_{n=1}^{\infty} T_n.$$

Then,

$$
\begin{aligned}
X + Y &= \lim_{n \to \infty} \left[(S_1 + T_1) + (S_2 + T_2) + \cdots + (S_n + T_n) \right] \\
&= \lim_{n \to \infty} \left[(A_0 - A_1) + (A_1 - A_2) + \cdots + (A_{n-1} - A_n) \right] \\
&= \lim_{n \to \infty} (A_0 - A_n) = A_0 = A. \quad \blacksquare \quad [14, 86, 107]
\end{aligned}
$$

4.6.65.   Suppose that $\mathfrak{A}$ is a C*-algebra and $\delta : \mathfrak{A} \to \mathfrak{A}$ is a linear mapping such that

$$\delta(AB) = A\delta(B) + \delta(A)B \qquad (A, B \in \mathfrak{A}).$$

(Such a mapping δ is called a *derivation* of $\mathfrak{A}$.) Let $\mathcal{I}$ be the set of all elements A in $\mathfrak{A}$ for which the linear mapping

$$T \to \delta(AT): \quad \mathfrak{A} \to \mathfrak{A}$$

is continuous.

    (i)   Show that if $A \in \mathfrak{A}$, then $A \in \mathcal{I}$ if and only if the linear mapping

$$T \to A\delta(T): \quad \mathfrak{A} \to \mathfrak{A}$$

is continuous.

    (ii)   Show that $\mathcal{I}$ is a closed two-sided ideal in $\mathfrak{A}$.

    (iii)   Show that the restriction $\delta|\mathcal{I}$ is continuous. [*Hint.* If $\delta|\mathcal{I}$ is discontinuous, there is a sequence $\{A_1, A_2, \ldots\}$ in $\mathcal{I}$ such that $\sum \|A_j\|^2 < 1$ and $\|\delta(A_j)\| \to \infty$. Use the result of Exercise 4.6.40 to obtain a contradiction.]

    (iv)   Show that the quotient C*-algebra $\mathfrak{A}/\mathcal{I}$ is finite dimensional. [*Hint.* Suppose the contrary, and deduce from Exercises 4.6.13 and 4.6.20(ii) that the unit ball of $\mathfrak{A}$ contains a sequence $\{S_1, S_2, \ldots\}$ of positive elements not in $\mathcal{I}$ such that $S_j S_k = 0$ when $j \ne k$. Prove that there is a sequence $\{T_1, T_2, \ldots\}$ in $\mathfrak{A}$ such that

$$\|T_j\| \le 2^{-j}, \qquad \|\delta(S_j^2 T_j)\| \ge j + \|\delta(S_j)\| \qquad (j = 1, 2, \ldots).$$

Obtain a contradiction by considering $S_j \delta(C)$, where $C = \sum S_j T_j$.]

    (v)   Deduce that $\delta : \mathfrak{A} \to \mathfrak{A}$ is continuous.

    Solution.   Compare the solution to 4.6.66. $\blacksquare$ [73, 88, 94]

4.6.66. Suppose that $\mathfrak{A}$ is a C*-algebra and $\mathfrak{X}$ is a Banach space. We describe $\mathfrak{X}$ as a *Banach $\mathfrak{A}$-module* if there are bounded bilinear mappings

$$(A, x) \to Ax, \qquad (A, x) \to xA : \quad \mathfrak{A} \times \mathfrak{X} \to \mathfrak{X}$$

such that $Ix = xI = x$ for each x in $\mathfrak{X}$, and the associative law holds for each type of triple product $A_1 A_2 x$, $A_1 x A_2$, $x A_1 A_2$. By a *derivation* from $\mathfrak{A}$ into a Banach $\mathfrak{A}$-module $\mathfrak{X}$, we mean a linear mapping $\delta : \mathfrak{A} \to \mathfrak{X}$ such that

$$\delta(AB) = A\delta(B) + \delta(A)B \qquad (A, B \in \mathfrak{A}).$$

Adapt the program set out in Exercise 4.6.65 to prove that every derivation from a C*-algebra $\mathfrak{A}$ into a Banach $\mathfrak{A}$-module is continuous.

Solution. Let $\mathcal{I}$ be the set of all elements A in $\mathfrak{A}$ for which the linear mapping $T \to \delta(AT) : \mathfrak{A} \to \mathfrak{X}$ is continuous. Since

$$\delta(AT) = A\delta(T) + \delta(A)T,$$

and (from boundedness of the bilinear mapping $(T, x) \to xT$) the linear mapping $T \to \delta(A)T : \mathfrak{A} \to \mathfrak{X}$ is continuous, it follows that $A \in \mathcal{I}$ if and only if the linear mapping $T \to A\delta(T) : \mathfrak{A} \to \mathfrak{X}$ is continuous.

It is clear that $\mathcal{I}$ is a linear subspace of $\mathfrak{A}$. Suppose that $A \in \mathcal{I}$ and $B \in \mathfrak{A}$. Then the four linear mappings

$$T \to BT : \mathfrak{A} \to \mathfrak{A}, \qquad T \to \delta(AT) : \mathfrak{A} \to \mathfrak{X},$$
$$T \to A\delta(T) : \mathfrak{A} \to \mathfrak{X}, \qquad x \to Bx : \mathfrak{X} \to \mathfrak{X},$$

are all continuous (the fourth, from boundedness of the bilinear mapping $(T, x) \to Tx$). By composing the first two, and then the last two, of these four mappings, we deduce that the linear mappings

$$T \to \delta(ABT), \qquad T \to BA\delta(T) : \mathfrak{A} \to \mathfrak{X}$$

are both continuous. Thus $AB, BA \in \mathcal{I}$, and $\mathcal{I}$ is a two-sided ideal in $\mathfrak{A}$.

Suppose that $A_0 \in \mathfrak{A}$ and $A_0 = \lim A_n$, where $A_1, A_2, \ldots \in \mathcal{I}$. Define linear mappings $L_0, L_1, L_2, \ldots$ from $\mathfrak{A}$ into $\mathfrak{X}$ by $L_n(T) = A_n \delta(T)$. Then $L_1, L_2, \ldots$ are continuous and (from continuity of the mapping $S \to S\delta(T) : \mathfrak{A} \to \mathfrak{X}$)

$$L_0(T) = A_0 \delta(T) = \lim A_n \delta(T) = \lim L_n(T)$$

for each T in $\mathfrak{A}$. It now follows from the principle of uniform boundedness (see Exercise 1.9.39) that L_0 is bounded. Thus $A_0 \in \mathcal{I}$, and $\mathcal{I}$ is closed in $\mathfrak{A}$.

If the restriction $\delta | \mathcal{I}$ is discontinuous, we can choose $A_1, A_2, \ldots$ in $\mathcal{I}$ so that $\|\delta(A_n)\| > 2n^2 \|A_n\|$. Upon replacing A_n by the scalar multiple $(2n)^{-1} \|A_n\|^{-1} A_n$, we may assume that

$$\|A_n\| = \frac{1}{2n}, \qquad \|\delta(A_n)\| > n.$$

Then, $\sum \|A_n\|^2 < 1$, and it follows from the result of Exercise 4.6.40 that there exist elements $B, C_1, C_2, \ldots$ of $\mathcal{I}$ such that $A_n = BC_n$ and $\|C_n\| \leq 1$, for $n = 1, 2, \ldots$. Since $\|C_n\| \leq 1$ and $\|\delta(BC_n)\| > n$, the mapping $T \to \delta(BT) : \mathfrak{A} \to \mathfrak{X}$ is not continuous, a contradiction (since $B \in \mathcal{I}$). Hence $\delta | \mathcal{I}$ is continuous.

If the C^*-algebra $\mathfrak{A}/\mathcal{I}$ is infinite dimensional, it contains a sequence $\{B_1, B_2, \ldots\}$ of non-zero positive elements such that $B_j B_k = 0$ when $j \neq k$ (Exercise 4.6.13). However, $B_j^2 \neq 0$, since $\|B_j^2\| = \|B_j\|^2 > 0$. By the result of Exercise 4.6.20(ii), with $\varphi : \mathfrak{A} \to \mathfrak{A}/\mathcal{I}$ the quotient mapping, there is a sequence $\{S_1, S_2, \ldots\}$ of elements of $\mathfrak{A}^+$ such that $\varphi(S_j) = B_j$ and $S_j S_k = 0$ when $j \neq k$. Upon replacing S_j (and B_j) by an appropriate positive multiple, we may assume that $\|S_j\| \leq 1$.

Since $\varphi(S_j^2) = B_j^2 \neq 0$, it follows that $S_j^2 \notin \mathcal{I}$. Accordingly, the mapping $T \to \delta(S_j^2 T) : \mathfrak{A} \to \mathfrak{X}$ is not continuous, and we can choose T_j in $\mathfrak{A}$ so that

$$\|T_j\| \leq 2^{-j}, \quad \|\delta(S_j^2 T_j)\| \geq j + M\|\delta(S_j)\|,$$

where M is the bound of the bilinear mapping $(T, x) \to xT : \mathfrak{A} \times \mathfrak{X} \to \mathfrak{X}$. Since $\|S_j T_j\| \leq 2^{-j}$, the series $\sum S_j T_j$ converges to an element C of $\mathfrak{A}$, and $S_j C = S_j^2 T_j$, $\|C\| \leq 1$. Thus

$$\begin{aligned}
\|S_j \delta(C)\| &= \|\delta(S_j C) - \delta(S_j)C\| \\
&\geq \|\delta(S_j C)\| - \|\delta(S_j)C\| \\
&= \|\delta(S_j^2 T_j)\| - \|\delta(S_j)C\| \\
&\geq j + M\|\delta(S_j)\| - M\|\delta(S_j)\|\,\|C\| \geq j,
\end{aligned}$$

for all $j = 1, 2, \ldots$. However, this is impossible, since $\|S_j\| \le 1$ and the mapping $T \to T\delta(C) : \mathfrak{A} \to \mathfrak{X}$ is bounded. Hence $\mathfrak{A}/\mathcal{I}$ is finite dimensional.

Choose $A_1, A_2, \ldots, A_r$ in $\mathfrak{A}$ so that $\varphi(A_1), \varphi(A_2), \ldots, \varphi(A_r)$ is a basis for the finite-dimensional Banach space $\mathfrak{A}/\mathcal{I}$, and let $\tau_1, \tau_2, \ldots, \tau_r$ be linear functionals (which are necessarily continuous) on $\mathfrak{A}/\mathcal{I}$ such that $\tau_j(\varphi(A_k))$ is 1 when $j = k$ and 0 when $j \ne k$. For each A in $\mathfrak{A}$, $\varphi(A)$ has the form $\sum_{k=1}^{r} c_k \varphi(A_k)$; and the scalars $c_1, c_2, \ldots, c_r$ are determined by $c_j = \tau_j(\varphi(A)) = \rho_j(A)$, where ρ_j is the continuous linear functional $\tau_j \circ \varphi$ on $\mathfrak{A}$. Since

$$\varphi(A) = \sum_{j=1}^{r} \rho_j(A)\varphi(A_j)$$

we have $\varphi(A - \sum_{j=1}^{r} \rho_j(A)A_j) = 0$; so

$$A \to A - \sum_{j=1}^{r} \rho_j(A)A_j$$

is a continuous mapping from $\mathfrak{A}$ into $\mathcal{I}$. From this, and since $\delta|\mathcal{I}$ and $\rho_1, \rho_2, \ldots, \rho_r$ are continuous, so is the mapping

$$A \to \left[\delta(A) - \sum_{j=1}^{r} \rho_j(A)\delta(A_j)\right] + \sum_{j=1}^{r} \rho_j(A)\delta(A_j) = \delta(A).$$

Thus δ is continuous on $\mathfrak{A}$. ∎ [88]

4.6.67.   Show that the set $\mathcal{P}$ of pure states of $\mathcal{B}(\mathcal{H})$ is weak * closed when $\mathcal{H}$ is finite dimensional.

Solution.   From Exercise 4.6.18(iv), the mapping $x \to \omega_x :$ $(\mathcal{H})_1 \to \mathcal{S}$ carries the boundary $(\mathcal{H})_1'$ of $(\mathcal{H})_1$ *onto* $\mathcal{P}$. If $A \in (\mathcal{B}(\mathcal{H}))_1$ and $x, y \in (\mathcal{H})_1$, then

$$|(\omega_x - \omega_y)(A)| = |\langle Ax, x \rangle - \langle Ay, y \rangle|$$
$$\le |\langle Ax, x - y \rangle| + |\langle A(x - y), y \rangle|$$
$$\le 2\|x - y\|.$$

Thus the mapping $x \to \omega_x$ is continuous from $(\mathcal{H})_1'$ (in its metric topology) to $\mathcal{S}$ in the norm topology on $\mathcal{B}(\mathcal{H})^\#$. But from Proposition 1.2.16, the norm and weak* topologies on $\mathcal{B}(\mathcal{H})^\#$ coincide. Since $(\mathcal{H})_1'$ is compact, its continuous image $\mathcal{P}$ is compact and, hence, closed in $\mathcal{B}(\mathcal{H})^\#$. ∎

4.6.68. Let $\mathcal{H}$ be a Hilbert space. Show that each vector state ω_x of $\mathcal{B}(\mathcal{H})$ is pure.

Solution. Suppose $0 < \eta_0 \leq \omega_x$ for some η_0 in $\mathcal{B}(\mathcal{H})^{\#}$. Let E be the projection in $\mathcal{B}(\mathcal{H})$ with range $[x]$. Then $0 \leq \eta_0(I - E) \leq \omega_x(I - E) = 0$. Now $\eta_0(I)^{-1}\eta_0$ is a state η of $\mathcal{B}(\mathcal{H})$ and $\eta(I - E) = 0$. Thus $\eta(E) = 1$ and $I - E$ is in the left and right kernels of η. It follows that

$$\eta(B) = \eta(EBE) = \eta(\langle Bx, x\rangle E) = \langle Bx, x\rangle\eta(E) = \langle Bx, x\rangle = \omega_x(B),$$

and $\omega_x = \eta$. Hence $\eta_0 = \eta_0(I)\omega_x$, and ω_x is a pure state of $\mathcal{B}(\mathcal{H})$. ∎

4.6.69. Suppose $\mathcal{H}$ is an infinite-dimensional Hilbert space, $\mathcal{K}$ is the ideal of compact operators in $\mathcal{B}(\mathcal{H})$, $\mathcal{P}$ is the set of pure states of $\mathcal{B}(\mathcal{H})$, and $\mathcal{S}_0$ is the set of vector states of $\mathcal{B}(\mathcal{H})$.

(i) Use Exercises 4.6.57 and 4.6.23 to show that there is a pure state ρ of $\mathcal{B}(\mathcal{H})$ that is 0 on $\mathcal{K}$.

(ii) With a in $[0, 1]$, x a unit vector in $\mathcal{H}$, and ρ in $\mathcal{P}^-$ and 0 on $\mathcal{K}$, let ω be the state $a\omega_x + (1 - a)\rho$ of $\mathcal{B}(\mathcal{H})$. Show that ω is in $\mathcal{P}^-$, the pure state space of $\mathcal{B}(\mathcal{H})$, and that ω is in $\mathcal{S}_0^-$. [*Hint.* Use Corollary 4.3.10 to approximate ρ by a vector state $\omega_{y'}$ of $\mathcal{B}(\mathcal{H})$. With $A_1, \ldots, A_n$ self-adjoint operators in $(\mathcal{B}(\mathcal{H}))_1$ and E the projection with range $[x, A_1 x, \ldots, A_n x]$, estimate $|(\omega - \omega_z)(A_j)|$ where $z = a^{1/2}x + (1 - a)^{1/2}y$ and $y = \|(I - E)y'\|^{-1}(I - E)y'$.]

(iii) Conclude that $\mathcal{P}$ is not weak* closed (that is, $\mathcal{P} \neq \mathcal{P}^-$) when $\mathcal{H}$ is infinite dimensional.

Solution. (i) From Exercise 4.6.57(iv) and (v), there is a representation π of $\mathcal{B}(\mathcal{H})$ with kernel $\mathcal{K}$. If ρ_0 is a pure state of $\pi(\mathcal{B}(\mathcal{H}))$, then $\rho \circ \pi$ is a pure state ρ of $\mathcal{B}(\mathcal{H})$ and ρ is 0 on $\mathcal{K}$ (from Exercise 4.6.23(iii)).

(ii) From Exercise 4.6.68, each vector state of $\mathcal{B}(\mathcal{H})$ is pure. We show that ω is in the weak* closure of $\mathcal{S}_0$ by following the program of the hint. Since the self-adjoint operators generate $\mathcal{B}(\mathcal{H})$ linearly, it will suffice to show that for $\varepsilon \, (> 0)$ and a given set $\{A_1, A_2, \ldots, A_n\}$ of self-adjoint operators in $(\mathcal{B}(\mathcal{H}))_1$, there is a unit vector z such that $|(\omega - \omega_z)(A_j)| < \varepsilon$ for j in $\{1, 2, \ldots, n\}$. By Corollary 4.3.10, $\mathcal{P}^-$ is contained in $\mathcal{S}_0^-$ (and, hence, in this case, $\mathcal{P}^- = \mathcal{S}_0^-$). Therefore, given a positive $\varepsilon\,(<1)$, there is a unit vector y' such that

$$|(\rho - \omega_{y'})(A_j)| < \frac{\varepsilon}{2}, \quad |(\rho - \omega_{y'})(E)| < \frac{\varepsilon^2}{100},$$

where E is the (finite-dimensional) projection in the algebra $\mathcal{B}(\mathcal{H})$ with range $[x, A_1 x, \ldots, A_n x]$. Since ρ is 0 on $\mathcal{K}$,

$$\|Ey'\|^2 = |\langle Ey', y'\rangle| = |(\rho - \omega_{y'})(E)| < \frac{\varepsilon^2}{100}.$$

Let c be $\|(I-E)y'\|$, y be $c^{-1}(I-E)y'$, and z be $a^{1/2}x + (1-a)^{1/2}y$. Then $1 - \frac{\varepsilon}{10} \le 1 - \|Ey'\| \le \|y' - Ey'\| = c \le 1$ and

$$\|y - y'\| = c^{-1}\|(I-E)y' - cy'\| = c^{-1}\|(1-c)y' - Ey'\|$$
$$\le c^{-1}[(1-c) + \|Ey'\|] < \frac{2\varepsilon}{10 - \varepsilon} < \frac{\varepsilon}{4}.$$

As in the solution to Exercise 4.6.67, $\|\omega_y - \omega_{y'}\| \le 2\|y - y'\| < \varepsilon/2$. Since $0 = \langle A_j x, y\rangle = \langle x, A_j y\rangle = \langle x, y\rangle$ for j in $\{1, \ldots, n\}$, $\|z\| = 1$ and

$$\begin{aligned}
&|(\omega - \omega_z)(A_j)| \\
&\quad = |a\langle A_j x, x\rangle + (1-a)\rho(A_j) - \langle A_j z, z\rangle| \\
&\quad = (1-a)|(\rho - \omega_y)(A_j)| \\
&\quad \le (1-a)[|(\rho - \omega_{y'})(A_j)| + |(\omega_{y'} - \omega_y)(A_j)|] < \varepsilon.
\end{aligned}$$

Hence $\omega \in \mathcal{S}_0^- = \mathcal{P}^-$.

(iii) Since $\rho \ne \omega_x$, ω is not pure when $a \in (0,1)$. It follows that $\mathcal{P} \ne \mathcal{P}^-$. ∎

4.6.70. Let $\mathfrak{A}$ be a C*-algebra and $\mathcal{B}$ be a self-adjoint subalgebra of $\mathfrak{A}$ containing I. Suppose that for each pair ρ_1, ρ_2 of distinct states of $\mathfrak{A}$ there is a B in $\mathcal{B}$ such that $\rho_1(B) \ne \rho_2(B)$ (that is, $\mathcal{B}$ *separates* the states of $\mathfrak{A}$). Show that $\mathcal{B}$ is norm dense in $\mathfrak{A}$.

Solution. If $\mathcal{B}$ separates the states of $\mathfrak{A}$, so does the norm closure of $\mathcal{B}$. We assume that $\mathcal{B}$ is norm closed and prove that $\mathcal{B} = \mathfrak{A}$. If $\mathcal{B} \ne \mathfrak{A}$, then $\mathcal{B}_h \ne \mathfrak{A}_h$. From Corollary 1.6.3, there is a bounded linear functional η' on $\mathfrak{A}_h$ such that $\|\eta'\| = 2$ and

η' is 0 on $\mathcal{B}_h$. Extend η' to a hermitian linear functional η on $\mathfrak{A}$. From the discussion preceding Proposition 4.3.1, $\|\eta\| = \|\eta'\| = 2$. From Theorem 4.3.6, $\eta = \eta^+ - \eta^-$, where η^+ and η^- are positive linear functionals on $\mathfrak{A}$ and $2 = \|\eta\| = \|\eta^+\| + \|\eta^-\|$. Since $I \in \mathcal{B}$, $0 = \eta(I) = \eta^+(I) - \eta^-(I)$. Thus $\|\eta^+\| = \eta^+(I) = \eta^-(I) = \|\eta^-\|$, and $1 = \|\eta^+\| = \eta^+(I) = \eta^-(I) = \|\eta^-\|$. As η is 0 on $\mathcal{B}$, $\eta^+|\mathcal{B} = \eta^-|\mathcal{B}$. But $0 \neq \eta = \eta^+ - \eta^-$. Hence η^+ and η^- are distinct states of $\mathfrak{A}$ that coincide on $\mathcal{B}$ — a contradiction. Thus $\mathcal{B} = \mathfrak{A}$.   ■

CHAPTER 5

ELEMENTARY VON NEUMANN
ALGEBRA THEORY

5.7.   Exercises

5.7.1.   Show that the mapping $A \to A^*$ of $\mathcal{B}(\mathcal{H})$ into $\mathcal{B}(\mathcal{H})$ is weak-operator continuous.

Solution.   Given A in $\mathcal{B}(\mathcal{H})$, a positive ε, and sets of vectors $x_1, \ldots, x_n$ and $y_1, \ldots, y_n$,

$$|\langle (A^* - B^*)x_j, y_j \rangle| < \varepsilon \qquad (j \in \{1, \ldots, n\})$$

provided

$$|\langle (A - B)y_j, x_j \rangle| < \varepsilon \qquad (j \in \{1, \ldots, n\})$$

since

$$|\langle (A^* - B^*)x_j, y_j \rangle| = |\langle (A - B)y_j, x_j \rangle| \qquad (j \in \{1, \ldots, n\}).$$

Thus $A \to A^*$ $(\mathcal{B}(\mathcal{H}) \to \mathcal{B}(\mathcal{H}))$ is weak-operator continuous.   ∎

5.7.2.   Show that the weak-operator topology on $\mathcal{B}(\mathcal{H})$ is *strictly* weaker (coarser) than the strong-operator topology on $\mathcal{B}(\mathcal{H})$.

Solution.   From the discussion following Definition 5.1.1, the weak-operator topology is coarser than the strong-operator topology. From Exercise 2.8.32, the adjoint operation is not strong-operator continuous on (the unit ball of) $\mathcal{B}(\mathcal{H})$. However, from Exercise 5.7.1, the adjoint operation is weak-operator continuous on $\mathcal{B}(\mathcal{H})$. It follows that the weak-operator topology is strictly coarser than the strong-operator topology.   ∎

5.7.3.   Let $\mathcal{P}$ denote the set of all projections from a Hilbert space $\mathcal{H}$ onto its closed subspaces, and suppose that $F \in \mathcal{P}$ and $0 \neq F \neq I$. Prove that the mappings

$$E \to E \wedge F, \qquad E \to E \vee F \qquad (\mathcal{P} \to \mathcal{P})$$

are not continuous from $\mathcal{P}$ with the norm topology to $\mathcal{P}$ with the weak-operator topology. (Compare Exercise 2.8.17.)

Solution.   See the solution to Exercise 2.8.17.   ∎

5.7.4.   Let $\mathcal{H}$ be a Hilbert space and $\mathcal{T}_{\mathrm{s}}, \mathcal{T}_{\mathrm{w}}$ be the restrictions of the strong- and weak-operator topologies to the set $\mathcal{P}$ of projections in $\mathcal{B}(\mathcal{H})$. Show that $\mathcal{T}_{\mathrm{s}} = \mathcal{T}_{\mathrm{w}}$. Conclude that if a sequence of projections tends to a *projection* in the weak-operator topology, it tends to that projection in the strong-operator topology (but compare Exercise 5.7.8(ii)).

Solution.   We must show that the identity mapping of $\{\mathcal{P}, \mathcal{T}_{\mathrm{s}}\}$ onto $\{\mathcal{P}, \mathcal{T}_{\mathrm{w}}\}$ is a homeomorphism. Since each weak-operator open subset of $\mathcal{B}(\mathcal{H})$ is strong-operator open, this mapping is continuous. (The inverse image of each set in $\mathcal{T}_{\mathrm{w}}$ is in $\mathcal{T}_{\mathrm{s}}$, since $\mathcal{T}_{\mathrm{w}} \subseteq \mathcal{T}_{\mathrm{s}}$.) It remains to establish the continuity of this mapping in the reverse direction.

Let E_0 in $\mathcal{P}$, a unit vector u in $\mathcal{H}$, and a positive ε be given. It will suffice, for the desired continuity, to find a positive δ and finite sets of vectors $x_1, \ldots, x_n$ and $y_1, \ldots, y_n$ in $\mathcal{H}$ such that if $E \in \mathcal{P}$, then $\|(E - E_0)u\| < \varepsilon$ provided $|\langle (E - E_0)x_j, y_j \rangle| < \delta$ for all j in $\{1, \ldots, n\}$. Toward this end, suppose E (in $\mathcal{P}$) satisfies:

$$|\langle (E - E_0)u, u \rangle| < \frac{\varepsilon^2}{4}, \qquad |\langle (E - E_0)u, E_0 u \rangle| < \frac{\varepsilon^2}{4}.$$

Then

$$\begin{aligned}
\|(E - E_0)u\|^2 &= \langle Eu, u \rangle + \langle E_0 u, u \rangle - 2Re\langle Eu, E_0 u \rangle \\
&= \langle (E - E_0)u, u \rangle + 2\langle E_0 u, E_0 u \rangle - 2Re\langle Eu, E_0 u \rangle \\
&\leq |\langle (E - E_0)u, u \rangle| + 2|\langle (E_0 - E)u, E_0 u \rangle| < \varepsilon^2.
\end{aligned}$$

A second solution proceeds as follows. Let u be $\sum_j a_j e_j + \sum_k b_k f_k$ where $\{e_j\}$ and $\{f_k\}$ are orthonormal bases for $E_0(\mathcal{H})$ and

$(I - E_0)(\mathcal{H})$, respectively. Let u_0 be $\sum_{j=1}^{N} a_j e_j + \sum_{k=1}^{N} b_k f_k$, where N is a positive integer such that $\|u - u_0\| < \frac{\varepsilon}{4}$ (here, e_j and f_k are taken to be 0 when j or k exceed the dimensions of $E_0(\mathcal{H})$ and $(I - E_0)(\mathcal{H})$, respectively). Suppose E in $\mathcal{P}$ is such that

$$|\langle (E - E_0)e_j, e_j\rangle| \ (= |\langle (I - E)e_j, e_j\rangle| = \|(I - E)e_j\|^2) < \frac{\varepsilon^2}{16N^2}$$

and

$$|\langle (E - E_0)f_k, f_k\rangle| \ (= |\langle Ef_k, f_k\rangle| = \|Ef_k\|^2) < \frac{\varepsilon^2}{16N^2}$$

for j in $\{1, \ldots, N\}$. Then

$$\|(E - E_0)u_0\| \le \|\sum_{j=1}^{N} a_j(E - E_0)e_j\| + \|\sum_{k=1}^{N} b_k(E - E_0)f_k\|$$

$$\le \sum_{j=1}^{N} \|(I - E)e_j\| + \sum_{k=1}^{N} \|Ef_k\| < \frac{\varepsilon}{2},$$

while

$$\|(E - E_0)(u - u_0)\| \le 2\|u - u_0\| < \frac{\varepsilon}{2}.$$

Hence $\|(E - E_0)u\| < \varepsilon$. ∎

5.7.5. Let $\mathcal{H}$ be a Hilbert space and $\mathcal{T}_s'$, $\mathcal{T}_w'$ be the restrictions of the strong- and weak-operator topologies to the set $\mathcal{U}$ of unitary operators in $\mathcal{B}(\mathcal{H})$. Show that $\mathcal{T}_s' = \mathcal{T}_w'$. Conclude that if a sequence of unitary operators tends to a *unitary operator* in the weak-operator topology, it tends to that unitary operator in the strong-operator topology.

Solution. As in Exercise 5.7.4, it suffices to show that, given a unitary operator U_0 in $\mathcal{U}$, a unit vector u in $\mathcal{H}$, and a positive ε, there is a positive δ and finite sets of vectors $x_1, \ldots, x_n$ and $y_1, \ldots, y_n$ in $\mathcal{H}$ such that if $U \in \mathcal{U}$, then $\|(U - U_0)u\| < \varepsilon$ provided that we have $|\langle (U - U_0)x_j, y_j\rangle| < \delta$ for all j in $\{1, \ldots, n\}$. Toward this end, suppose U in $\mathcal{U}$ satisfies: $|\langle (U - U_0)u, U_0 u\rangle| < \frac{\varepsilon^2}{2}$. Then

$$\|(U - U_0)u\|^2 = 2\langle u, u\rangle - 2\mathbf{Re}\langle Uu, U_0 u\rangle$$

$$= 2Re\langle (U_0 - U)u, U_0 u\rangle$$

$$\le 2|\langle (U - U_0)u, U_0 u\rangle| < \varepsilon^2. \qquad \blacksquare$$

5.7.6.   Let $\mathcal{H}$ be a Hilbert space, $\{y_a\}_{a \in \mathbf{A}}$ be an orthonormal basis for $\mathcal{H}$, and $\mathcal{S}$ be a bounded subset of $\mathcal{B}(\mathcal{H})$. With $\mathbf{F}$ a finite subset of $\mathbf{A}$, ε a positive number, and S_0 in $\mathcal{S}$,

$$\{S : |\langle (S - S_0)y_a, y_{a'} \rangle| < \varepsilon, \ a, a' \in \mathbf{F}, \ S \in \mathcal{S}\} \ (= \mathcal{V}_{\mathbf{F},\varepsilon})$$

is a weak-operator open neighborhood of S_0. Show that $\{\mathcal{V}_{\mathbf{F},\varepsilon}\}$ is a base for the weak-operator open neighorhoods of S_0 in $\mathcal{S}$.

Solution.   Note that

$$\mathcal{V}_{\mathbf{F},\varepsilon} \subseteq \mathcal{V}_{\mathbf{F}_1,\varepsilon_1} \cap \cdots \cap \mathcal{V}_{\mathbf{F}_n,\varepsilon_n},$$

where $\varepsilon = \min\{\varepsilon_1, \ldots, \varepsilon_n\}$ and $\mathbf{F} = \mathbf{F}_1 \cup \cdots \cup \mathbf{F}_n$. It suffices, therefore, to show that each subbasic weak-operator open neighborhood

$$\{S : |\langle (S - S_0)u, u \rangle| < \varepsilon, \ S \in \mathcal{S}\} \ (= \mathcal{V})$$

of S_0, where $\varepsilon > 0$ and u is a unit vector in $\mathcal{H}$, contains some $\mathcal{V}_{\mathbf{F},\varepsilon'}$. For this, choose a finite subset $\mathbf{F}$ of $\mathbf{A}$ such that $\|u - u_0\| < \frac{\varepsilon}{4M}$, where $u_0 = \sum_{a \in \mathbf{F}} \langle u, y_a \rangle y_a$ and $\|S\| \leq \frac{1}{2}M$ for S in $\mathcal{S}$. Let ε' be $\frac{\varepsilon}{2k^2}$, where k is the number of elements in $\mathbf{F}$. With these choices,

$$|\langle (S - S_0)u, u \rangle - \langle (S - S_0)u_0, u_0 \rangle| < 2M \cdot \frac{\varepsilon}{4M} = \frac{\varepsilon}{2},$$

for S in $\mathcal{S}$. If $S \in \mathcal{V}_{\mathbf{F},\varepsilon'}$, then

$$|\langle (S - S_0)u_0, u_0 \rangle| = \Big| \sum_{a,a' \in \mathbf{F}} \langle u, y_a \rangle \langle y_{a'}, u \rangle \langle (S - S_0)y_a, y_{a'} \rangle \Big|$$

$$\leq \sum_{a,a' \in \mathbf{F}} |\langle (S - S_0)y_a, y_{a'} \rangle| < k^2 \varepsilon' = \frac{\varepsilon}{2},$$

and $|\langle (S - S_0)u, u \rangle < \varepsilon$. Thus $\mathcal{V}_{\mathbf{F},\varepsilon'} \subseteq \mathcal{V}$.   ■

5.7.7.   Let $\mathcal{H}$ be a separable Hilbert space and $\{y_1, y_2, \ldots\}$ be an orthonormal basis for $\mathcal{H}$. Show that the equation

$$d(S, T) = \sum_{n,m=1}^{\infty} 2^{-(n+m)} |\langle (S - T)y_n, y_m \rangle|$$

defines a translation-invariant metric d on $\mathcal{B}(\mathcal{H})$ (that is, $d(S + R, T + R) = d(S, T)$ for each R in $\mathcal{B}(\mathcal{H})$, and the associated metric topology coincides on bounded subsets of $\mathcal{B}(\mathcal{H})$ with the weak-operator topology.

Solution. Since $|\langle (S-T)y_n, y_m \rangle| < \|S-T\|$, the series

$$\sum_{n,m=1}^{\infty} 2^{-(n+m)} |\langle (S-T)y_n, y_m \rangle|$$

converges. Clearly $d(S,T) \geq 0$ and $d(S,S) = 0$. If $d(S,T) = 0$, then $\langle (S-T)y_n, y_m \rangle = 0$ for all n and m. Since

$$\|(S-T)y_n\|^2 = \sum_{m=1}^{\infty} |\langle (S-T)y_n, y_m \rangle|^2 = 0,$$

$(S-T)y_n = 0$ for all n. Hence $S = T$. Of course $d(S,T) = d(T,S)$ and $d(S+R, T+R) = d(T,S)$. Finally,

$$\begin{aligned}
d(S,T) &= \sum_{n,m=1}^{\infty} 2^{-(n+m)} |\langle (S-R+R-T)y_n, y_m \rangle| \\
&\leq \sum_{n,m=1}^{\infty} 2^{-(n+m)} |\langle (S-R)y_n, y_m \rangle| \\
&\quad + \sum_{n,m=1}^{\infty} 2^{-(n+m)} |\langle (R-T)y_n, y_m \rangle| \\
&= d(S,R) + d(R,T).
\end{aligned}$$

Thus d is a translation-invariant metric on $\mathcal{B}(\mathcal{H})$.

Now let $\mathcal{S}$ be a bounded subset of $\mathcal{B}(\mathcal{H})$ and choose $M \,(> 0)$ so that $\|S\| \leq M$ whenever $S \in \mathcal{S}$. Given S_0 in $\mathcal{S}$ and $\varepsilon \,(> 0)$, we can choose a positive integer k such that $k^{-1} + M2^{2-k} < \varepsilon$. Then the metric ball $\{S \in \mathcal{S} : d(S, S_0) < \varepsilon\}$ contains the weak-operator neighborhood

$$\{S \in \mathcal{S} : |\langle (S-S_0)y_j, y_{j'} \rangle| < k^{-1} \ \ (j,j' = 1,\ldots,k)\} \ (= \mathcal{V})$$

of S_0 in $\mathcal{S}$, since

$$\begin{aligned}
d(S, S_0) &\leq \sum_{n,m=1}^{k} k^{-1} 2^{-(n+m)} + 2 \sum_{n=k+1}^{\infty} \sum_{m=1}^{\infty} 2M 2^{-(n+m)} \\
&< \frac{1}{k} + \frac{M}{2^{k-2}} < \varepsilon.
\end{aligned}$$

Conversely, the metric ball $\{S \in \mathcal{S} : d(S, S_0) < 2^{-2k}\varepsilon\}$ is contained in the weak-operator neighborhood

$$\{S \in \mathcal{S} : |\langle (S - S_0)y_j, y_{j'}\rangle| < \varepsilon \quad (j, j' = 1, \ldots, k)\}.$$

From Exercise 5.7.6, these form a base of weak-operator neigborhoods of S_0 in $\mathcal{S}$. Thus the weak-operator topology and the metric topology associated with d coincide on $\mathcal{S}$.   ∎

5.7.8.   With the notation of Exercise 5.7.4, assume that $\mathcal{H}$ is infinite dimensional.

(i)   Show that $\mathcal{P}$ is weak-operator dense in $(\mathcal{B}(\mathcal{H}))_1^+$, the set of positive operators in the unit ball of $\mathcal{B}(\mathcal{H})$. [*Hint.* Note that

$$\begin{bmatrix} A & (A - A^2)^{\frac{1}{2}} \\ (A - A^2)^{\frac{1}{2}} & I - A \end{bmatrix}$$

is a projection in $\mathcal{B}(\mathcal{K} \oplus \mathcal{K})$ when $A \in (\mathcal{B}(\mathcal{K}))_1^+$.]

(ii)   Show that $\mathcal{P}$ is strong-operator closed in $\mathcal{B}(\mathcal{H})$, and conclude that there is a sequence of projections tending to an operator in $(\mathcal{B}(\mathcal{H}))_1^+$ in the weak-operator topology, when $\mathcal{H}$ is separable, that does not tend to it in the strong-operator topology. (Compare Exercises 5.7.2 and 5.7.4 and Theorem 5.1.2.)

Solution. (i)   Let $x_1, \ldots, x_n$ be a finite set of vectors in $\mathcal{H}$ and let E be the projection with range $[x_1, \ldots, x_n]$. Since $\mathcal{H}$ is infinite dimensional, there is a projection F orthogonal to E with range of the same dimension as $[x_1, \ldots, x_n]$. Let B be an operator in $(\mathcal{B}(\mathcal{H}))_1^+$. By the methods of subsection 2.6, *Matrix representations*, we may view $\mathcal{H}_0$, the subspace of $\mathcal{H}$ spanned by $[x_1, \ldots, x_n]$ and $F(\mathcal{H})$, as the direct sum of a Hilbert space (namely, $E(\mathcal{H})$) with itself, and represent the operators in $\mathcal{B}(\mathcal{H}_0)$ by 2×2 matrices with entries from $\mathcal{B}(E(\mathcal{H}))$. Let A be $EBE|E(\mathcal{H})$ and let G be the operator on $\mathcal{H}$ that annihilates each vector orthogonal to $\mathcal{H}_0$ and acts on $\mathcal{H}_0$ as the operator with matrix as in the hint. Then $G = G^*$ and $G^2 = G$, from the matrix rules established in subsection 2.6, *Matrix representations*. Moreover, $E|\mathcal{H}_0$ is represented by the matrix

$$\begin{bmatrix} I & 0 \\ 0 & 0 \end{bmatrix},$$

so that $EGE = EBE$.

Let two finite sets of vectors $u_1, \ldots, u_m$ and $v_1, \ldots, v_m$ in $\mathcal{H}$ be given and let M be the projection in the algebra $\mathcal{B}(\mathcal{H})$ with range $[u_1, \ldots, u_m, v_1, \ldots, v_m]$. With B in $(\mathcal{B}(\mathcal{H}))_1^+$, we can, from the foregoing, find a projection N in $\mathcal{B}(\mathcal{H})$ such that $MNM = MBM$. Thus, for each j in $\{1, \ldots, m\}$,

$$\begin{aligned}
\langle Nu_j, v_j \rangle &= \langle MNMu_j, v_j \rangle \\
&= \langle MBMu_j, v_j \rangle \\
&= \langle Bu_j, v_j \rangle.
\end{aligned}$$

It follows that $\mathcal{P}$ is weak-operator dense in $(\mathcal{B}(\mathcal{H}))_1^+$.

(ii) Suppose A is in the strong-operator closure of $\mathcal{P}$. Then A is in the weak-operator closure of $\mathcal{P}$ and, from Exercise 5.7.1, A is self-adjoint. With x a unit vector in $\mathcal{H}$ and ε a positive number, there is an E in $\mathcal{P}$ such that $\|(A - E)x\| \le \varepsilon$. Thus $\|Ax\| \le \|Ex\| + \varepsilon \le 1 + \varepsilon$. Hence $\|Ax\| \le 1$ and $A \in (\mathcal{B}(\mathcal{H}))_1$. From Remark 2.5.10, the mapping $(S, T) \to ST$ of $(\mathcal{B}(\mathcal{H}))_1 \times (\mathcal{B}(\mathcal{H}))_1$ into $(\mathcal{B}(\mathcal{H}))_1$ is strong-operator continuous. Thus $\{S : S^2 = S, \, S \in (\mathcal{B}(\mathcal{H}))_1\}$ is strong-operator closed. This set contains $\mathcal{P}$, so that $A^2 = A$. Hence $A \in \mathcal{P}$ and $\mathcal{P}$ is strong-operator closed.

Assume $\mathcal{H}$ is separable and infinite dimensional. From Exercise 5.7.7, the weak-operator topology on $(\mathcal{B}(\mathcal{H}))_1$ is metrizable. From (i), each A in $(\mathcal{B}(\mathcal{H}))_1^+$ is in the weak-operator closure of $\mathcal{P}$ and hence, A is the weak-operator limit of a sequence $\{E_n\}$ of elements of $\mathcal{P}$. Choosing A not in $\mathcal{P}$, $\{E_n\}$ does not tend to A in the strong-operator topology(since $\mathcal{P}$ is strong-operator closed and $A \notin \mathcal{P}$). $\blacksquare$

5.7.9. Let $\mathcal{H}$ be a Hilbert space.

(i) Show that the mapping $A \to AB$ of $\mathcal{B}(\mathcal{H})$ into $\mathcal{B}(\mathcal{H})$ is weak-operator continuous for each B in $\mathcal{B}(\mathcal{H})$.

(ii) Show that the mapping $A \to BA$ of $\mathcal{B}(\mathcal{H})$ into $\mathcal{B}(\mathcal{H})$ is weak-operator continuous for each B in $\mathcal{B}(\mathcal{H})$.

(iii) When $\mathcal{H}$ is infinite dimensional, show that the mapping $(A, B) \to AB$ of $(\mathcal{B}(\mathcal{H}))_1^+ \times (\mathcal{B}(\mathcal{H}))_1^+$ into $\mathcal{B}(\mathcal{H})$ is not weak-operator continuous.

Solution. (i) To establish continuity at A_0, it suffices to note that the (subbasic) weak-operator open neighborhood of A_0 given by

$\{A : |\langle (A - A_0)Bu, v\rangle| < \varepsilon, A \in \mathcal{B}(\mathcal{H})\}$ maps, under the given mapping, into the (subbasic) weak-operator open neighborhood $\{S : |\langle (S - A_0 B)u, v\rangle| < \varepsilon, S \in \mathcal{B}(\mathcal{H})\}$ of the image $A_0 B$ of A_0.

(ii)   In this case, the (subbasic) weak-operator open neighborhood $\{A : |\langle (A - A_0)u, B^*v\rangle| < \varepsilon, A \in \mathcal{B}(\mathcal{H})\}$ of A_0 maps, under the given mapping, into the (subbasic) weak-operator open neighborhood $\{S : |\langle (S - BA_0)u, v\rangle| < \varepsilon, S \in \mathcal{B}(\mathcal{H})\}$ of the image BA_0 of A_0.

(iii)   We saw, in Exercise 5.7.8(i), that $\mathcal{P}$, the set of projections in $\mathcal{B}(\mathcal{H})$, is weak-operator dense in $(\mathcal{B}(\mathcal{H}))_1^+$ when $\mathcal{H}$ is infinite dimensional. Let $\mathcal{V}_0$ be a weak-operator open neighborhood of $\frac{1}{4}I$ in $(\mathcal{B}(\mathcal{H}))_1$ with closure not containing $\frac{1}{2}I$ and let $\mathcal{V}$ be a weak-operator open neighborhood of $\frac{1}{2}I$. There is a weak-operator open neighborhood $\mathcal{V}'$ of $\frac{1}{2}I$ contained in $\mathcal{V}$ and disjoint from $\mathcal{V}_0$ and a projection E in $\mathcal{V}'$. But then $E^2 = E \in \mathcal{V}'$ so that $E^2 \notin \mathcal{V}_0$. Hence each neighborhood $\mathcal{V}$ of $\frac{1}{2}I$ contains elements E (in $(\mathcal{B}(\mathcal{H}))_1^+$) whose square does not lie in the given neighborhood $\mathcal{V}_0$ of $\frac{1}{4}I$. Thus the mapping $(A, B) \to AB$ is not weak-operator continuous on $(\mathcal{B}(\mathcal{H}))_1^+ \times (\mathcal{B}(\mathcal{H}))_1^+$.   ■

5.7.10.   Let $\mathcal{H}$ be a Hilbert space and $(\mathcal{B}(\mathcal{H}))_r$ be $\{T \in \mathcal{B}(\mathcal{H}) : \|T\| \le r\}$.

(i)   Show that the mapping

$$(A, B) \to AB : \quad (\mathcal{B}(\mathcal{H}))_r \times \mathcal{B}(\mathcal{H}) \to \mathcal{B}(\mathcal{H})$$

is continuous when $(\mathcal{B}(\mathcal{H}))_r \times \mathcal{B}(\mathcal{H})$ is provided with the product of the weak-operator topology on $(\mathcal{B}(\mathcal{H}))_r$ and the strong-operator topology on $\mathcal{B}(\mathcal{H})$, and the range $\mathcal{B}(\mathcal{H})$ is provided with the weak-operator topology. (Compare Exercise 2.8.33.)

(ii)   Let $\mathcal{H}$ be infinite dimensional and separable. Show that the mapping

$$(A, B) \to AB : \quad (\mathcal{B}(\mathcal{H}))_1 \times (\mathcal{B}(\mathcal{H}))_1 \to \mathcal{B}(\mathcal{H})$$

is *not* continuous when the first factor of $(\mathcal{B}(\mathcal{H}))_1 \times (\mathcal{B}(\mathcal{H}))_1$ is provided with the strong-operator topology and the second is provided with the weak-operator topology.

Solution. (i)   The stated continuity follows from the inequality:

$$|\langle (AB - A_0 B_0)x, y \rangle| \leq |\langle (AB - AB_0)x, y \rangle| + |\langle (AB_0 - A_0 B_0)x, y \rangle|$$
$$\leq \|(B - B_0)x\| \, \|A^* y\| + |\langle (A - A_0)B_0 x, y \rangle|$$
$$\leq r\|y\| \, \|(B - B_0)x\| + |\langle (A - A_0)B_0 x, y \rangle|,$$

when $A \in (\mathcal{B}(\mathcal{H}))_r$.

(ii)   Let $\{e_1, e_2, \ldots\}$ be an orthonormal basis for $\mathcal{H}$ and let V_n be the isometry of $\mathcal{H}$ into $\mathcal{H}$ determined by the equation, $V_n e_j = e_{j+n}$, $j = 1, 2, \ldots$. Then, for each x in $\mathcal{H}$ and each e_j, $\langle V_n e_j, x \rangle = \langle e_{j+n}, x \rangle \to 0$ as $n \to \infty$, from which we conclude that $\{V_n\}$ tends to 0 in the weak-operator topology. At the same time, $\langle V_n^* e_j, e_k \rangle = \langle e_j, e_{k+n} \rangle = 0$ if $j \neq k + n$. Thus $V_n^* e_j = 0$ if $j < n + 1$ and $V_n^* e_j = e_{j-n}$ if $j \geq n + 1$. It follows that, for each x in $\mathcal{H}$, $\{V_n^* x\}$ tends to 0 as n tends to ∞. Thus $\{V_n^*\}$ tends to 0 in the strong-operator topology. But $V_n^* V_n e_j = e_j$ for $j = 1, 2, \ldots$, so that $V_n^* V_n = I$ for each n, and the mapping described in (ii) fails to have the continuity noted.   ∎

5.7.11.   Let $\mathcal{H}$ be a Hilbert space and $\mathcal{R}$ be a von Neumann algebra on $\mathcal{H}$.

(i)   Show that $(\mathcal{R})_1$ is weak-operator compact.

(ii)   Show that $(\mathcal{R})_1^+$ is weak-operator compact.

(iii)   With $\mathcal{H}$ infinite dimensional, show that $(\mathcal{B}(\mathcal{H}))_1$ and $(\mathcal{B}(\mathcal{H}))_1^+$ are not strong-operator compact.

Solution. (i)   Since $(\mathcal{R})_1 = \mathcal{R} \cap (\mathcal{B}(\mathcal{H}))_1$, $(\mathcal{R})_1$ is a weak-operator closed subset of $(\mathcal{B}(\mathcal{H}))_1$. From Theorem 5.1.3, $(\mathcal{B}(\mathcal{H}))_1$ is weak-operator compact. Hence $(\mathcal{R})_1$ is weak-operator compact.

(ii)   Since

$$(\mathcal{B}(\mathcal{H}))_1^+ = \bigcap_{x \in \mathcal{H}} \{A \in (\mathcal{B}(\mathcal{H}))_1 : \langle Ax, x \rangle \geq 0\}$$

and $\{A \in (\mathcal{B}(\mathcal{H}))_1 : \langle Ax, x \rangle \geq 0\}$ is a weak-operator closed subset of $(\mathcal{B}(\mathcal{H}))_1$ for each x in $\mathcal{H}$, $(\mathcal{B}(\mathcal{H}))_1^+$ is a weak-operator closed subset of $(\mathcal{B}(\mathcal{H}))_1$. Hence $(\mathcal{B}(\mathcal{H}))_1^+$ is weak-operator compact. As $(\mathcal{R})_1^+ = \mathcal{R} \cap (\mathcal{B}(\mathcal{H}))_1^+$, $(\mathcal{R})_1^+$ is a weak-operator closed subset of $(\mathcal{B}(\mathcal{H}))_1^+$ (and of $(\mathcal{B}(\mathcal{H}))_1$). Hence $(\mathcal{R})_1^+$ is weak-operator compact.

(iii)   Since $(\mathcal{B}(\mathcal{H}))_1$ and $(\mathcal{B}(\mathcal{H}))_1^+$ are weak-operator compact and the identity mappings are one–to–one and continuous from each in its strong-operator topology to each in its weak-operator topology, if either were strong-operator compact, the corresponding identity mapping would be a homeomorphism. But with $\mathcal{H}$ infinite dimensional, the strong- and weak-operator topologies are different on $(\mathcal{B}(\mathcal{H}))_1^+$ (and, hence, on $(\mathcal{B}(\mathcal{H}))_1$), from Exercise 5.7.8(ii). Thus neither $(\mathcal{B}(\mathcal{H}))_1$ nor $(\mathcal{B}(\mathcal{H}))_1^+$ is strong-operator compact. (However, see Proposition 2.5.11.)   ■

5.7.12.   With the notation of Exercise 5.7.5, assume that $\mathcal{H}$ is infinite dimensional.

(i)   Show that $\mathcal{U}$ is weak-operator dense in $(\mathcal{B}(\mathcal{H}))_1$. [*Hint.* Follow the pattern of the solution to Exercise 5.7.8(i) and use the facts that each operator on a finite-dimensional Hilbert space has the form VH, where V is unitary and H is positive, and that such an operator R_0 is unitary if *either* of $R_0^* R_0$ or $R_0 R_0^*$ is I.]

(ii)   Show that the set of isometries in $\mathcal{B}(\mathcal{H})$ is strong-operator closed.

(iii)   Conclude that, with $\mathcal{H}$ separable, there is a sequence of unitary operators in $\mathcal{B}(\mathcal{H})$ that tends to an operator in $(\mathcal{B}(\mathcal{H}))_1$ in the weak-operator topology but not in the strong-operator topology.

(iv)   Find a sequence of unitary operators that converges in the strong-operator topology to an operator (isometry) that is not a unitary.

Solution. (i)   With the notation of the solution to Exercise 5.7.8(i), let T be in $(\mathcal{B}(\mathcal{H}))_1$ and let S be $ETE|E(\mathcal{H})$. As noted in the hint, $S = VH$, where V is a unitary operator and H is a positive operator on the finite-dimensional Hilbert space $E(\mathcal{H})$. Let R_0 be the operator on $\mathcal{H}_0$ with corresponding 2×2 matrix

$$\begin{bmatrix} VH & V(I - H^2)^{\frac{1}{2}} \\ -V(I - H^2)^{\frac{1}{2}} & VH \end{bmatrix}.$$

Then $R_0^* R_0$ is the identity operator on $\mathcal{H}_0$. Since $\mathcal{H}_0$ is finite dimensional, R_0 is unitary.

Let R be the operator on $\mathcal{H}$ that agrees with R_0 on $\mathcal{H}_0$ and leaves fixed each vector orthogonal to $\mathcal{H}_0$. Then R is a unitary operator on $\mathcal{H}$ and $ERE = ETE$. As in the last paragraph of the solution to Exercise 5.7.8(i), $\mathcal{U}$ is weak-operator dense in $(\mathcal{B}(\mathcal{H}))_1$.

(ii)   Suppose W is in the strong-operator closure of the set $\mathcal{I}$ of isometries of $\mathcal{H}$ into itself. Given a positive ε and a vector x in $\mathcal{H}$, there is an isometry V of $\mathcal{H}$ into itself such that $\|(W - V)x\| < \varepsilon$. Thus

$$\|x\| - \varepsilon = \|Vx\| - \varepsilon \leq \|Wx\| \leq \|Vx\| + \varepsilon = \|x\| + \varepsilon.$$

It follows that $\|Wx\| = \|x\|$ and $W \in \mathcal{I}$.

(iii)   If $\mathcal{H}$ is separable, the weak-operator topology on $(\mathcal{B}(\mathcal{H}))_1$ is metrizable (from Exercise 5.7.7). From (i), then, 0 is the weak-operator limit of a sequence $\{U_n\}$ of unitary operators. But 0 is not an isometry, so that 0 is not the strong-operator limit of $\{U_n\}$.

(iv)   Let $\{e_1, e_2, \ldots\}$ and $\{f_0, f_1, \ldots\}$ be orthonormal bases for $\mathcal{H}$. Let $U_n e_j$ be f_j if $j \leq n$, $U_n e_{n+1}$ be f_0, and $U_n e_k$ be f_{k-1} if $k > n + 1$, with U_n in $\mathcal{B}(\mathcal{H})$. Then $U_n \in \mathcal{U}$ and $\{U_n e_j\}$ tends to f_j as n tends to ∞, for each j. Thus $\{U_n\}$ tends in the strong-operator topology to W, where $W e_j = f_j$ for all j. Since $\langle f_0, Wx \rangle = 0$ for all x in $\mathcal{H}$, $W \notin \mathcal{U}$.   ∎

5.7.13.   Let $\mathcal{H}$ be an infinite-dimensional Hilbert space.
(i)   Show that $(\mathcal{H})_1$ is a weakly compact convex subset of $\mathcal{H}$ whose extreme points form a dense subset of it.
(ii)   Show that $(\mathcal{B}(\mathcal{H}))_1^+$ is a weak-operator compact convex subset of $\mathcal{B}(\mathcal{H})$ whose extreme points form a dense subset of it.
(iii)   Show that $(\mathcal{B}(\mathcal{H}))_1$ is a weak-operator compact convex subset of $\mathcal{B}(\mathcal{H})$ whose extreme points form a dense subset of it.

Solution.   (i)   With x in $(\mathcal{H})_1$, $x = cx_1$ for some orthonormal sequence $\{x_1, x_2, x_3, \ldots\}$ and some c in $[0, 1]$. We can define unit vectors $y_2, y_3, y_4, \ldots$ by $y_n = cx_1 + (1 - c^2)^{\frac{1}{2}} x_n$, and each is an extreme point of $(\mathcal{H})_1$ by Exercise 2.8.13(i). For each z in $\mathcal{H}$, $\sum_{n=1}^{\infty} |\langle z, x_n \rangle|^2 \leq \|z\|^2$, and thus $\langle z, x_n \rangle \to 0$. Hence, in the weak topology, $x_n \to 0$ and

$$y_n = cx_1 + (1 - c^2)^{\frac{1}{2}} x_n \to cx_1 = x.$$

(ii)   It follows from Exercise 5.7.11(ii) that $(\mathcal{B}(\mathcal{H}))_1^+$ is weak-operator compact and from Exercise 5.7.8(i) that $\mathcal{P}$ is weak-operator dense in $(\mathcal{B}(\mathcal{H}))_1^+$. From Exercise 2.8.14, each element of $\mathcal{P}$ is an extreme point of $(\mathcal{B}(\mathcal{H}))_1^+$.

(iii)   It follows from Theorem 5.1.3 that $(\mathcal{B}(\mathcal{H}))_1$ is weak-operator compact and from Exercise 5.7.12(i) that $\mathcal{U}$ is weak-operator dense in $(\mathcal{B}(\mathcal{H}))_1$. From Exercise 2.8.13(ii), each element of $\mathcal{U}$ is an extreme point of $(\mathcal{B}(\mathcal{H}))_1$.   ∎

5.7.14.   Let X be an extremely disconnnected compact Hausdorff space, and let $\{f_a : a \in \mathbf{A}\}$ be a family of real-valued functions in $C(X)$ bounded above by some constant.

(i)   Suppose that each f_a is the characteristic function of some clopen subset X_a of X. Show that $[\cup_{a \in \mathbf{A}} X_a]^-$ is a clopen set whose characteristic function $\vee_{a \in \mathbf{A}} f_a$ is the least upper bound of $\{f_a\}$ and that the interior of $\cap_{a \in \mathbf{A}} X_a$ is a clopen set whose characteristic function $\wedge_{a \in \mathbf{A}} f_a$ is the greatest lower bound of $\{f_a\}$ in $C(X)$.

(ii)   Show that

$$X \setminus \left[\bigcup_{a \in \mathbf{A}} \{x \in X : f_a(x) > \lambda\} \right]^- \quad (= X_\lambda)$$

is a clopen subset of X and that if Y is a clopen subset of X with the property that $f_a(p) \leq \lambda$ for all a in $\mathbf{A}$ and all p in Y, then $Y \subseteq X_\lambda$.

(iii)   Let e_λ be the characteristic function of X_λ. Let k be a constant that bounds $\{f_a\}$ above and such that $-k \leq f_{a'}$ for some a' in $\mathbf{A}$. Show that
  (1) $e_\lambda = 0$ for $\lambda < -k$ and $e_\lambda = 1$ for $\lambda > k$;
  (2) $e_\lambda \leq e_{\lambda'}$ if $\lambda \leq \lambda'$;
  (3) $e_\lambda = \wedge_{\lambda' > \lambda} e_{\lambda'}$.

(iv)   Show that $\int_{-k}^{k} \lambda \, de_\lambda$ converges in norm (in the sense of approximating Riemann sums) to a function f in $C(X)$ and that X_λ is the largest clopen set on which f takes values not exceeding λ.

(v)   Show that f is the least upper bound of $\{f_a\}$, and conclude that $C(X)$ is a boundedly complete lattice.

(vi)   Deduce that $C(X)$ is a boundedly complete lattice if and only if X is extremely disconnected.

Solution. (i)   Since $\cup_{a \in \mathbf{A}} X_a$ is open, its closure Y_0 is clopen. If g is an upper bound for $\{f_a\}$, then $1 \leq g(p)$ for each p in $\cup_{a \in \mathbf{A}} X_a$. By continuity of g, the same is true for each p in Y_0. Thus $\vee_{a \in \mathbf{A}} f_a$ is the least upper bound of $\{f_a\}$. Now $1 - f_a$ is the characteristic function of $X \setminus X_a$. Hence the least upper bound $1 - f_0$ of the family

$\{1 - f_a\}$ is the characteristic function of $[\cup_{a\in\mathbb{A}}(X \setminus X_a)]^-$. Thus f_0, the characteristic function of

$$X \setminus \left[\bigcup_{a\in\mathbb{A}}(X \setminus X_a)\right]^{-} \quad (= X \setminus \left[X \setminus \bigcap_{a\in\mathbb{A}} X_a\right]^{-}),$$

which is the interior of $\cap_{a\in\mathbb{A}} X_a$, is the greatest lower bound of $\{f_a\}$.

(ii)   Since X_λ is the complement in X of the closure of a union of open subsets of X, X_λ is clopen. If $p \in X_\lambda$, then for each a in $\mathbb{A}$, $p \notin \{x \in X : f_a(x) > \lambda\}$; that is, $f_a(p) \le \lambda$. By assumption on Y, $Y \subseteq X \setminus f_a^{-1}((\lambda,\infty))$ so that $f_a^{-1}((\lambda,\infty)) \subseteq X \setminus Y$. As Y is open, $X \setminus Y$ is closed. Thus

$$\left[\bigcup_{a\in\mathbb{A}} f_a^{-1}((\lambda,\infty))\right]^{-} \subseteq X \setminus Y,$$

and $Y \subseteq X_\lambda$.

(iii)   If $\lambda < -k$ and $p \in X_\lambda$, then $p \notin \{x \in X : f_{a'}(x) > \lambda\}$. But $-k \le f_{a'}$, so $\{x \in X : f_{a'}(x) > \lambda\} = X$. Thus $X_\lambda = \emptyset$ and $e_\lambda = 0$.

If $\lambda \ge k$ and $p \in X$, then $f_a(p) \le \lambda$ for all a in $\mathbb{A}$. Thus X is a clopen set on which all f_a take values not exceeding λ. From (ii), $X \subseteq X_\lambda$; so $X = X_\lambda$ and $e_\lambda = 1$.

If $\lambda \le \lambda'$, then

$$\{x \in X : f_a(x) > \lambda'\} \subseteq \{x \in X : f_a(x) > \lambda\}$$

and $X_\lambda \subseteq X_{\lambda'}$. Hence $e_\lambda \le e_{\lambda'}$.

Since $e_\lambda \le e_{\lambda'}$ when $\lambda \le \lambda'$, $e_\lambda \le \wedge_{\lambda'>\lambda} e_{\lambda'}$. Thus $X_\lambda \subseteq Y_\lambda$, where Y_λ is the set whose characteristic function is $\wedge_{\lambda'>\lambda} e_{\lambda'}$. Now $Y_\lambda \subseteq X_{\lambda'}$ for each λ' greater than λ. Thus if $p \in Y_\lambda$, $f_a(p) \le \lambda'$ for each such λ' and each a in $\mathbb{A}$. Hence $f_a(p) \le \lambda$ for all a in $\mathbb{A}$, and Y_λ is a clopen set on which all f_a take values not exceeding λ. From (ii), $Y_\lambda \subseteq X_\lambda$. Hence $Y_\lambda = X_\lambda$, and $e_\lambda = \wedge_{\lambda'>\lambda} e_{\lambda'}$.

(iv)   The proof of Theorem 5.2.4 (transferred to $C(X)$) applies to $\{e_\lambda\}$ to yield the fact that $\int_{-k}^{k} \lambda \, de_\lambda$ converges in norm to some f in $C(X)$. From the second paragraph of the proof of Theorem 5.2.3 (transferred to $C(X)$), $fe_\lambda \le \lambda e_\lambda$ and $\lambda(1 - e_\lambda) \le f(1 - e_\lambda)$. It follows that X_λ is the largest clopen set on which f takes values not exceeding λ (as in the first paragrph of that proof).

(v)   If $f(p) < f_a(p)$ for some p in X and some a in $\mathbf{A}$, there is a λ and a clopen set Y containing p such that $f(q) \leq \lambda$ and $\lambda < f_a(q)$ for each q in Y. From (iv), $p \in Y \subseteq X_\lambda$. But $p \in \{x \in X : f_a(x) > \lambda\}$, so $p \notin X_\lambda$, a contradiction. Thus $f_a \leq f$ for each a in $\mathbf{A}$, and f is an upper bound for $\{f_a\}$.

If g is an upper bound for $\{f_a\}$ and $g(p) < f(p)$ for some p in X, then, again, there is a λ and a clopen set Y containing p such that $g(q) \leq \lambda < f(q)$ for each q in Y. Since $f_a \leq g$ for all a in $\mathbf{A}$, Y is a clopen set on which all f_a take values not exceeding λ. From (ii), $p \in Y \subseteq X_\lambda$. From (iv), $f(p) \leq \lambda$, a contradiction. Thus $f \leq g$, and f is the least upper bound of $\{f_a\}$ in $C(X)$. It follows that $C(X)$ is a boundedly complete lattice.

(vi)   From Theorem 3.4.16 and (v), X is extremely disconnected if and only if $C(X)$ is a boundedly complete lattice. ∎   [105]

5.7.15.   Let X be a compact Hausdorff space.

(i)   Show that X is extremely disconnected if and only if disjoint open subsets have disjoint closures.

(ii)   Show that X is extremely disconnected if and only if it satisfies the following two conditions:

(a) X is totally disconnected;

(b)   the family $\mathcal{C}$ of clopen subsets of X partially ordered by inclusion is a complete lattice.

Solution. (i)   Suppose $\mathcal{O}_1$ and $\mathcal{O}_2$ are disjoint open subsets of X and X is extremely disconnected. Since O_2 is open, $X \backslash \mathcal{O}_2$ is closed and contains $\mathcal{O}_1^-$. Thus $X \backslash \mathcal{O}_1^-$ contains $\mathcal{O}_2$. Since X is extremely disconnected, $\mathcal{O}_1^-$ is open so that $X \backslash \mathcal{O}_1^-$ is closed. Hence $X \backslash \mathcal{O}_1^-$ contains $\mathcal{O}_2^-$ — that is, $\mathcal{O}_1^- \cap \mathcal{O}_2^- = \emptyset$.

Suppose now that disjoint open subsets of X have disjoint closures and let $\mathcal{O}$ be an open subset of X. Then $\mathcal{O}$ and $X \backslash \mathcal{O}^-$ are disjoint open subsets of X. By assumption, $\mathcal{O}^-$ is disjoint from the closure F of $X \backslash \mathcal{O}^-$. But $F \cup \mathcal{O}^- = X$ (since $X = (X \backslash \mathcal{O}^-) \cup \mathcal{O}^-$). Hence $\mathcal{O}^-$ is the complement in X of F and $\mathcal{O}^-$ is open. Thus X is extremely disconnected.

(ii)   Assume that X is totally disconnected and $\mathcal{C}$ is a complete lattice. Since X is a compact Hausdorff space in which points can be separated by *clopen* subsets of X, a standard compactness argument (replacing "open" by "clopen") shows that a point can be separated from a closed ($=$ compact) subset of X by clopen sets. In particular, each open set in X is a union of clopen sets. Let $\mathcal{O}_1$ and $\mathcal{O}_2$ be disjoint open subsets of X and let $\mathcal{C}_j$ be $\{X_0 \in \mathcal{C} : X_0 \subseteq \mathcal{O}_j\}$ for j in $\{1,2\}$. By assumption, $\mathcal{C}_1$ has a least upper bound X_1 in $\mathcal{C}$. If $X_0 \in \mathcal{C}_2$, then $X\backslash X_0$ is a clopen set, contains $\mathcal{O}_1$, and, hence, contains each element of $\mathcal{C}_1$. Thus $X_1 \subseteq X\backslash X_0$ and $X_0 \subseteq X\backslash X_1$. Since $\mathcal{O}_2 = \cup\mathcal{C}_2$, $\mathcal{O}_2 \subseteq X\backslash X_1$. But $X\backslash X_1$ is clopen so that $\mathcal{O}_2^- \subseteq X\backslash X_1$. As $\mathcal{O}_1 \subseteq X_1$ and X_1 is clopen, $\mathcal{O}_1^- \subseteq X_1$. Thus $\mathcal{O}_1^- \cap \mathcal{O}_2^- = \emptyset$. From (i), X is extremely disconnected.

Assume, now, that X is extremely disconnected. From Exercise 5.7.14(i), $\mathcal{C}$ is a complete lattice; and, of course X is totally disconnected. ∎

5.7.16.   With the notation of Exercises 3.5.4, 3.5.5, and 3.5.6:

(i) show that l_∞ is isometrically isomorphic to the maximal abelian algebra $\mathcal{A}$ of multiplication operators M_f $(f \in l_\infty)$ on $l_2(\mathbf{N},\mathbf{C})$ via the mapping $f \to M_f$;

(ii) show that the pure state space of the quotient C*-algebra $\mathcal{A}/\mathcal{C}_0$ (see Exercise 4.6.60) is (naturally homeomorphic to) $\beta(\mathbf{N})\backslash\mathbf{N}$, where $\mathcal{C}_0$ is the image of c_0 under the mapping $f \to M_f$.

Solution. (i)   The mapping $f \to M_f$ of l_∞ onto the multiplication algebra $\mathcal{A}$ of $l_2(\mathbf{N},\mathbf{C})$ is linear and multiplicative on the Banach algebra l_∞ described in Exercise 3.5.4(i), from Example 2.4.11. This same example assures us that $\|f\|_\infty = \|M_f\|$ so that the mapping $f \to M_f$ is an isometric isomorphism.

(ii) The pure states of the (abelian) C^*-algebra $\mathcal{A}$ are the non-zero multiplicative linear functionals on $\mathcal{A}$ and correspond, via the mapping $f \to M_f$, to the set of such functionals on l_∞, by Proposition 4.4.1. From Exercise 3.5.6(iv), the pure states of $\mathcal{A}$ that are 0 on $\mathcal{C}_0$ correspond to $\beta(\mathbf{N}) \backslash \mathbf{N}$. Exercise 4.6.23(iii), (iv) applies now, and $\rho \to \rho \circ \varphi$ is a one-to-one mapping of the set $\mathcal{P}'$ of pure states of $\mathcal{A}/\mathcal{C}_0$ onto the set $\mathcal{P}_0$ of pure states of $\mathcal{A}$ corresponding to $\beta(\mathbf{N}) \backslash \mathbf{N}$, where φ is the quotient mapping of $\mathcal{A}$ onto $\mathcal{A}/\mathcal{C}_0$. Since

$$|(\rho \circ \varphi)(A) - (\rho' \circ \varphi)(A)| = |\rho(\varphi(A)) - \rho'(\varphi(A))|,$$

the mapping $\rho \to \rho \circ \varphi$ is continuous from $\mathcal{P}'$ with its weak* topology to $\mathcal{P}_0$ with its weak* topology. Since $\mathcal{P}_0$ and $\mathcal{P}'$ are weak* compact, this mapping is a homeomorphism. Thus $\mathcal{P}'$ is (naturally homeomorphic to) $\beta(\mathbf{N}) \backslash \mathbf{N}$. ∎

5.7.17.   With the notation of Exercise 5.7.16, let E_0 be a projection in the quotient C^*-algebra $\mathcal{A}/\mathcal{C}_0$.

(i) Show that there is a projection E in $\mathcal{A}$ such that E maps onto E_0 under the quotient mapping.

(ii) Let Y_0 be a subset of $\beta(\mathrm{N})\setminus\mathrm{N}$ clopen in the relative topology on $\beta(\mathrm{N})\setminus\mathrm{N}$. Show that there is a clopen subset Y of $\beta(\mathrm{N})$ such that $Y\cap(\beta(\mathrm{N})\setminus\mathrm{N}) = Y_0$.

Solution. (i) There is an element K in $\mathcal{A}$ such that K maps onto E_0. Thus $K^2 - K$ maps onto 0 and $K^2 - K \in \mathcal{C}_0$. Let k be the function in $l_\infty(\mathrm{N},\mathbb{C})$ corresponding to K. Then $k^2 - k \in c_0$ and $\{k^2(n) - k(n)\}$ tends to 0 as $n \to 0$. Now $|k(n)| \cdot |k(n) - 1| = |k(n)^2 - k(n)|$ $(= \varepsilon(n)^2)$, so at least one of $|k(n)|$ or $|k(n) - 1|$ does not exceed $\varepsilon(n)$. If $|k(n)| \le \varepsilon(n)$, let $e(n)$ be 0; otherwise, let $e(n)$ be 1. Then $|k(n) - e(n)| \le \varepsilon(n)$. Since $\varepsilon(n) \to 0$ as $n \to \infty$, $k - e \in c_0$ and $K - E \in \mathcal{C}_0$, where E is the projection in $\mathcal{A}$ corresponding to e. It follows that the quotient mapping takes E onto E_0.

(ii) From Exercise 5.7.16(ii), $C(\beta(\mathrm{N})\setminus\mathrm{N})$ is isomorphic to $\mathcal{A}/\mathcal{C}_0$. The characteristic function of Y_0 corresponds to a projection E_0 in $\mathcal{A}/\mathcal{C}_0$. From (i), there is a projection E in $\mathcal{A}$ and an idempotent e in $l_\infty(\mathrm{N},\mathbb{C})$ corresponding to it such that E maps onto E_0. Thus the restriction of $\hat{e}$ to $\beta(\mathrm{N})\setminus\mathrm{N}$ is the characteristic function of Y_0. Let Y be the (clopen) subset of $\beta(\mathrm{N})$ on which e takes the value 1. Then $Y\cap(\beta(\mathrm{N})\setminus\mathrm{N}) = Y_0$. ∎

5.7.18.   With the notation of Exercise 3.5.5:

(i) show that $\mathcal{O}^- = (\mathcal{O}\cap\mathrm{N})^-$ for each open subset $\mathcal{O}$ of $\beta(\mathrm{N})$;

(ii) show that a subset Y_0 of $\beta(\mathrm{N})\setminus\mathrm{N}$ is clopen in $\beta(\mathrm{N})\setminus\mathrm{N}$ if and only if it has the form $\mathrm{N}_0^-\cap(\beta(\mathrm{N})\setminus\mathrm{N})$ for some subset N_0 of N, and that Y_0 is non-empty if and only if N_0 is infinite.

Solution. (i) Clearly $(\mathcal{O}\cap\mathrm{N})^- \subseteq \mathcal{O}^-$. Suppose $p \in \mathcal{O}^-$ and $\mathcal{O}_0$ is an open subset of $\beta(\mathrm{N})$ containing p. Then $\mathcal{O}_0\cap\mathcal{O}$ is a non-empty open subset of $\beta(\mathrm{N})$. From Exercise 3.5.6(ii), N is dense in $\beta(\mathrm{N})$, so that $\mathcal{O}_0\cap\mathcal{O}\cap\mathrm{N}$ is non-empty. It follows that $p \in (\mathcal{O}\cap\mathrm{N})^-$. Thus $\mathcal{O}^- = (\mathcal{O}\cap\mathrm{N})^-$.

(ii) From Exercise 3.5.6(iii), N_0 is an open subset of $\beta(\mathrm{N})$ so that N_0^- is clopen and $\mathrm{N}_0^-\cap(\beta(\mathrm{N})\setminus\mathrm{N})$ is a clopen subset of $\beta(\mathrm{N})\setminus\mathrm{N}$. If Y_0 is a clopen subset of $\beta(\mathrm{N})\setminus\mathrm{N}$, then $Y_0 = Y\cap(\beta(\mathrm{N})\setminus\mathrm{N})$ for some clopen subset Y of $\beta(\mathrm{N})$, from Exercise 5.7.17(ii). From (i),

$Y = \mathbf{N}_0^-$, where $\mathbf{N}_0 = Y \cap \mathbf{N}$. Thus $Y_0 = \mathbf{N}_0^- \cap (\beta(\mathbf{N}) \setminus \mathbf{N})$.

If $\mathbf{N}_0$ is finite, then $\mathbf{N}_0$ is closed (since $\beta(\mathbf{N})$ is a Hausdorff space), and

$$\mathbf{N}_0^- \cap (\beta(\mathbf{N}) \setminus \mathbf{N}) = \mathbf{N}_0 \cap (\beta(\mathbf{N}) \setminus \mathbf{N}) = \emptyset.$$

If $\mathbf{N}_0$ is infinite, say $\mathbf{N}_0 = \{n_1, n_2, \ldots\}$, where $n_1 < n_2 < \ldots$, then the equation, $\rho_0(f) = \lim_j f(n_j)$, defines a pure state ρ_0 of the C^*-subalgebra of $l_\infty(\mathbf{N}, \mathbb{C})$ consisting of those f in l_∞ such that $\{f(n_j)\}$ tends to a limit as j tends to ∞. (Compare Exercise 4.6.56.) Let ρ be a pure state of l_∞ extending ρ_0. Then ρ is a point of $\beta(\mathbf{N}) \setminus \mathbf{N}$ (as in the solution to Exercise 4.6.56(ii)). Let $\mathcal{O}$ be an open subset of $\beta(\mathbf{N})$ containing ρ and let f be a funtion in l_∞ that is 1 at ρ and 0 outside of $\mathcal{O}$. Since $\rho(f) \neq 0$, $f(n_j) \neq 0$ for some j, and $n_j \in \mathcal{O}$. Thus $\rho \in \mathbf{N}_0^-$. Hence $\mathbf{N}_0^- \cap (\beta(\mathbf{N}) \setminus \mathbf{N})$ is non-empty. ∎

5.7.19.   Let φ be a one-to-one mapping of the set of rational numbers onto $\mathbf{N}$. For each real number t, choose a sequence $\{r_1, r_2, \ldots\}$ of distinct rational numbers tending to t. With the notation of Exercise 5.7.16, let $\mathbf{N}_t$ be $\{\varphi(r_1), \varphi(r_2), \ldots\}$ and let Y_t be $\mathbf{N}_t^- \cap (\beta(\mathbf{N}) \setminus \mathbf{N})$.

(i)   Show that Y_t and Y_s are disjoint, non-empty, clopen subsets of $\beta(\mathbf{N}) \setminus \mathbf{N}$ when $s \neq t$.

(ii)   Let S be a subset of $\mathbb{R}$ and let Y_S be the closure of $\cup_{s \in S} Y_s$. Show that if $t \notin S$, then $Y_t \cap Y_S = \emptyset$.

(iii)   Show that Y_S is not a clopen subset of $\beta(\mathbf{N}) \setminus \mathbf{N}$ for some subset S of $\mathbb{R}$. [*Hint.* "Count" subsets of $\mathbb{R}$ and of $\mathbf{N}$ and use Exercise 5.7.18(ii).]

(iv)   Deduce that $\beta(\mathbf{N}) \setminus \mathbf{N}$ is totally disconnected but not extremely disconnected and that $\mathcal{A}/\mathcal{C}_0$ is not a boundedly complete lattice.

Solution. (i)   Since each $\mathbf{N}_t$ is infinite, Y_t is a non-empty clopen subset of $\beta(\mathbf{N}) \setminus \mathbf{N}$, from Exercise 5.7.18(ii). When $t \neq s$, $\mathbf{N}_t \cap \mathbf{N}_s \ (= \mathbf{N}')$ is finite. Thus $(\mathbf{N}_t \setminus \mathbf{N}')^- = \mathbf{N}_t^- \setminus \mathbf{N}'$ and $(\mathbf{N}_s \setminus \mathbf{N}')^- = \mathbf{N}_s^- \setminus \mathbf{N}'$. It follows that

$$(\mathbf{N}_t \setminus \mathbf{N}')^- \cap (\beta(\mathbf{N}) \setminus \mathbf{N}) = (\mathbf{N}_t^- \setminus \mathbf{N}') \cap (\beta(\mathbf{N}) \setminus \mathbf{N})$$
$$= \mathbf{N}_t^- \cap (\beta(\mathbf{N}) \setminus \mathbf{N}) = Y_t$$

and, similarly,
$$(\mathbf{N}_s \setminus \mathbf{N}')^- \cap (\beta(\mathbf{N}) \setminus \mathbf{N}) = Y_s.$$

Now $N_t \setminus N'$ and $N_s \setminus N'$ are disjoint open subsets of $\beta(N)$ so that, from Exercise 5.7.15, their closures are disjoint. Hence Y_s and Y_t are disjoint, when $s \neq t$.

(ii)   From (i), Y_t is an open subset of $\beta(N) \setminus N$ disjoint from $\cup_{s \in S} Y_s$. Thus $Y_t \cap Y_S = \emptyset$.

(iii)   If Y_S is a clopen subset of $\beta(N) \setminus N$, then from Exercise 5.7.18(ii), there is a subsete N_S of N such that $N_S^- \cap (\beta(N) \setminus N) = Y_S$. From (ii), if S and T are distinct subsets of $\mathbb{R}$, $Y_S \neq Y_T$. Thus, if Y_S and Y_T are clopen, $N_S \neq N_T$. Since $\mathbb{R}$ has cardinality c greater than $\aleph_0$ and there are $2^c (> 2^{\aleph_0})$ subsets of $\mathbb{R}$, for some subset T of $\mathbb{R}$, Y_T is not clopen.

(iv)   Since $\cup_{t \in T} Y_t$ is an open subset of $\beta(N) \setminus N$ and its closure Y_T is not clopen (for some subset T of $\mathbb{R}$, from (iii)), $\beta(N) \setminus N$ is not extremely disconnected. From Theorem 3.4.16 and Exercise 5.7.16(ii), $\mathcal{A}/\mathcal{C}_0$ is not a boundedly complete lattice. Since $\beta(N)$ is totally disconnected, $\beta(N) \setminus N$ is totally disconnected.   ∎

5.7.20.   Let X be a complete metric space. We say that an open subset of X is *regular* when it coincides with the interior of its closure.

(i)   Show that the interiors of closed (hence of the closures and complements of open) sets in X are regular.

(ii)   Show that each open subset of X differs from a regular open subset on a meager set.

(iii)   Show that each Borel subset of X differs from a regular open subset on a meager (Borel) set. [*Hint.* Follow the pattern of the argument of the first paragraph of the proof of Lemma 5.2.10.]

(iv)   Show that there is a *unique* regular open subset of X that differs from a given Borel set on a meager (Borel) set.

(v)   Let $\mathcal{F}_0$ be the family of regular open subsets of X partially ordered by inclusion. Show that $\mathcal{F}_0$ is a complete lattice.

(vi)   Let $\mathcal{F}$ be the family of Borel subsets of X and $\mathcal{M}$ the σ-ideal of meager Borel subsets of X (a countable union of sets in $\mathcal{M}$ is in $\mathcal{M}$ and the intersection of a set of $\mathcal{M}$ with any set of $\mathcal{F}$ is in $\mathcal{M}$). Let $\mathcal{F}/\mathcal{M}$ be the family of equivalence classes of sets in $\mathcal{F}$ under the relation $S \sim S'$ when S and S' differ by a meager set. With $\mathcal{S}$ and $\mathcal{S}'$ in $\mathcal{F}/\mathcal{M}$, define $\mathcal{S} \preceq \mathcal{S}'$ when $S \subseteq S'$ for some S in $\mathcal{S}$ and S' in $\mathcal{S}'$. Show that $\preceq$ is a partial ordering of $\mathcal{F}/\mathcal{M}$ (the "quotient" of "inclusion" on $\mathcal{F}$ by the ideal $\mathcal{M}$), that each $\mathcal{S}$ in $\mathcal{F}/\mathcal{M}$ contains precisely one regular open set, and that the mapping that assigns to each $\mathcal{S}$ in $\mathcal{F}/\mathcal{M}$ the regular open set it contains is an order isomorphism of $\mathcal{F}/\mathcal{M}$ onto $\mathcal{F}_0$. Conclude that $\mathcal{F}/\mathcal{M}$ is a complete lattice.

(vii)   Show that the algebra $\mathcal{B}(X)$ of bounded Borel functions on X is a commutative C^*-algebra and that the family $\mathcal{M}_0$ of functions in $\mathcal{B}(X)$ that vanish on the completment of a meager Borel set is a closed ideal in $\mathcal{B}(X)$. Conclude that $\mathcal{B}(X)/\mathcal{M}_0$ is a commutative C^*-algebra.

(viii)   Let Y be the compact Hausdorff space such that $\mathcal{B}(X)/\mathcal{M}_0 \cong C(Y)$. Show that Y is totally disconnected and that the family of clopen subsets of Y, partially ordered by inclusion, form a complete lattice. Conclude that Y is extremely disconnected and that $C(Y)$ (and $\mathcal{B}(X)/\mathcal{M}_0$) are boundedly complete lattices.

Solution.   (i)   Let Y be a closed subset of X and $\mathcal{O}$ be its interior. Since $\mathcal{O}$ is an open subset of X contained in $\mathcal{O}^-$, $\mathcal{O}$ is contained in the interior $\mathcal{O}_0$ of $\mathcal{O}^-$. Since $\mathcal{O}_0 \subseteq \mathcal{O}^- \subseteq Y$ and $\mathcal{O}_0$ is an open subset of X, $\mathcal{O}_0$ is contained in the interior $\mathcal{O}$ of Y. Thus $\mathcal{O} = \mathcal{O}_0$, $\mathcal{O}$ is the interior of $\mathcal{O}^-$, and $\mathcal{O}$ is regular.

(ii)   Let $\mathcal{O}$ be an open subset of X and $\mathcal{O}_0$ be the interior of $\mathcal{O}^-$. Then

$$(\mathcal{O}_0 \setminus \mathcal{O}) \cup (\mathcal{O} \setminus \mathcal{O}_0) = \mathcal{O}_0 \setminus \mathcal{O} \subseteq \mathcal{O}^- \setminus \mathcal{O}.$$

But $\mathcal{O}^- \setminus \mathcal{O}$ is a (closed) nowhere-dense set. Hence $\mathcal{O}$ and $\mathcal{O}_0$ differ on a meager subset of X (that is, $\mathcal{O} \sim \mathcal{O}_0$). From (i), $\mathcal{O}_0$ is regular.

(iii)   Let $\mathcal{F}'$ be the family of Borel subsets of X that differ from a regular open set by a meager (Borel) set. If $S \in \mathcal{F}'$ and $\mathcal{O}_0$ is a regular open set such that $S \sim \mathcal{O}_0$, then $(S \setminus \mathcal{O}_0) \cup (\mathcal{O}_0 \setminus S)$ $(= [(X \setminus S) \setminus (X \setminus \mathcal{O}_0)] \cup [(X \setminus \mathcal{O}_0) \setminus (X \setminus S)])$ is meager. Thus $X \setminus S \sim X \setminus \mathcal{O}_0$. From (i), the interior $\mathcal{O}_1$ of $X \setminus \mathcal{O}_0$ is regular, and $\mathcal{O}_1 \sim X \setminus \mathcal{O}_0 \sim X \setminus S$. Thus $X \setminus S \in \mathcal{F}'$.

Suppose $S_1, S_2, \ldots$ are in $\mathcal{F}'$. Let $\mathcal{O}_j$ be a regular open set such that $S_j \sim \mathcal{O}_j$. As in the proof of Lemma 5.2.10, $\cup_{j=1}^{\infty} S_j \sim \cup_{j=1}^{\infty} \mathcal{O}_j$. From (i), the interior $\mathcal{O}_0$ of $(\cup_{j=1}^{\infty} \mathcal{O}_j)^-$ is regular and $\mathcal{O}_0 \sim (\cup_{j=1}^{\infty} \mathcal{O}_j)^- \sim \cup_{j=1}^{\infty} S_j$. Thus $\cup_{j=1}^{\infty} S_j \in \mathcal{F}'$ and $\mathcal{F}'$ is a σ-algebra containing the open sets and contained in $\mathcal{F}$. It follows that $\mathcal{F}' = \mathcal{F}$.

(iv)   If $S \in \mathcal{F}$, $S \sim \mathcal{O}_1$, $S \sim \mathcal{O}_2$, and $\mathcal{O}_1$ and $\mathcal{O}_2$ are regular open sets, then $\mathcal{O}_1 \sim \mathcal{O}_2$. Since $\mathcal{O}_2^-$ is closed, if some p in $\mathcal{O}_1$ is not in $\mathcal{O}_2^-$, then some open set $\mathcal{O}$ containing p does not meet $\mathcal{O}_2^-$, and $\mathcal{O} \cap \mathcal{O}_1 \subseteq \mathcal{O}_1 \setminus \mathcal{O}_2$. But $\mathcal{O}_1 \setminus \mathcal{O}_2$ is meager and meager sets in a complete metric space have null interior. Thus $\mathcal{O}_1 \subseteq \mathcal{O}_2^-$ and $\mathcal{O}_1$ is contained in the interior $\mathcal{O}_2$ of $\mathcal{O}_2^-$. Symmetrically, $\mathcal{O}_2 \subseteq \mathcal{O}_1$ and $\mathcal{O}_1 = \mathcal{O}_2$.

(v)   Suppose $\mathcal{O}_a \in \mathcal{F}_0$ for a in $\mathbf{A}$. Let $\mathcal{O}_1$ be the interior of $(\cup_{a\in\mathbf{A}}\mathcal{O}_a)^-$. Then $\mathcal{O}_1 \in \mathcal{F}_0$, from (i), since $\cup_{a\in\mathbf{A}}\mathcal{O}_a$ is open. Clearly $\mathcal{O}_1$ is an upper bound for $\{\mathcal{O}_a : a \in \mathbf{A}\}$. If $\mathcal{O}$ is another upper bound, then $(\cup_{a\in\mathbf{A}}\mathcal{O}_a)^- \subseteq \mathcal{O}^-$ and the interior $\mathcal{O}$ of $\mathcal{O}^-$ contains the interior $\mathcal{O}_1$ of $(\cup_{a\in\mathbf{A}}\mathcal{O}_a)^-$. Thus $\mathcal{O}_1$ is the least upper bound of $\{\mathcal{O}_a : a \in \mathbf{A}\}$ and $\mathcal{F}_0$ is a complete lattice.

(vi)   As defined, the relation $\precsim$ is clearly reflexive and transitive. Suppose $\mathcal{S} \precsim \mathcal{S}'$ and $\mathcal{S}' \precsim \mathcal{S}$, with $\mathcal{S}$ and $\mathcal{S}'$ in $\mathcal{F}/\mathcal{M}$. Then there are S_1 and S_2 in $\mathcal{S}$ and S_1' and S_2' in $\mathcal{S}'$ such that $S_1 \subseteq S_1'$ and $S_2' \subseteq S_2$. Let M be $(S_1 \setminus S_2) \cup (S_2 \setminus S_1)$ and M' be $(S_1' \setminus S_2') \cup (S_2 \setminus S_1')$. Since $S_1 \sim S_2$ and $S_1' \sim S_2'$; M, M', and $M \cup M'$ are meager, and $S_1 \cup M = S_2 \cup M$, $S_1' \cup M' = S_2' \cup M'$. Thus

$$S_1' \cup M' \cup M = S_2' \cup M' \cup M \subseteq S_2 \cup M' \cup M$$
$$= S_1 \cup M' \cup M \subseteq S_1' \cup M' \cup M,$$

so that $S_1 \sim S_1'$ and $\mathcal{S} = \mathcal{S}'$. Hence $\precsim$ is a partial ordering of $\mathcal{F}/\mathcal{M}$.

From (iii) and (iv), each S in $\mathcal{F}$ differs from a unique regular open set $\mathcal{O}$ by a meager set. Thus the equivalence class $\mathcal{S}$ of S contains $\mathcal{O}$ and no other regular open set. If $\mathcal{S}'$ is another equivalence class and $\mathcal{O}'$ is the regular open set it contains, then $\mathcal{S} \precsim \mathcal{S}'$ if $\mathcal{O} \subseteq \mathcal{O}'$, by definition of $\precsim$. Conversely, if $\mathcal{S} \precsim \mathcal{S}'$, then $\mathcal{O} \cup M \subseteq \mathcal{O}' \cup M'$, where M and M' are in $\mathcal{M}$. Thus $\mathcal{O} \subseteq \mathcal{O}' \cup M'$ so that $\mathcal{O} \setminus \mathcal{O}'^- \subseteq \mathcal{O} \setminus \mathcal{O}' \subseteq M'$. Since $\mathcal{O} \setminus \mathcal{O}'^-$ is open and M' is meager, $\mathcal{O} \setminus \mathcal{O}'^- = \emptyset$, that is $\mathcal{O} \subseteq \mathcal{O}'^-$. Hence $\mathcal{O}$ is contained in the interior $\mathcal{O}'$ of $\mathcal{O}'^-$. It follows that the mapping $\mathcal{S} \to \mathcal{O}$ of $\mathcal{F}/\mathcal{M}$ onto $\mathcal{F}_0$ is an order isomorphism and, from (v), $\mathcal{F}/\mathcal{M}$ is a complete lattice.

(vii)   If $\mathcal{B}(X)$ is provided with the supremum norm it becomes a Banach algebra. The operation of complex conjugation of functions is an adjoint operation on $\mathcal{B}(X)$. Since $\|\bar{f}f\| = \||f|^2\| = \||f|\|^2 = \|f\|^2$, $\mathcal{B}(X)$ with the given norm and adjoint operation is a C^*-algebra. If f_1 and f_2 in $\mathcal{M}_0$ vanish outside of the meager sets M_1 and M_2, respectively, then $f_1 + f_2$ vanishes outside $M_1 \cup M_2$, a meager subset of X, and ff_1 vanishes outside M_1 for each f in $\mathcal{B}(X)$. Thus $\mathcal{M}_0$ is an ideal in $\mathcal{B}(X)$. If $f_n \in \mathcal{M}_0$ and $\|f - f_n\| \to 0$, then f vanishes outside $\cup_{n=1}^{\infty} M_n$, a meager set, where M_n is a meager set on the complement of which f_n vanishes. Thus $f \in \mathcal{M}_0$ and $\mathcal{M}_0$ is a closed (two-sided) ideal in $\mathcal{B}(X)$. From Exercise 4.6.60, $\mathcal{B}(X)/\mathcal{M}_0$ is a C^*-algebra.

(viii)   Let $\mathcal{S}$ be in $\mathcal{F}/\mathcal{M}$ and e be the characteristic function of a set in $\mathcal{S}$. Define $\eta(\mathcal{S})$ to be the projection in $\mathcal{B}(X)/\mathcal{M}_0$ that is the image of e under the quotient mapping. If e' corresponds to another set in $\mathcal{S}$, then $e - e' \in \mathcal{M}_0$ so that e and e' have the same image in $\mathcal{B}(X)/\mathcal{M}_0$ and $\eta(\mathcal{S})$ is well defined. If $\mathcal{S} \precsim \mathcal{S}'$, there are sets S in $\mathcal{S}$ and S' in $\mathcal{S}'$ such that $S \subseteq S'$. With e and e' the characteristic functions of S and S', respectively, $e \leq e'$ so that $\eta(\mathcal{S}) \leq \eta(\mathcal{S}')$.

Let E be a projection in $\mathcal{B}(X)/\mathcal{M}_0$ and f be an element of $\mathcal{B}(X)$ mapping onto E. Then $f^2 - f$ maps onto 0 and $f^2 - f$ vanishes outside some meager Borel set M. Let $e(p)$ be $f(p)$ for p in $X \setminus M$ and 0 for p in M. Then e is an idempotent in $\mathcal{B}(X)$ so that e is the characteristic function of a set S in $\mathcal{F}$. If $\mathcal{S}$ in $\mathcal{F}/\mathcal{M}$ is the equivalence class of S, then $\eta(\mathcal{S}) = E$. Hence η is an order-preserving mapping of $\mathcal{F}/\mathcal{M}$ *onto* the set $\mathcal{P}$ of projections in $\mathcal{B}(X)/\mathcal{M}_0$.

If E and E' are in $\mathcal{P}$ and $E \leq E'$, there are $\mathcal{S}$ and $\mathcal{S}'$ in $\mathcal{F}/\mathcal{M}$ such that $\eta(\mathcal{S}) = E$ and $\eta(\mathcal{S}') = E'$. By definition of η, there are sets S and S' in $\mathcal{S}$ and $\mathcal{S}'$ whose characteristic functions e and e' map onto E and E', respectively. Thus $e - ee'$ maps onto $E - EE' (= 0)$ and $e - ee'$ is 0 on $X \setminus M'$ for some meager set M'. It follows that $S \setminus S' \subseteq M'$ so that $S \subseteq S' \cup M'$. Since $S' \cup M' \in \mathcal{S}'$, $\mathcal{S} \precsim \mathcal{S}'$. Hence η is one-to-one, for if $\eta(\mathcal{S}) = \eta(\mathcal{S}') = E$, then $\mathcal{S} \precsim \mathcal{S}'$ and $\mathcal{S}' \precsim \mathcal{S}$ so that $\mathcal{S} = \mathcal{S}'$ from (vi). It follows, too, that η^{-1} is order preserving.

If φ is the isomorphism of $\mathcal{B}(X)/\mathcal{M}_0$ onto $C(Y)$, then $\varphi \circ \eta$ is an order isomorphism of $\mathcal{F}/\mathcal{M}$ with the set of idempotents $\mathcal{P}'$ in $C(Y)$. From (vi), $\mathcal{P}'$ is a complete lattice. Each function in $\mathcal{B}(X)$ is approximable in norm as closely as we wish by step functions. Thus linear combinations of idempotents lie dense in $\mathcal{B}(X)$, $\mathcal{B}(X)/\mathcal{M}_0$, and in $C(Y)$. Hence Y is totally disconnected. From Exercise 5.7.15(ii), Y is extremely disconnected. Thus $C(Y)$ and $\mathcal{B}(X)/\mathcal{M}_0$ are boundedly complete lattices, by Exercise 5.7.14(vi). ∎ [13,26]

5.7.21.   With the notation of Exercise 5.7.20, assume that X is $[0, 1]$ and let ρ be a state of $C(Y)$.

(i)   Suppose $\rho(\vee_{n=1}^{\infty} e_n) = \sum_{n=1}^{\infty} \rho(e_n)$ whenever $\{e_n\}$ is a countable family of idempotents in $C(Y)$ such that $e_n \cdot e_{n'} = 0$ unless $n = n'$. (We say that ρ is a *normal* state in this case.) Show that $\rho(\vee_{n=1}^{\infty} f_n) \leq \sum_{n=1}^{\infty} \rho(f_n)$ for each countable set $\{f_n\}$ of idempotents f_n in $C(Y)$ (where "$a \leq +\infty$" is envisaged in the inequality of this assertion).

(ii)   Enumerate the open intervals in $[0,1] (= X)$ with rational endpoints and let $f_1, f_2, \ldots$ be the idempotents in $C(Y)$ that are the images of their characteristic functions (in $\mathcal{B}(X)$) under the composition of the quotient mapping of $\mathcal{B}(X)$ onto $\mathcal{B}(X)/\mathcal{M}_0$ and the isomorphism of $\mathcal{B}(X)/\mathcal{M}_0$ with $C(Y)$. For each j in $\{1, 2, \ldots\}$, let e_j be an idempotent in $C(Y)$ such that $0 < e_j \leq f_j$. Show that $\bigvee_{j=1}^{\infty} e_j = 1$.

(iii)   With the notation of (ii) and given a positive ε, show that e_j can be chosen such that $\rho(e_j) \leq 2^{-j}\varepsilon$. Conclude that $C(Y)$ has no normal states.

(iv)   Deduce that $C(Y)$ is isomorphic to no abelian von Neumann algebra although Y is extremely disconnected.

Solution. (i)   Let f_1' be f_1 and f_n' be $f_1 \vee \cdots \vee f_n - f_1 \vee \cdots \vee f_{n-1}$ for n in $\{2, 3, \ldots\}$. If $m < n$, then $f_m' \leq f_1 \vee \cdots \vee f_{n-1}$ so that $f_m' \cdot f_n' = 0$. Moreover, $f_1' + \cdots + f_n' = f_1 \vee \cdots \vee f_n$ for each n in $\{1, 2, \ldots\}$ so that $\bigvee_{n=1}^{\infty} f_n = \sum_{n=1}^{\infty} f_n'$ and

$$\rho\left(\bigvee_{n=1}^{\infty} f_n\right) = \rho\left(\sum_{n=1}^{\infty} f_n'\right) = \sum_{n=1}^{\infty} \rho(f_n').$$

From Exercise 2.8.31(i), $f_1 \vee \cdots \vee f_n \leq f_1 + \cdots + f_n$ so that

$$\sum_{n=1}^{m} \rho(f_n') = \rho(f_1 \vee \cdots \vee f_m) \leq \sum_{n=1}^{m} \rho(f_n) \leq \sum_{n=1}^{\infty} \rho(f_n).$$

Hence

$$\rho\left(\bigvee_{n=1}^{\infty} f_n\right) = \sum_{n=1}^{\infty} \rho(f_n') \leq \sum_{n=1}^{\infty} \rho(f_n).$$

(ii)   With the notation (and results) of the solution to Exercise 5.7.20(viii), let $\mathcal{S}_j$ be $(\varphi \circ \eta)^{-1}(e_j)$. From Exercise 5.7.20(iii), $\mathcal{S}_j$ contains a regular open set $\mathcal{O}_j$. Let $\mathcal{O}$ be $\cup_{j=1}^{\infty} \mathcal{O}_j$. If $p \in [0,1]\backslash\mathcal{O}^-$, then some open interval (a, b) with rational endpoints contains p and does not meet $\mathcal{O}$. Let $\mathcal{S}$ be the equivalence class of (a, b) in $\mathcal{F}/\mathcal{M}$ and f_j be $(\varphi \circ \eta)(\mathcal{S})$. Since $0 < e_j \leq f_j$, $\mathcal{S}_j \precsim \mathcal{S}$. Now (a, b) is regular and from Exercise 5.7.20(iv), (a, b) is the only regular set in $\mathcal{S}$. From Exercise 5.7.20(vi), $\mathcal{O}_j \subseteq (a, b)$, contradicting the choice of (a, b) (not meeting $\mathcal{O}$). Thus $\mathcal{O}^- = [0,1]$, $\bigvee_{j=1}^{\infty} \mathcal{O}_j = [0,1]$, and $\bigvee_{j=1}^{\infty} e_j = 1$.

(iii)   We note, first, that no non-zero idempotent f in $C(Y)$ is minimal. If f were minimal, $(\varphi \circ \eta)^{-1}(f)\,(= \mathcal{S})$ would be minimal in $\mathcal{F}/\mathcal{M}$ and the regular open set $\mathcal{O}$ in $\mathcal{S}$ would be non-empty and minimal in $\mathcal{F}_0$. But $\mathcal{O}$ contains some open interval (a, b) as a proper subset and as noted in (ii), (a, b) is regular. Thus we can choose an idempotent f' in $C(Y)$ such that $0 < f' < f$. One of $\rho(f')$ and $\rho(f - f')$ is not greater than $\frac{1}{2}\rho(f)$. Continuing this "division" process, we find an idempotent f'' in $C(Y)$ such that $0 < f'' < f$ and $\rho(f'') < \varepsilon$. Applying this to f_j, we find e_j as described.

(iv)   From (i), and with the notation of (iii),

$$1 = \rho(1) = \rho\Big(\bigvee_{j=1}^{\infty} e_j \Big) \leq \sum_{j=1}^{\infty} \rho(e_j) \leq \sum_{j=1}^{\infty} 2^{-j}\varepsilon = \varepsilon,$$

a contradiction. Thus $C(Y)$ has no normal states. Since vector states of an abelian von Neumann algebra are normal, $C(Y)$ is isomorphic to no such algebra. ■   [13(Thm. 12,Cor. 1, p.186),26(Sect. 7)]

5.7.22.   Let A be an abelian von Neumann algebra acting on a Hilbert space $\mathcal{H}$, and let A be an operator in $\mathcal{A}$. Suppose $\mathcal{A} \cong C(X)$, where X is an extremely disconnected compact Hausdorff space, and f in $C(X)$ represents A. Show that $Ax = \lambda x$ for some unit vector x in $\mathcal{H}$ and some complex number λ if and only if $f^{-1}(\lambda)$ contains a non-empty clopen subset of X.

Solution.   Let E be the projection on the subspace $\mathcal{V}$ of all vectors y in $\mathcal{H}$ such that $Ay = \lambda y$. Suppose $T \in \mathcal{A}'$. Then $TA = AT$ and $TA^* = A^*T$ so that $T^*A = AT^*$ and $T^*A^* = A^*T^*$. With y in $\mathcal{V}$, $ATy = TAy = \lambda Ty$ and $Ty \in \mathcal{V}$. Similarly $T^*y \in \mathcal{V}$. Thus $TE = ETE$ and $T^*E = ET^*E$. It follows that

$$TE = ETE = (ET^*E)^* = (T^*E)^* = ET,$$

so that $E \in (\mathcal{A}')'$. From the double commutant theorem, $E \in \mathcal{A}$. Of course, $AE = \lambda E$ and E corresponds to the characteristic function e of a clopen subset X_0 of X such that $fe = \lambda e$.

If there are x and λ as described, then $E \neq 0$, $X_0 \neq \emptyset$, and $X_0 \subseteq f^{-1}(\lambda)$. If $f^{-1}(\lambda)$ contains a non-null clopen set X_0, the characteristic function of that set corresponds to a non-zero projection E in $\mathcal{A}$ such that $AE = \lambda E$. Thus $Ax = \lambda x$ for each unit vector x in the range of E.   ■

5.7.23.   Let $\mathcal{R}$ be a von Neumann algebra and $\mathcal{P}$ be its family of projections. Show that, with ρ_0 in $\mathcal{R}^\#$, the family of all sets,

$$\{\rho \in \mathcal{R}^\# : |(\rho - \rho_0)(E_j)| < \varepsilon,\ j \in \{1,\ldots,n\}\}\ (= \mathcal{V}(E_1,\ldots,E_n,\varepsilon)),$$

where $\varepsilon > 0$ and $\{E_1,\ldots,E_n\} \subseteq \mathcal{P}$, is a base for the open neighborhoods of ρ_0 in the weak* topology on a bounded subset of $\mathcal{R}^\#$.

Solution. Since

$$\mathcal{V}(E_1,\ldots,E_n,F_1,\ldots,F_m,\varepsilon'') \subseteq \mathcal{V}(E_1,\ldots,E_n,\varepsilon) \cap \mathcal{V}(F_1,\ldots,F_m,\varepsilon'),$$

where $\varepsilon'' = \min\{\varepsilon,\varepsilon'\}$, it will suffice to show that each subbasic weak* open neighborhood of ρ_0,

$$\{\rho \in (\mathcal{R}^\#)_k : |(\rho - \rho_0)(A)| < 1\},$$

where $A \in \mathcal{R}$ and $(\mathcal{R}^\#)_k$ is the closed ball in $\mathcal{R}^\#$ with center 0 and radius k, contains some $\mathcal{V}(E_1,\ldots,E_n,\varepsilon) \cap (\mathcal{R}^\#)_k$. Let A be $A_1 + iA_2$, with A_1 and A_2 self-adjoint operators in $\mathcal{R}$. From Theorem 5.2.2(v), we can approximate A_1 and A_2 as closely as we wish in norm by finite linear combinations of elements of $\mathcal{P}$. Thus, there is such a combination $A_0\ (= a_1 E_1 + \cdots + a_n E_n)$ that approximates A in norm to within $\frac{1}{4k}$. Suppose $\rho \in \mathcal{V}(E_1,\ldots,E_n,\varepsilon) \cap (\mathcal{R}^\#)_k$, where $\varepsilon = \frac{1}{2an}$ and $a = \max\{|a_1|,\ldots,|a_n|\}$. Then

$$|(\rho - \rho_0)(A)| \leq |(\rho - \rho_0)(A - A_0)| + |(\rho - \rho_0)(A_0)|$$

$$\leq 2k(\frac{1}{4k}) + \sum_{j=1}^{n} |a_j|\,|(\rho - \rho_0)(E_j)|$$

$$< \frac{1}{2} + \frac{na}{2na} = 1. \qquad \blacksquare$$

5.7.24.   Let $\mathcal{R}$ be a von Neumann algebra acting on a Hilbert space $\mathcal{H}$.

(i)   Let φ be a representation of a C^*-algebra $\mathfrak{A}$ with image $\mathcal{R}$, and let V be a unitary operator in $\mathcal{R}$. Show that there is a unitary operator U in $\mathfrak{A}$ such that $\varphi(U) = V$.

(ii)   Show that the unitary group $\mathcal{R}_u$ of $\mathcal{R}$ is (pathwise) connected (in its norm topology).

Solution. (i)   From Theorem 5.2.5, the abelian von Neumann algebra generated by V and V^* contains a positive operator H such that $V = \exp iH$ (and $H \in \mathcal{R}$).   From Exercise 4.6.2, there is a unitary operator U in $\mathfrak{A}$ such that $\varphi(U) = V$.

(ii)   With the notation of (i), the mapping $t \to \exp itH$ of $[0,1]$ into $\mathcal{R}_u$ is norm continuous by uniform continuity of the mapping $s \to \exp(is)$ on $[0,2\pi]$. This mapping is a path in $\mathcal{R}_u$ with endpoints I and V. Thus $\mathcal{R}_u$ is pathwise connected.   ■

5.7.25.   Let S be a locally compact topological space, $\mathcal{S}$ the σ-algebra of Borel sets, and m a σ-finite regular Borel measure on S. Let $\mathfrak{A}$ be the algebra of multiplications by bounded continuous functions on $L_2(S, m)$ and $\mathcal{A}$ be its weak-operator closure. Show that $\mathcal{A}$ is the algebra of multiplications by essentially bounded measurable functions on S.

Solution.   From the discussion of Example 5.1.6, we know that the algebra of multiplications by essentially bounded measurable functions on S is weak-operator closed in $\mathcal{B}(L_2(S, m))$. It remains to show that each such multiplication operator M_f is a weak-operator limit of multiplications by continuous functions. It follows from Lusin's theorem [R, p 53] that f is the limit almost everywhere of a sequence $\{f_n\}$ of bounded continuous functions (with uniform bound). As in the last paragraph of Example 2.5.12, M_f is the strong (hence, weak) -operator limit of $\{M_{f_n}\}$.   ■

5.7.26.   Let $(S, \mathcal{S}, m)$ be a σ-finite measure space, f be a measurable function on S, and A be a bounded operator on $L_2(S, m)$ such that $f \cdot g = Ag$ almost everywhere for each essentially bounded measurable function g in $L_2(S, m)$. Show that f is essentially bounded and that $M_f = A$.

Solution.   Let $\{S_n\}$ be a countable family of measurable subsets of S with finite measure such that $S_n \subseteq S_{n+1}$ and $\cup_{n=1}^\infty S_n = S$. Let S_{nk} be $\{s \in S_n : \|A\| + 1 \leq |f(s)| \leq k\}$. Since f is finite almost everywhere (our hypothesis entails this), if $m(S_{nk}) = 0$ for each n and k, then $\|A\| + 1$ is an essential bound for f. Suppose that some S_{nk} has a positive measure. Since $m(S_{nk}) \leq m(S_n) < \infty$, the characteristic function x of S_{nk} is a non-zero element of

$L_\infty(S, m) \cap L_2(S, m)$, and

$$\|A\|^2 \|x\|^2 \geq \|Ax\|^2 = \int_{S_{nk}} |f(s)|^2 \, dm(s)$$

$$\geq (\|A\| + 1)^2 \int_{S_{nk}} dm(s) = (\|A\| + 1)^2 \|x\|^2,$$

a contradiction.

Since f is essentially bounded, M_f is a bounded operator on $L_2(S, m)$. Now the bounded operators M_f and A agree on the dense subspace of $L_2(S, m)$ consisting of the essentially bounded measurable functions (in $L_2(S, m)$). Thus $A = M_f$.   ∎

5.7.27.   Let $(S, \mathcal{S}, m)$ be a σ-finite measure space and g be an essentially bounded measurable function on S.

(i)   Show that $m(g^{-1}(D)) = 0$ for each closed disk D contained in $\mathbb{C} \setminus \mathrm{sp}\,(g)$, where $\mathrm{sp}(g)$ is the essential range of g (defined and studied in Example 3.2.16).

(ii)   Note that each open subset of $\mathbb{C}$ is the union of a countable family of closed disks and conclude that $m(g^{-1}(\mathbb{C} \setminus \mathrm{sp}\,(g))) = 0$.

(iii)   Let λ be some point in $\mathrm{sp}(g)$ and define $g_0(s)$ to be $g(s)$ for s in $g^{-1}(\mathrm{sp}\,(g))$ and λ for s in $g^{-1}(\mathbb{C} \setminus \mathrm{sp}\,(g))$. Show that $\mathrm{sp}\,(g_0) = \mathrm{sp}\,(g)$.

(iv)   Let f be a bounded Borel function on $\mathrm{sp}\,(g)$. Show that $f(M_g) = f(M_{g_0}) = M_{f \circ g_0}$. [*Hint.* Use uniqueness of the Borel function calculus.]

Solution. (i)   From Example 3.2.16, $\mathrm{sp}(g) = \mathrm{sp}(M_g)$, so that $\mathrm{sp}(g)$ is closed and $\mathbb{C} \setminus \mathrm{sp}(g)$ is open. By definition of $\mathrm{sp}(g)$, each point of $\mathbb{C} \setminus \mathrm{sp}(g)$ is contained in an open disk whose inverse image under g has measure 0 (in (S, m)). Such open disks cover D. Select a finite subcovering $\{D_1, \ldots, D_n\}$. Then

$$m(g^{-1}(D)) \leq m(g^{-1}(D_1 \cup \cdots \cup D_n)) = m\Big(\bigcup_{j=1}^{n} g^{-1}(D_j)\Big) = 0.$$

(ii)   Let $\mathcal{O}$ be an open subset of $\mathbb{C}$. Enumerate the closed disks contained in $\mathcal{O}$ whose centers are rational complex numbers and whose radii are rational as $D_1, D_2, \ldots$. If $p \in \mathcal{O}$, then p is the center of some open disk D contained in $\mathcal{O}$. There is a rational complex number q near p such that some closed disk D' of rational radius contains p and is contained in D. But then the disk D' is one of $\{D_n : n = 1, 2, \ldots\}$. Hence $\mathcal{O} = \cup_{n=1}^\infty D_n$. If we take $\mathbb{C} \setminus \mathrm{sp}(g)$ as $\mathcal{O}$, then $m(g^{-1}(D_n)) = 0$ for each n, from (i). Thus

$$m(g^{-1}(\mathbb{C} \setminus \mathrm{sp}(g))) = m(\bigcup_{n=1}^\infty g^{-1}(D_n)) = 0.$$

(iii)   From (ii) and the definition of g_0, $g = g_0$ almost everywhere on S. Thus $M_{g_0} = M_g$. From Example 3.2.16, $\mathrm{sp}(g_0) = \mathrm{sp}(M_{g_0}) = \mathrm{sp}(M_g) = \mathrm{sp}(g)$.

(iv)   With h in $\mathcal{B}(\mathrm{sp}(g))$, define $\varphi(h)$ to be $M_{h \circ g_0}$. Then, $\varphi(1) = M_{1 \circ g_0} = M_1 = I$ and $\varphi(\iota) = M_{\iota \circ g_0} = M_{g_0} = M_g$. Let $\{h_n\}$ be an increasing sequence of bounded Borel functions tending pointwise to the bounded Borel function h. Then $\{h_n \circ g_0\}$ is an increasing sequence of bounded measurable functions tending pointwise to $h \circ g_0$. From the properties of multiplication operators described in Example 2.4.11, $\varphi(h_n)(= M_{h_n \circ g_0})$ is an increasing sequence of operators bounded above by $\varphi(h)(= M_{h \circ g_0})$. Moreover, from the monotone convergence theorem,

$$\langle \varphi(h_n)x, x \rangle = \int_S h_n(g_0(s))|x(s)|^2 \, dm(s)$$
$$\rightarrow \int_S h(g_0(s))|x(s)|^2 \, dm(s) = \langle \varphi(h)x, x \rangle,$$

for each x in $\mathcal{H}$. Thus $\varphi(h)$ is the least upper bound of $\{\varphi(h_n)\}$ and φ is a σ-normal homomorphism of $\mathcal{B}(\mathrm{sp}(g))$ into the multiplication algebra of $L_2(S, m)$. Theorem 5.2.9 applies and $M_{f \circ g_0} = \varphi(f) = f(M_g)$.   ■

5.7.28.   Let $\mathcal{R}$ be a von Neumann algebra acting on a Hilbert space $\mathcal{H}$, and let A be a self-adjoint operator in $\mathcal{B}(\mathcal{H})$ such that $UA + AU \leq 2A$ for each self-adjoint unitary operator U in $\mathcal{R}$. Show that $A \in \mathcal{R}'$.

Solution.   With E a projection in $\mathcal{R}$ and $I - 2E$ in place of U, we have

$$(I - 2E)A + A(I - 2E) \leq 2A;$$

so that $0 \leq AE + EA$. With x in $E(\mathcal{H})$ and y in $(I - E)(\mathcal{H})$, we have

$$0 \leq \langle AE(x - y), x - y \rangle + \langle EA(x - y), x - y \rangle$$
$$= \langle Ax, x \rangle - \langle Ax, y \rangle + \langle Ax, x \rangle - \langle Ay, x \rangle.$$

For a suitable choice of a of modulus 1 and t real,

$$0 \leq \langle Ax, aty \rangle = \langle A(aty), x \rangle \to +\infty$$

as $t \to +\infty$ unless $\langle Ax, y \rangle = 0$. Hence $\langle Ax, y \rangle = 0$ and Ax is orthogonal to $(I - E)(\mathcal{H})$. Thus $Ax \in E(\mathcal{H})$, $AE = EAE = (EAE)^* = (AE)^* = EA$. From Theorem 5.2.2(v), $A \in \mathcal{R}'$.   ∎

5.7.29.   Let $\mathcal{S}$ and $\mathcal{T}$ be two families of bounded operators on a Hilbert space $\mathcal{H}$, and suppose that $\mathcal{S} \subseteq \mathcal{T}$.
(i)   Show that $\mathcal{T}' \subseteq \mathcal{S}'$.
(ii)   Show that $\mathcal{S}' = (\mathcal{S}')'' (= \mathcal{S}''')$. (Compare Theomrem 5.3.1.)

Solution. (i) If $T' \in \mathcal{T}'$, then $T'T = TT'$ for each T in $\mathcal{T}$. Since $\mathcal{S} \subseteq \mathcal{T}$, $T'S = ST'$ for each S in $\mathcal{S}$. Thus $T' \in \mathcal{S}'$ and $\mathcal{T}' \subseteq \mathcal{S}'$.
(ii) If $S \in \mathcal{S}$ and $S' \in \mathcal{S}'$, then $SS' = S'S$. Hence $S \in (\mathcal{S}')' (= \mathcal{S}'')$ and $\mathcal{S} \subseteq \mathcal{S}''$. In particular $\mathcal{S}' \subseteq (\mathcal{S}')'' (= \mathcal{S}''')$. From (i), $\mathcal{S}''' = (\mathcal{S}'')' \subseteq \mathcal{S}'$, since $\mathcal{S} \subseteq \mathcal{S}''$. Thus $\mathcal{S}' = \mathcal{S}'''$.   ∎

5.7.30.   Let $\mathcal{H}$ be a Hilbert space of dimension greater than 1. Find a weak-operator closed subalgebra $\mathcal{B}$ of $\mathcal{B}(\mathcal{H})$ such that $\mathcal{B} \neq \mathcal{B}''$.

Solution. Let x_0 be a unit vector in $\mathcal{H}$ and let $\mathcal{B}$ be the subalgebra of $\mathcal{B}(\mathcal{H})$ consisting of those operators for which x_0 is an eigenvector. Let E_0 be the projection with range $[x_0]$. Then $B \in \mathcal{B}$ if and only if $E_0 B E_0 = B E_0$. Since both of the mappings $T \to E_0 T E_0$ and $T \to T E_0$ are weak-operator continuous on $\mathcal{B}(\mathcal{H})$ and $\mathcal{B}(\mathcal{H})$ with the weak-operator topology is a Hausdorff space, $\mathcal{B}$ is weak-operator closed. We note that $\mathcal{B}' = \{aI : a \in \mathbb{C}\}$ so that $\mathcal{B}'' = \mathcal{B}(\mathcal{H}) (\neq \mathcal{B}$ since $\mathcal{H}$ has dimension greater than one). For this, choose y_0 in $\mathcal{H}$ a unit vector orthogonal to x_0. Let E be the projection with range $[x_0, y_0]$ and V be the operator defined by $V y_0 = x_0$, $V x_0 = 0$, and $V(I - E) = 0$. Then E_0, E, and V are in $\mathcal{B}$. Thus, if $B' \in \mathcal{B}'$, we have that x_0 and y_0 are eigenvectors for B', say $B' x_0 = a x_0$ and $B' y_0 = b y_0$. Since $B'V = VB'$,

$$b x_0 = bV y_0 = V B' y_0 = B' V y_0 = B' x_0 = a x_0,$$

and $a = b$. As this applies to each y_0 orthogonal to x_0, $B' = aI$; and $B' = \{aI : a \in \mathbb{C}\}$. ∎

5.7.31.   Let $\mathcal{H}$ be a Hilbert space and $\mathfrak{A}_0$ be a self-adjoint subalgebra of $\mathcal{B}(\mathcal{H})$. Assume that $\mathfrak{A}_0(\mathcal{H})$ is dense in $\mathcal{H}$ but *not* that $I \in \mathfrak{A}_0$. Show that $\mathfrak{A}_0''$ is the strong-operator closure of $\mathfrak{A}_0$ (and that $I \in \mathfrak{A}_0''$).

Solution.   Let $\mathfrak{A}_0^-$ be the strong-operator closure of $\mathfrak{A}_0$. From Theorem 5.1.2, $\mathfrak{A}_0^-$ is also the weak-operator closure of $\mathfrak{A}_0$. From Exercise 5.7.1, $\mathfrak{A}_0^-$ is stable under the adjoint operation. Thus $\mathfrak{A}_0^-$ is a strong (and weak) -operator closed, self-adjoint operator algebra acting on $\mathcal{H}$. From Proposition 5.1.8, $\mathfrak{A}_0^-$ has a unit element P, a projection in $\mathfrak{A}_0^-$ that contains the range projection of each operator in $\mathfrak{A}_0$. Thus $Py = y$ for each y in $\mathcal{H}$ such that $y = Ax$ for some A in $\mathfrak{A}_0$. Since $\mathfrak{A}_0(\mathcal{H})$ is dense in $\mathcal{H}$, $P = I$. Now $\mathfrak{A}_0 \subseteq \mathfrak{A}_0^-$, so that $(\mathfrak{A}_0^-)' \subseteq \mathfrak{A}_0'$ and $\mathfrak{A}_0'' \subseteq (\mathfrak{A}_0^-)'' = \mathfrak{A}_0^-$, by Theorem 5.3.1. But $\mathfrak{A}_0''$ is strong-operator closed and $\mathfrak{A}_0 \subseteq \mathfrak{A}_0''$. Thus $\mathfrak{A}_0^- \subseteq \mathfrak{A}_0''$, and $\mathfrak{A}_0^- = \mathfrak{A}_0''$.
∎ [79]

5.7.32.   Use the double commutant theorem to re-prove (compare Proposition 5.1.8), in the special case of a von Neumann algebra $\mathcal{R}$ (so $\mathcal{R}$ is assumed to contain I), that the union and intersection of each family of projections in $\mathcal{R}$ lie in $\mathcal{R}$.

Solution.   Let $\{E_a\}$ be a family of projections in $\mathcal{R}$, E be its union and F its intersection. If $T' \in \mathcal{R}'$, then $T'E_a = E_aT'$ for each a. Thus the linear span of the ranges of E_a and the intersection of these ranges are stable under T'. By continuity of T', the closures of these linear spaces are stable under T'; that is, $T'E = ET'E$ and $T'F = FT'F$ for each T' in $\mathcal{R}'$. Since $T'^* \in \mathcal{R}'$,

$$T'E = ET'E = (ET'^*E)^* = (T'^*E)^* = ET'.$$

Similarly, $T'F = FT'$, and $E, F \in (\mathcal{R}')' = \mathcal{R}$. ∎

5.7.33.   Let $\mathfrak{A}$ be a simple C^*-algebra (that is, $\mathfrak{A}$ has no proper two-sided ideals) acting on a Hilbert space $\mathcal{H}$.
  (i)   Show that the center of $\mathfrak{A}$ is $\{aI : a \in \mathbb{C}\}$.
  (ii)   Suppose $\mathfrak{A}$ contains a maximal abelian subalgebra of $\mathcal{B}(\mathcal{H})$. Show that $\mathfrak{A}$ acts irreducibly on $\mathcal{H}$.

Solution. (i)   Let C be the center of $\mathfrak{A}$. From Theorem 4.4.3, $C \cong C(X)$, where X is the pure state space of C. Suppose C is not $\{aI : a \in \mathbb{C}\}$. Then $C(X)$ is not one-dimensional and X has two distinct points p and q. Let $\mathcal{O}$ and $\mathcal{O}'$ be disjoint open sets containing p and q, respectively. Let f and g be functions in $C(X)$ taking the value 1 at p and q and vanishing outside of $\mathcal{O}$ and $\mathcal{O}'$, respectively. Suppose A and B in C correspond to f and g, respectively. Then $\mathfrak{A}A$ is a two-sided ideal in $\mathfrak{A}$ (since $A \in C$) containing $A\,(\neq 0)$ but not I (since $\mathfrak{A}AB = (0)$ and $B \neq 0$). Thus $\mathfrak{A}A$ is a proper two-sided ideal, contradicting the assumption that $\mathfrak{A}$ is simple. Thus X consists of a single point and $C = \{aI : a \in \mathbb{C}\}$.

(ii)   Let $\mathcal{A}$ be a maximal abelian algebra on $\mathcal{B}(\mathcal{H})$ contained in $\mathfrak{A}$. If $T \in \mathfrak{A}'$, then $T \in \mathcal{A}' = \mathcal{A} \subseteq \mathfrak{A}$. Thus $T \in \mathfrak{A} \cap \mathfrak{A}' = C$. From (i), $T = aI$ for some scalar a. It follows that $\mathfrak{A}' = \{aI : a \in \mathbb{C}\}$ and $\mathfrak{A}$ acts irreducibly on $\mathcal{H}$.   ∎

5.7.34.   Find a von Neumann algebra $\mathcal{R}$ and a strong-operator dense, self-adjoint subalgebra $\mathfrak{A}_0$ (containing I) such that *no* unitary operator in $\mathcal{R}$ other than a scalar is the strong-operator limit of unitary operators in $\mathfrak{A}_0$. [*Hint.* Consider polynomials on $[0,1]$.)

Solution.   Let $\mathcal{H}$ be $L_2([0,1])$ (relative to Lebesgue measure), $\mathcal{R}$ be the multiplication algebra of $L_2([0,1])$, and $\mathfrak{A}_0$ be the subalgebra of multiplications by polynomials. We show that $\mathfrak{A}_0$ has no unitary operators other than scalars so that no unitary operator in $\mathcal{R}$ other than a scalar is the strong-operator limit of unitary operators in $\mathfrak{A}_0$.

If p is a polynomial such that M_p is a unitary operator, then $p = p_1 + ip_2$, where p_1 and p_2 are polynomials with real coefficients and $|p|^2 = p_1^2 + p_2^2 = 1$. Let ax^n and bx^m be the highest order terms in p_1 and p_2, respectively. Then $a^2 x^{2n}$ and $b^2 x^{2m}$ are the highest order terms in p_1^2 and p_2^2, respectively. Since a^2 and b^2 are positive, $p_1^2 + p_2^2 - 1$ is a polynomial with non-constant term *unless n and m are 0*. As $p_1^2 + p_2^2 - 1$ vanishes on $[0,1]$, $p_1, p_2,$ and p are constant polynomials.

From the Weierstrass approximation theorem, $\mathfrak{A}_0$ has norm closure the algebra of multiplications by continuous functions on $[0,1]$ and this last algebra has strong-operator closure $\mathcal{R}$. Thus $\mathfrak{A}_0$ is strong-operator dense in $\mathcal{R}$.   ∎

5.7.35.   Let $\mathcal{H}$ be a Hilbert space, S a closed subset of $\mathbb{R}$, $\mathcal{S}$ the set of self-adjoint operators on $\mathcal{H}$ with spectrum in S, and h a real-valued, bounded, continuous function defined on S. With A_0 in $\mathcal{S}$, let k be a continuous function on $\mathbb{R}$ that takes the value 1 at each point of $\mathrm{sp}(A_0)$ and vanishes outside of $[-(\|A_0\|+1), \|A_0\|+1]$. Let p be hk and q be $1 - k + p$.

(i)   Show that $p(A_0) = q(A_0) = h(A_0)$, $h = (1-h)p + hq$, and

$$h(A) - h(A_0) = (I - h(A))(p(A) - p(A_0)) + h(A)(q(A) - q(A_0)).$$

(ii)   Show that the mapping $A \to h(A)$ of $\mathcal{S}$ into (the set of normal operators in) $\mathcal{B}(\mathcal{H})$ is strong-operator continuous.

Solution. (i)   Since k takes the value 1 on $\mathrm{sp}(A_0)$, $k(A_0) = I$ and $p(A_0) = h(A_0)I = h(A_0)$. At the same time, we have that $q(A_0) = I - k(A_0) + p(A_0) = p(A_0)$. By definition of p and q,

$$(1-h)p + hq = p - hp + h - hk + hp = h.$$

Finally, with the aid of the preceding identities,

$$
\begin{aligned}
(I - h(A))(p(A) &- p(A_0)) + h(A)(q(A) - q(A_0)) \\
&= h(A) - ((I - h(A))p(A_0) + h(A)q(A_0)) \\
&= h(A) - ((I - h(A))h(A_0) + h(A)h(A_0)) \\
&= h(A) - h(A_0).
\end{aligned}
$$

(ii)   Since p and $q - 1$ extend to continuous functions that vanish outside a compact set, Theorem 5.3.4 applies, and they are strong-operator continuous on $\mathcal{S}$. Thus given a vector x in $\mathcal{H}$ and a positive ε, there are vectors $y_1, \ldots, y_n$ in $\mathcal{H}$ and a positive δ such that

$$\|(p(A) - p(A_0))x\| < \frac{\varepsilon}{2(1+c)}, \quad \|(q(A) - q(A_0))x\| < \frac{\varepsilon}{2c},$$

when $A \in \mathcal{S}$ and $\|(A - A_0)y_j\| < \delta$ for j in $\{1, \ldots, n\}$, where c is a bound for $|h|$. From the last identity of (i),

$$\|(h(A) - h(A_0))x\|$$

$$\leq \|I - h(A)\|\,\|(p(A) - p(A_0))x\| + \|h(A)\|\,\|(q(A) - q(A_0))x\|$$

$$< \frac{(1+c)\varepsilon}{2(1+c)} + \frac{c\varepsilon}{2c} = \varepsilon,$$

when $A \in \mathcal{S}$ and $\|(A - A_0)y_j\| < \delta$ for j in $\{1, \ldots, n\}$. Thus h is strong-operator continuous on $\mathcal{S}$. ∎   [72]

5.7.36.   Let $\mathcal{H}$ be a Hilbert space, S a closed subset of $\mathbb{R}$, and $\mathcal{S}$ the set of self-adjoint operators in $\mathcal{B}(\mathcal{H})$ with spectrum in S.

(i)   Show that the mapping $A \to |A|$ is strong-operator continuous on $\mathcal{S}$.

(ii)   Let h be a real-valued, continuous function on S (so h is bounded on bounded subsets of S). Suppose S_0 is a bounded subset of S such that g is bounded on $S \setminus S_0$, where $g(t) = h(t)/|t|$ for t in $S \setminus S_0$. Show that the mapping $A \to h(A)$ is strong-operator continuous on $\mathcal{S}$.

(iii)   Deduce that $A \to A^{1/n}$ is a strong-operator continuous mapping on $\mathcal{B}(\mathcal{H})^+$ for each positive integer n.

Solution. (i)   Let $r(\lambda) = \lambda$ for λ in $[-1,1]$ and $|\lambda|/\lambda$ for λ in $\mathbb{R} \setminus [-1,1]$. Let $s(\lambda)$ be $r(\lambda) \cdot \lambda$ and $t(\lambda)$ be $|\lambda| - s(\lambda)$ for λ in $\mathbb{R}$. Since t is continuous on $\mathbb{R}$ and vanishes outside of $[-1,1]$, $A \to t(A)$ is strong-operator continuous on $\mathcal{S}$ from Theorem 5.3.4. Note that with A and B in $\mathcal{S}$ and x in $\mathcal{H}$,

$$\|[s(A) - s(B)]x\| \le \|r(A)\| \, \|(A - B)x\| + \|[r(A) - r(B)]Bx\|.$$

Since r is bounded and continuous on $\mathbb{R}$, the mapping $A \to r(A)$ is strong-operator continuous on $\mathcal{S}$ from Exercise 5.7.35. It follows from this and the preceding inequality that the mapping $A \to s(A)$ is strong-operator continuous on $\mathcal{S}$. We can now conclude that the mapping $A \to s(A) + t(A) = |A|$ is strong-operator continuous on $\mathcal{S}$.

(ii)   Let $g_0(t)$ be $h(t)/(1 + |t|)$ for t in S. For each x in $\mathcal{H}$, with A, B in $\mathcal{S}$,

$$\|[h(A) - h(A_0)]x\| = \|(g_0(A)[(I + |A|) - g_0(A_0)(I + |A_0|)]x\|$$

$$(*) \qquad \le \|g_0(A)\| \, \|[(I + |A|) - (I + |A_0|)]x\|$$

$$+ \|[g_0(A) - g_0(A_0)](I + |A_0|)x\|.$$

Since g_0 is bounded and continuous on S, $A \to g_0(A)$ is strong-operator continuous on $\mathcal{S}$ from Exercise 5.7.35. From (i), the strong-operator continuity of g_0, and the inequality $(*)$, the mapping $A \to h(A)$ is continuous.

(iii)   Let S be $\{t \in \mathbb{R} : t \ge 0\}$ and $h(t)$ be $t^{1/n}$. Since $h(t)/|t|$ is continuous and bounded outside of $[0,1]$, the conditions of (ii) apply and the mapping $A \to A^{1/n}$ is strong-operator continuous on $\mathcal{B}(\mathcal{H})^+$. ■ [62]

5.7.37.   Let S be a subset of $\mathbb{R}$. Suppose f is a function defined on S such that the mapping $A \to f(A)$ is strong-operator continuous on $\mathcal{S}$ for each Hilbert space $\mathcal{H}$, where $\mathcal{S}$ is the set of self-adjoint operators in $\mathcal{B}(\mathcal{H})$ with spectrum in S. Show that f is continuous on S, bounded on bounded subsets of S, and that there is a bounded subset S_0 of S such that $\{f(t)/|t| \,:\, t \in S \setminus S_0\}$ is bounded.

Solution.   Let $\mathcal{H}$ be $L_2(0,1)$ relative to Lebesgue measure. With x a unit vector in $\mathcal{H}$, the function $aI \to \langle f(aI)x, x \rangle = f(a)$ is a continuous function on S, since $A \to f(A)$ is strong-operator continuous on $\mathcal{S}$. Hence f is continuous on S.

Suppose f is not bounded on some bounded subset of S. Then there is a sequence $\{s_n\}$ of points s_n in S that tend to some s in $\mathbb{R}$ such that $n \le |f(s_n)|$ for each n. Choose b in S. We show that the mapping $A \to f(A)$ is not strong-operator continuous at bI. Suppose $x_1, \ldots, x_m$ are in $L_2(0,1)$ and let x be $\sum_{k=1}^m |x_k|$. There is a subset X of $(0,1)$ with positive Lebesgue measure a such that $\int_X |x(t)|^2 \, dt \le (|b|+|s|+1)^{-2}$. Choose j such that $a|f(s_j) - f(b)|^2 \ge 1$ and $|s_j| < |s|+1$. Let g be s_j on X and b on $(0,1) \setminus X$. From Exercise 5.7.27, $f(M_g) = M_{f \circ g}$, so that

$$\|[f(M_g) - f(bI)]1\|^2 = \int |f(g(t)) - f(b)|^2 \, dt = a|f(s_j) - f(b)|^2 \ge 1.$$

But for each k in $\{1, \ldots, m\}$,

$$\|(bI - M_g)x_k\|^2 = \int |(b - g(t))x_k(t)|^2 \, dt = \int_X |(b - s_j)x_k(t)|^2 \, dt$$
$$\le (|b| + |s| + 1)^2 \int_X |x(t)|^2 \, dt \le 1.$$

Thus the mapping $A \to f(A)$ is not strong-operator continuous at bI on $\mathcal{S}$.

Suppose, next, that there is no bounded subset S_0 as described in the exercise. Then there is a sequence $\{s_n\}$ in S such that $|s_n| \to \infty$ and $n|s_n| \le |f(s_n)|$ for each n. Again, we show that the mapping $A \to f(A)$ is not strong-operator continuous at bI, where $b \in S$. Suppose $x_1, \ldots, x_m$ in $L_2(0,1)$ are given and $x = \sum_{k=1}^m |x_k|$. Let X_n be the subset of $(0,1)$ at which x does not exceed n, for n in $\{1, 2, \ldots\}$. Since $x \in L_2(0,1)$, some X_n has positive measure a.

Choose j larger than n such that $(c^{-1} =)\,|s_j - b|^2 an^2 \ge 1$, $(j|s_j| - |f(b)|)^2 ac \ge 1$, and $j|s_j| > |f(b)|$. Then $0 < c \le 1$, and there is a subset X of X_n with measure ac. Let g be s_j on X and b on $(0,1) \setminus X$. Then

$$\|(bI - M_g)x_k\|^2 = \int_X |s_j - b|^2 |x_k(t)|^2 \, dt \le |s_j - b|^2 acn^2 = 1,$$

while

$$\begin{aligned}
\|[f(M_g) - f(bI)]1\|^2 &= \int |f(g(t)) - f(b)|^2 \, dt \\
&= \int_X |f(s_j) - f(b)|^2 \, dt \\
&\ge (|f(s_j)| - |f(b)|)^2 ac \\
&\ge (j|s_j| - |f(b)|)^2 ac \ge 1.
\end{aligned}$$

Thus the mapping $A \to f(A)$ is not strong-operator continuous at bI on $\mathcal{S}$. ■ [62]

5.7.38.   Let $\mathcal{R}$ be a von Neumann algebra acting on a Hilbert space $\mathcal{H}$, and let E be a projection in $\mathcal{R}$. Find the central carrier of a projection F in $E\mathcal{R}E$ relative to $E\mathcal{R}E$ in terms of its central carrier C_F relative to $\mathcal{R}$.

Solution.   From Propositions 5.5.6 and 5.5.2, the range of the central carrier of F in $E\mathcal{R}E$ is

$$[E\mathcal{R}EFE(\mathcal{H})] = [E\mathcal{R}F(\mathcal{H})] = [EC_F(\mathcal{H})] = EC_F(\mathcal{H}).$$

Thus EC_F is the central carrier of F relative to $E\mathcal{R}E$.   ■

5.7.39.   Let $\mathcal{R}$ be a von Neumann algebra of infinite linear dimension. Show that there is an orthogonal infinite family of non-zero projections with sum I.

Solution.   From Exercise 4.6.12 (iv), each maximal abelian self-adjoint subalgebra of $\mathcal{R}$ has infinite linear dimension. Let $\mathcal{A}$ be such a subalgebra of $\mathcal{R}$. Since $\mathcal{R}$ is a von Neumann algebra, $\mathcal{A}$ is a von Neumann algebra. From Theorem 5.2.1, $\mathcal{A} \cong C(X)$, where X is an extremely disconnected compact Hausdorff space. Let Y be the set of points q in X with the property that $\{q\}$ is clopen. If Y is an infinite set, we can select from it a countably infinite set of points. The projections E_n corresponding to the characteristic functions of each of the points in the countably infinite set form a countably infinite orthogonal family $\{E_1, E_2, \ldots\}$ of non-zero projections in $\mathcal{A} (\subseteq \mathcal{R})$. If $\sum_{n=1}^{\infty} E_n \neq I$, adjoin $I - \sum_{n=1}^{\infty} E_n$ to this family and the resulting family has sum I.

We may suppose that Y is finite. Since $C(X)$ has infinite linear dimension, there is a point p of X not contained in Y. Suppose we have found clopen sets $X_1, \ldots, X_n$ such that X_{j+1} is a proper subset of X_j and each X_j contains p. Since $p \notin Y$, there is a q in X_n distinct from p. Let X_0 be a clopen subset of X containing p but not q and let X_{n+1} be $X_0 \cap X_n$. Then X_{n+1} is a clopen subset of X containing p and properly contained in X_n. In this way, we construct an infinite sequence $X_1, X_2, \ldots$ of clopen sets containing p such that X_{n+1} is properly contained in X_n. Let E_n be the projection in $\mathcal{A}$ corresponding to the characteristic function of the clopen set $X_n \setminus X_{n+1}$. Then $\{E_1, E_2, \ldots\}$ is an infinite orthogonal family of non-zero projections in $\mathcal{A}$ and, again, we complete the solution by adjoining $I - \sum_{n=1}^{\infty} E_n$ to this family if $I - \sum_{n=1}^{\infty} E_n \neq 0$. ∎

5.7.40.   Let $\mathfrak{A}$ be a C^*-algebra acting on a Hilbert space $\mathcal{H}$, and let $\{E_n'\}$ be an orthogonal family of non-zero projections in the commutant $\mathfrak{A}'$ of $\mathfrak{A}$ with sum I. Suppose x_0 is a generating vector for $\mathfrak{A}$ and $E_n' x_0 = a_n x_n$, where $|a_n| = \|E_n' x_0\|$ and $\|x_n\| = 1$.

(i)   Show that $a_n \neq 0$ and that $x_0 = \sum_n a_n x_n$.

(ii)   Suppose $\{E_n'\}$ is a (countably) infinite family (indexed by positive integers). Choose $n(j)(> n(j-1))$ such that $|a_{n(j)}| < j^{-2}$ for j in $\mathbb{N}$. Let x' be $\sum_{j=1}^{\infty} j^{-1} x_{n(j)}$. Show that for each A in the strong-operator closure $\mathfrak{A}^-$ of $\mathfrak{A}$, $A x_0 \neq x'$.

(iii)   Conclude that $\mathfrak{A}'$ is finite dimensional if there is a vector x_0 in $\mathcal{H}$ such that $\{A x_0 : A \in \mathfrak{A}^-\} = \mathcal{H}$.

Solution. (i)   From Corollary 5.5.12, x_0 is separating for $\mathfrak{A}' (= (\mathfrak{A}^-)')$ since x_0 is generating for $\mathfrak{A}^-$. Thus $E_n' x_0 \neq 0$ and $a_n \neq 0$. Since $\sum_n E_n' = I$, $x_0 = (\sum_n E_n') x_0 = \sum_n a_n x_n$.

(ii)   Since $\sum_{n=1}^{\infty} a_n x_n$ converges (to x_0), $\|a_n x_n\| = |a_n| \to 0$ as $n \to \infty$. Thus $n(j)$ can be chosen as indicated. Note that for A in $\mathfrak{A}^-$, $A x_0 = \sum_{n=1}^{\infty} a_n A x_n$ and that $E_n' A x_n = A E_n' x_n = A x_n$ for each n. Thus if $A x_0 = x'$, then $a_{n(j)} A x_{n(j)} = j^{-1} x_{n(j)}$, since $\{E_n'\}$ is an orthogonal family. It follows that

$$\|Ax_{n(j)}\| = \|a_{n(j)}^{-1} j^{-1} x_{n(j)}\| \geq j^2 j^{-1} = j.$$

Thus $\|A\| \geq j$ for each positive integer j, contradicting the fact that $A \in \mathcal{B}(\mathcal{H})$. It follows that $Ax_0 = x'$ for no A in $\mathfrak{A}^-$.

(iii)   If $\{Ax_0 : A \in \mathfrak{A}^-\} = \mathcal{H}$, there is no vector x', and every orthogonal family $\{E_n'\}$ of projections with sum I in $\mathfrak{A}'$ is finite. From Exercise 5.7.39, $\mathfrak{A}'$ is finite dimensional. ■ [44]

5.7.41.   Let $\mathfrak{A}$ be a C^*-algebra that acts topologically irreducibly on a Hilbert space $\mathcal{H}$, and let $\{x_1, \ldots, x_n\}$ and $\{y_1, \ldots, y_n\}$ be sets of vectors in $\mathcal{H}$.

(i)   Let H be a self-adjoint operator in $\mathcal{B}(\mathcal{H})$ such that $Hx_j = y_j$ for j in $\{1, \ldots, n\}$. Show that there is a self-adjoint operator K in $\mathfrak{A}$ such that $Kx_j = y_j$ for j in $\{1, \ldots, n\}$ and $\|K\| \leq \|H\|$. [*Hint.* Use a diagonalizing orthonormal basis for EHE restricted to $[x_1, \ldots, x_n, y_1, \ldots, y_n]$, where E is the projection in $\mathcal{B}(\mathcal{H})$ with this subspace as range, and apply Remark 5.6.32.]

(ii)   Let B be an operator in $\mathcal{B}(\mathcal{H})$ such that $Bx_j = y_j$ for j in $\{1, \ldots, n\}$. Show that there is an operator A in $\mathfrak{A}$ such that $Ax_j = y_j$ for j in $\{1, \ldots, n\}$ and $\|A\| \leq \|B\|$. [*Hint.* With E as in the hint to (i), use the fact that $EBE|E(\mathcal{H})$ has the form VH with V a unitary operator and H a positive operator on $E(\mathcal{H})$.]

Solution.   Since $E(\mathcal{H})$ is finite dimensional, there is an orthonormal basis $z_1, \ldots, z_m$ for $E(\mathcal{H})$ such that $EHz_j = \lambda_j z_j$ for some real scalars $\lambda_1, \ldots, \lambda_m$. From Theorem 5.4.3, there is a self-adjoint operator K_0 in $\mathfrak{A}$ such that $K_0 z_j = \lambda_j z_j$. Let $f(t)$ be t for t in $[-\|H\|, \|H\|]$, $\|H\|$ if $t > \|H\|$, and $-\|H\|$ if $t < -\|H\|$. Then f is continuous and $f(K_0)$ is a self-adjoint operator K in $\mathfrak{A}$ such that $Kz_j = f(\lambda_j)z_j = \lambda_j z_j$, from Remark 5.6.32 and the fact that $|\lambda_j| \leq \|H\|$. But then $Kz = EHz$ for each z in $E(\mathcal{H})$. In particular, $Kx_j = EHx_j = y_j$ for j in $\{1, \ldots, n\}$.

(ii)   With V and H as in the hint, $\|H\| = \|EBE\| \leq \|B\|$. From (i), there is a self-adjoint K in $\mathfrak{A}$ such that $Kz = Hz$ for each z in $E(\mathcal{H})$ and $\|K\| \leq \|H\| \leq \|B\|$. Of course V extends to a unitary operator V' on $\mathcal{H}$. From Theorem 5.4.5, there is a unitary operator U in $\mathfrak{A}$ such that $UKx_j = V'Kx_j = VHx_j = y_j$ for j in $\{1, \ldots, n\}$. Let A be UK. Then $A \in \mathfrak{A}$, $Ax_j = y_j$ for j in $\{1, \ldots, n\}$ and $\|A\| = \|K\| \leq \|B\|$. ■ [61(pp.69-70)]

5.7.42.   Let $\mathcal{H}_0$ be a Hilbert space and $\mathcal{H}$ be the direct sum $\sum_n \oplus \mathcal{H}_n$ of countably many copies $\mathcal{H}_n$ of $\mathcal{H}_0$.

(i)   With the notation of Subsection 2.6, *Matrix representations*, let $\mathcal{R}$ be the subalgebra of $\mathcal{B}(\mathcal{H})$ consisting of operators whose matrix has the same element of $\mathcal{B}(\mathcal{H}_0)$ at each diagonal entry and 0 at each off-diagonal entry. (With $\mathcal{H}$ viewed as a tensor product $\mathcal{H}_0 \otimes \mathcal{K}$ of $\mathcal{H}_0$ with a separable, infinite-dimensional Hilbert space $\mathcal{K}$, $\mathcal{R}$ is $\{T \otimes I : T \in \mathcal{B}(\mathcal{H}_0)\}$.) Show that $\mathcal{R}'$ consists of those operators whose matrix representations have scalar multiples of I at each entry. (In tensor product form, $\mathcal{R}' = \{I \otimes S : S \in \mathcal{B}(\mathcal{K})\}$.) Show that $\mathcal{R}'' = \mathcal{R}$ and conclude that $\mathcal{R}$ is a von Neumann algebra (as well as $\mathcal{R}'$).

(ii)   Let $x_1, x_2, \ldots$ be a sequence of vectors in $\mathcal{H}_0$. Call this sequence $l_2 - independent$ when $\sum_{j=1}^{\infty} \|x_j\|^2 < \infty$ and $\sum_{j=1}^{\infty} a_j x_j = 0$ for a sequence $\{a_j\}$ in $l_2(\mathbb{N}, \mathbb{C})$ only if $a_j = 0$ for all j. Note that an l_2-independent sequence is linearly independent and find a linearly independent sequence that is not l_2-independent when $\mathcal{H}_0$ is infinite dimensional.

(iii)   Show that $\{x_1, x_2, \ldots\}$ is a generating vector for $\mathcal{R}$ if and only if $x_1, x_2, \ldots$ is l_2-independent.

(iv)   Show that $\{x_1, x_2, \ldots\}$ is separating for $\mathcal{R}$ if and only if $[x_1, x_2, \ldots] = \mathcal{H}_0$.

Solution. (i)   Suppose $A \in \mathcal{B}(\mathcal{H}_0)$ and T is the operator in $\mathcal{B}(\mathcal{H})$ with matrix $[T_{j,k}]$, where $T_{j,j} = A$ for all j and $T_{j,k} = 0$ when $j \neq k$. Let T', with matrix $[T'_{j,k}]$, lie in $\mathcal{R}'$. Then TT' has matrix $[AT'_{j,k}]$, and $T'T$ has matrix $[T'_{j,k}A]$. Since $TT' = T'T$, we have that $AT'_{j,k} = T'_{j,k}A$ for all A in $\mathcal{B}(\mathcal{H}_0)$. Thus each $T'_{j,k} \in \mathcal{B}(\mathcal{H}_0)'$ and $T'_{j,k} = t'_{j,k}I$ for all j and k. Conversely, if S' in $\mathcal{B}(\mathcal{H})$ has matrix $[s'_{j,k}I]$, then $TS' = S'T$ and $S' \in \mathcal{R}'$.

Suppose S, with matrix $[S_{j,k}]$, lies in $\mathcal{R}''$. If T' has matrix whose only non-zero entry occurs at the (j,j) position and is I, then $T'S$ has all entries 0 other than those occurring in the jth row, and the jth row of $T'S$ coincides with the jth row of S. Moreover, ST' has all entries 0 other than those occurring in the jth column, and the jth column of ST' coincides with the jth column of S. Since $T' \in \mathcal{R}'$ and $S \in \mathcal{R}''$, $ST' = T'S$ and $S_{j,k} = 0$ unless $j = k$. If S' has matrix whose only non-zero entry is I in the (j,k) position, then from the equality of SS' and $S'S$, we have that $S_{j,j} = S_{k,k}$. Thus $S \in \mathcal{R}$.

(ii)   A linear dependence relation, $a_1 x_1 + \cdots + a_n x_n = 0$, is an

l_2-dependence relation, so that $a_1 = a_2 = \cdots = a_n = 0$ if $x_1, x_2, \ldots$ is an l_2-independent sequence. Hence an l_2-independent sequence is linearly independent.

Let $\{e_n\}$ be a countably-infinite, orthonormal set in $\mathcal{H}_0$, y_n be $2^{-n}e_n$, x_0 be $-y_1$, and x_n be $2y_n - y_{n+1}$ with n in $\{1, 2, \ldots\}$. Then $x_n = 2^{1-n}e_n - 2^{-(1+n)}e_{n+1}$ and $\|x_n\|^2 = 2^{2-2n} + 2^{-2(n+1)}$. Thus $\sum_{n=0}^{\infty} \|x_n\|^2 < \infty$. If $\sum_{j=0}^{n} b_j x_j = 0$, then $b_n = 0$ since the coefficient of e_{n+1} in $\sum_{j=0}^{n} b_j x_j$ is $-b_n 2^{-(n+1)}$. Thus $x_0, x_1, \ldots$ is a linearly independent sequence. Let a_n be 2^{-n}. Then $\sum_{n=0}^{\infty} |a_n|^2 < \infty$ and

$$\sum_{j=0}^{n} a_j x_j = -y_1 + \sum_{j=1}^{n} 2^{-j}[2^{1-j}e_j - 2^{-j-1}e_{j+1}]$$

$$= -y_1 + \sum_{j=1}^{n} 2^{1-2j}e_j - \sum_{j=2}^{n+1} 2^{1-2j}e_j = 2^{-(2n+1)}e_{n+1} \to 0.$$

Thus $\sum_{j=0}^{\infty} a_j x_j = 0$ and $\{x_0, x_1, \ldots\}$ is not l_2-independent.

(iii) Suppose $\{x_1, x_2, \ldots\}$ is an l_2-independent sequence. Since $\sum_{j=1}^{\infty} \|x_j\|^2 < \infty$, $\{x_1, x_2, \ldots\}(=x) \in \mathcal{H}$. From Corollary 5.5.12, x is generating for $\mathcal{R}$ if and only if it is separating for $\mathcal{R}'$. From (i), if $T' \in \mathcal{R}'$, then T' has matrix $[t'_{j,k}I]$. Let u be a unit vector in $\mathcal{H}_0$ and u' be the vector in $\mathcal{H}$ with u at the j-coordinate and all other coordinates 0. Then $\|T'^*u'\|^2 = \sum_{k=1}^{\infty} |t'_{j,k}|^2 < \infty$. Now

$$T'x = \left\{ \sum_{k=1}^{\infty} t'_{1,k}x_k, \sum_{k=1}^{\infty} t'_{2,k}x_k, \ldots \right\},$$

so that $\sum_{k=1}^{\infty} t'_{j,k}x_k = 0$ for all j if $T'x = 0$. Since $\{x_1, x_2, \ldots\}$ is l_2-independent, all $t'_{j,k}$ are 0 in this case and $T' = 0$. It follows that x is separating for $\mathcal{R}'$ and generating for $\mathcal{R}$ when $\{x_1, x_2, \ldots\}$ is l_2-independent.

Suppose $\sum_{j=1}^{\infty} \|x_j\|^2 < \infty$ and $x_1, x_2, \ldots$ is not l_2-independent. Let $\{a_1, a_2, \ldots\}$ be a non-zero element of $l_2(\mathbb{N}, \mathbb{C})$ such that

$$\sum_{j=1}^{\infty} a_j x_j = 0.$$

Let T' be the operator in $\mathcal{R}'$ with matrix $[t'_{j,k}I]$, where $t'_{j,k} = 0$ unless $j = 1$ and $t'_{1,k} = a_k$. With $\{y_1, y_2, \ldots\}(= y)$ in $\mathcal{H}$, $T'y =$

$\{\sum_{k=1}^{\infty} a_k y_k, 0, 0, \ldots\}$ so that

$$\|T'y\|^2 = \Big\| \sum_{k=1}^{\infty} a_k y_k \Big\|^2 \le \Big(\sum_{k=1}^{\infty} |a_k| \|y_k\| \Big)^2 \le \Big(\sum_{k=1}^{\infty} |a_k|^2 \Big) \Big(\sum_{k=1}^{\infty} \|y_k\|^2 \Big)$$
$$= \Big(\sum_{k=1}^{\infty} |a_k|^2 \Big) \|y\|^2,$$

which establishes that T' is bounded $(\|T'\| \le (\sum_{k=1}^{\infty} |a_k|^2)^{\frac{1}{2}})$, so that $T' \in \mathcal{R}'$ and that $T'x = 0$. Thus x is not separating for $\mathcal{R}'$ and, hence, not generating for $\mathcal{R}$.

(iv) Suppose $(x =) \{x_1, x_2, \ldots\} \in \mathcal{H}$. With T in $\mathcal{R}$, there is an A in $\mathcal{B}(\mathcal{H}_0)$ such that $Ty = \{Ay_1, Ay_2, \ldots\}$, where $(y =) \{y_1, y_2, \ldots\}$ is in $\mathcal{H}$. Thus if $[x_1, x_2, \ldots] = \mathcal{H}_0$, $Tx = 0$ implies that $T = 0$ and x is separating for $\mathcal{R}$. Conversely, if $[x_1, x_2, \ldots] \ne \mathcal{H}_0$ and E is the (non-zero) projection on the orthogonal complement of $[x_1, x_2, \ldots]$ in $\mathcal{H}_0$, then with E for A, $Tx = 0$ and T is a non-zero element of $\mathcal{R}$ so that x is not separating for $\mathcal{R}$. $\blacksquare$

5.7.43.   Let $\mathcal{H}$ be $L_2([0,1])$ relative to Lebesgue measure. With $\mathcal{A}$ the multiplication algebra of $\mathcal{H}$ and $\mathfrak{A}$ the C^*-algebra of multiplications by continuous functions, find a unit vector u that is separating for $\mathfrak{A}$ (if $A \in \mathfrak{A}$ and $Au = 0$, then $A = 0$) but not for $\mathcal{A}$. [*Hint.* Use the "Cantor process" to find an open dense subset of $[0,1]$ that has measure $\frac{1}{2}$.]

Solution.   Let $\mathcal{O}_1$ be $(\frac{3}{8}, \frac{5}{8})$, $\mathcal{O}_2$ be the union of $(\frac{5}{32}, \frac{7}{32})$ and $(\frac{25}{32}, \frac{27}{32}), \ldots$. (Let $\mathcal{O}_n$ be the union of the open intervals of lengths 2^{-2n} centered in the closed intervals whose union is $[0,1] \setminus \cup_{j=1}^{n-1} \mathcal{O}_j$.) Then $\mathcal{O}_n$ has measure $2^{-(n+1)}$ and $\{\mathcal{O}_n\}$ is a family of mutually disjoint open subsets of $[0,1]$. Thus $\cup_{n=1}^{\infty} \mathcal{O}_n (= \mathcal{O})$ has measure $\frac{1}{2}$. The closed intervals with union $[0,1] \setminus \cup_{j=1}^{n} \mathcal{O}_j$ have lengths tending to 0 as n tends to ∞. Thus $[0,1] \setminus \mathcal{O}$ contains no open interval and $\mathcal{O}$ is dense in $[0,1]$.

Let $u(t)$ be 0 for t in $[0,1] \setminus \mathcal{O}$ and $\sqrt{2}$ for t in $\mathcal{O}$. Then $\|u\| = 1$. If $f \in C([0,1])$ and $M_f u = 0$, then $f(t) = 0$ for t in $\mathcal{O}$. Hence $f(t) = 0$ for t in $\mathcal{O}^- (= [0,1])$ and f and M_f are 0. Let g be the characteristic function of $[0,1] \setminus \mathcal{O}$. Then $M_g u = 0$; but $M_g \ne 0$ since $[0,1] \setminus \mathcal{O}$ has measure $\frac{1}{2}$. Thus u is separating for $\mathfrak{A}$ but not for $\mathcal{A}$. $\blacksquare$

5.7.44.   Let $\mathcal{H}$ be $L_2([0,1])$ relative to Lebesgue measure and $\mathcal{A}$ be the multiplication algebra of $\mathcal{H}$.

(i)   Describe the vectors in $\mathcal{H}$ that are generating for $\mathcal{A}$.

(ii)   Show that the set of generating vectors for $\mathcal{A}$ is dense in $\mathcal{H}$.

(iii)   Deduce that a norm limit of generating vectors need not be a generating vector.

Solution. (i)   Since $\mathcal{A}$ is maximal abelian, a vector is separating for $\mathcal{A}$ if and only if it is generating for $\mathcal{A}$. If $f \in \mathcal{H}$ and f vanishes on a set of positive measure, then $M_\chi \in \mathcal{A}$ and $M_\chi \neq 0$, where χ is the characteristic function of that set. But $M_\chi f = 0$, so that f is not separating (hence, not generating) for $\mathcal{A}$. If $g \in \mathcal{H}$ and the set of points at which g is 0 has measure 0, then $hg = 0$ almost everywhere for some h in $L_\infty([0,1])$ only if h is 0 almost everywhere. That is $M_h(g) = 0$ only if $M_h = 0$, and g is separating (and generating) for $\mathcal{A}$ in this case. Thus the generating vectors for $\mathcal{A}$ are precisely those g in $L_2([0,1])$ that are non-zero on the complement of a set of measure 0.

(ii)   Suppose $g \in L_2([0,1])$ and S is the set of points at which g vanishes. Given a positive ε, let $g_0(t)$ be $g(t)$ if $t \notin S$ and ε if $t \in S$. Then $g_0 \in L_2$ and

$$\|g - g_0\|^2 = \int |g(t) - g_0(t)|^2 \, dt = \int_S \varepsilon^2 \, dt = a\varepsilon^2 \leq \varepsilon^2,$$

where a is the measure of S. Now g_0 vanishes nowhere on $[0,1]$. From (i), g_0 is generating for $\mathcal{A}$. It follows that the set of generating vectors for $\mathcal{A}$ is norm dense in $\mathcal{H}$.

(iii)   If g is the characteristic function of $[0, \frac{1}{2}]$, then g is not generating for $\mathcal{A}$ from (i). From (ii), g is a norm limit of generating vectors.   ∎

5.7.45.   Let $\mathcal{R}$ be a von Neumann algebra and $\{E_1, E_2, \ldots\}$ be a countable family of countably decomposable projections in $\mathcal{R}$. Show that $\bigvee_{n=1}^{\infty} E_n$ is a countably decomposable projection in $\mathcal{R}$.

Solution.   From Proposition 5.5.9, each E_j is the union of an orthogonal family of cyclic projections in $\mathcal{R}$. Since each E_j is countably decomposable, each orthogonal family of cyclic projections is countable. It follows that E is the union of a countable family of cyclic projections in $\mathcal{R}$. From Proposition 5.5.19, E is countably decomposable.   ∎

5.7.46.   Let $\mathcal{R}$ be a countably decomposable von Neumann algebra acting on a Hilbert space $\mathcal{H}$. Define a metric on $\mathcal{R}$ with the property that its associated metric topology coincides with the strong-operator topology on bounded subsets of $\mathcal{R}$.

Solution.   From Proposition 5.5.9 and the assumption that $\mathcal{R}$ is countably decomposable, there are cyclic projections $E_1, E_2, \ldots$ in $\mathcal{R}$ with sum I. Suppose $y_1, y_2, \ldots$ are generating unit vectors for $E_1, E_2, \ldots$, respectively. With S and T in $\mathcal{R}$ define:

$$d(S,T) = \sum_{n=1}^{\infty} 2^{-n}\|(S-T)y_n\|.$$

Suppose $d(S,T) = 0$. Then $(S-T)y_n = 0$ for each n, from which we have that $(S-T)A'y_n = A'(S-T)y_n = 0$ for each A' in $\mathcal{R}'$. Thus $(S-T)E_n = 0$ for each n and

$$S - T = (S-T)\sum_{n=1}^{\infty} E_n = \sum_{n=1}^{\infty}(S-T)E_n = 0.$$

The remaining properties of a metric are established for d precisely as in the solution to Exercise 2.8.35.

Let $\mathcal{S}$ be a bounded subset of $\mathcal{R}$, and choose $M(>0)$ so that $\|S\| \leq M$ for S in $\mathcal{S}$. As in the solution to Exercise 2.8.35, the metric ball $\{S \in \mathcal{S} : d(S,S_0) < \varepsilon\}$ contains the strong-operator neighborhood

$$\{S \in \mathcal{S} : \|Sy_j - S_0y_j\| < k^{-1} \ (j = 1, \ldots, k)\},$$

where $S_0 \in \mathcal{S}$, $\varepsilon > 0$, and k is a positive integer such that $k^{-1} + M2^{1-k} < \varepsilon$. Neighborhoods of the form

$$\{S \in \mathcal{S} : \|(S - S_0)A_j'y_j\| < \varepsilon, A_j' \in \mathcal{R}', \ (j = 1, \ldots, k)\} \ (= \mathcal{V})$$

constitute a base of strong-operator neighborhoods of S_0 in $\mathcal{S}$ from the discussion preceding Remark 2.5.9 since the closed linear span of the set of vectors $\{A'y_j : A' \in \mathcal{R}, j = 1, 2, \ldots\}$ is $\mathcal{H}$. We note that $\{S \in \mathcal{S} : d(S,S_0) < M_0^{-1}2^{-k}\varepsilon\}$ is contained in $\mathcal{V}$, where $M_0 = \max\{\|A_j'\| : j = 1, \ldots, k\}$; for if $d(S,S_0) < M_0^{-1}2^{-k}\varepsilon$, then $2^{-j}\|(S-S_0)y_j\| < M_0^{-1}2^{-k}\varepsilon$ for all j and

$$\|(S-S_0)A_j'y_j\| = \|A_j'(S-S_0)y_j\| \leq M_0\|(S-S_0)y_j\| < 2^{j-k}\varepsilon \leq \varepsilon$$

when $j = 1, \ldots, k$. Thus the metric topology and the strong-operator topology coincide on $\mathcal{S}$.   ∎

5.7.47. Let $(S, \mathcal{S}, m)$ be a σ-finite measure space, $\mathcal{H}$ be $L_2(S, m)$, $\mathcal{A}$ be the multiplication algebra, and f and g be measurable functions on S finite almost everywhere. Show that:

   (i) $M_f = M_g$ if and only if $f = g$ almost everywhere;

   (ii) $M_{af+g} = aM_f \hat{+} M_g$ for each scalar a;

   (iii) $M_{f \cdot g} = M_f \hat{\cdot} M_g$;

   (iv) $M_f \geq 0$ if and only if $f \geq 0$ almost everywhere.

Solution. (i) Suppose $f = g$ almost everywhere. With h in $L_2(S, m)$, $fh = gh$ almost everywhere so that $fh \in L_2(S, m)$ if and only if $gh \in L_2(S, m)$. Thus M_f and M_g have the same domain. If h is in this domain, $M_f h = fh = gh = M_g h$, whence $M_f = M_g$.

Suppose $M_f = M_g$. Let S_n be the subset of S on which $|f|$ and $|g|$ take values not exceeding n and at each point of which f and g differ. Let χ be the characteristic function of some subset S' of S_n of finite measure. Then $f\chi$ and $g\chi$ are in $L_2(S, m)$ so that $f\chi = g\chi$ almost everywhere. Since $f(s') \neq g(s')$ for each s' in S', $m(S') = 0$. Thus $m(S_n) = 0$ and $m(\cup_{n=1}^{\infty} S_n) = 0$. Hence $f = g$ almost everywhere.

(ii) If h is in the domain of $aM_f + M_g$, then fh and gh are in $L_2(S, m)$ so that $(af + g)h \in L_2(S, m)$ and h is in the domain of M_{af+g}. Moreover, $M_{af+g} h = afh + gh = (aM_f + M_g)h$. Thus M_{af+g} is a closed extension of $aM_f + M_g$ and hence of its closure $aM_f \hat{+} M_g$. From Theorem 5.6.15(vii), $M_{af+g} = aM_f \hat{+} M_g$.

(iii) If h is in the domain of $M_f M_g$, then gh and fgh are in $L_2(S, m)$. Hence h is in the domain of M_{fg}. Moreover, $M_f M_g h = fgh = M_{fg} h$. Thus M_{fg} is a closed extension of $M_f M_g$ and hence of its closure $M_f \hat{\cdot} M_g$. From Theorem 5.6.15(vii), $M_{fg} = M_f \hat{\cdot} M_g$.

(iv) Suppose $f \geq 0$ almost everywhere and h is in the domain of M_f. Then $fh \in L_2(S, m)$ and

$$0 \leq \int_S f(s)|h(s)|^2 \, dm(s) = \langle fh, h \rangle = \langle M_f h, h \rangle.$$

Hence $M_f \geq 0$.

Suppose $M_f \geq 0$. For each positive integer n, let S_n be the set $\{s \in S : -n \leq f(s) \leq -\frac{1}{n}\}$ and let χ be the characteristic function of a subset S' of S_n of finite measure. Then χ is in the domain of M_f and

$$0 \leq \langle M_f \chi, \chi \rangle = \int_S f(s)\chi(s) \, ds \leq -m(S')/n \leq 0.$$

Thus $m(S') = 0$, $m(S_n) = 0$, and $m(\cup_{n=1}^{\infty} S_n) = 0$. Hence f is nonnegative almost everywhere. ∎

5.7.48.   Let $(S, \mathcal{S}, m)$ be a σ-finite measure space, $\mathcal{H}$ be $L_2(S, m)$, $\mathcal{A}$ be the multiplication algebra, g be a measurable function on S finite almost everywhere, and f be a Borel function on $\mathrm{sp}(M_g)$.

(i)   Define a concept of "essential range" $\mathrm{sp}(g)$ analogous to that of Example 3.2.16 and show that $\mathrm{sp}(g) = \mathrm{sp}(M_g)$.

(ii)   Show that there is a measurable g_0 on S equal to g almost everywhere such that the range of g_0 is contained in $\mathrm{sp}(g)$.

(iii)   With the notation of (ii), show that $f(M_g) = M_{f \circ g_0}$.

Solution. (i)   Precisely as in Example 3.2.16, we say that λ is in the essential range of g, when $0 < m(g^{-1}(D))$ for each open disk D in $\mathbb{C}$ containing λ. We denote by $\mathrm{sp}(g)$ the subset of $\mathbb{C}$ consisting of points in the essential range of g.

Suppose $\lambda \in \mathrm{sp}(g)$. Let D_n be the open disk of radius $\frac{1}{n}$ with center λ and let x_n be the charateristic function of S_n, a subset of $g^{-1}(D_n)$ of finite positive measure a_n. Let h_n be $a_n^{-\frac{1}{2}} x_n$. Then $\|h_n\| = 1$ and

$$\|(M_g - \lambda I)h_n\|^2 = \int_S |g(s) - \lambda|^2 |h(s)|^2 \, dm(s)$$
$$= a_n^{-1} \int_{S_n} |g(s) - \lambda|^2 \, dm(s) \le n^{-2}.$$

If B in $\mathcal{A}$ is an inverse to $M_g - \lambda I$ (in $\mathcal{N}(\mathcal{A})$), then $B(M_g - \lambda I)h_n = h_n$; and $n \le \|B\|$ for each n. Hence $M_g - \lambda I$ has no inverse in $\mathcal{N}(\mathcal{A})$ that lies in $\mathcal{A}$. From the discussion preceding Proposition 5.6.20, $\lambda \in \mathrm{sp}(M_g)$.

If $\lambda \notin \mathrm{sp}(g)$, there is an open disk D with center λ of radius $\frac{1}{n}$ for some positive integer n such that $g^{-1}(D) (= S_0)$ has measure 0. Let $h(s)$ be $1/(g(s) - \lambda)$ when $s \notin S_0$ and 0 when $s \in S_0$. Then h has essential supremum not exceeding n so that $\|M_h\| \le n$ and $M_h \in \mathcal{A}$. Now $(g - \lambda)h = h(g - \lambda) = 1$ almost everywhere; and from Exercise 5.7.47, M_h is an inverse to $M_g - \lambda I$ in $\mathcal{N}(\mathcal{A})$ that lies in $\mathcal{A}$. Consequently, $\lambda \notin \mathrm{sp}\, M_g$ and $\mathrm{sp}(g) = \mathrm{sp}(M_g)$.

(ii)   We may assume that g is defined and finite everywhere. Choose λ_0 in $\mathrm{sp}(g)$. For each s in S and each positive integer n, let

$g_n(s)$ be $g(s)$ if $|g(s)| < n$ and λ_0 otherwise. Then $g_n \in L_\infty(S, m)$. From Exercise 5.7.27, there is a g_{n0} in $L_\infty(S, m)$ equal to g_n on the complement of a set S_n of measure 0 such that g_{n0} has range in $\mathrm{sp}(g_n)$. For each s in S, there is a unique positive integer n such that $n - 1 \leq |g(s)| < n$. At such an s, let $g_0(s)$ be $g_{n0}(s)$. For such an s, $g(s) = g_n(s)$ and if, in addition, $s \notin S_n$, then $g(s) = g_n(s) = g_{n0}(s) = g_0(s)$. Thus g and g_0 agree on $S \setminus \cup_{n=1}^\infty S_n$ and $m(\cup_{n=1}^\infty S_n) = 0$.

If $n-1 \leq |g(s)| < n$, then $g_0(s) = g_{n0}(s) \in \mathrm{sp}(g_n)$. If $g_0(s) = \lambda_0$, then $g_0(s) \in \mathrm{sp}(g)$. Assume $g_0(s) \neq \lambda_0$. If D is an open disk in $\mathbb{C}$ with center $g_0(s)$, $0 < m(g_n^{-1}(D))$ and $0 < m(g_n^{-1}(D_0))$, where $D_0 \subseteq D$, D_0 has center $g_0(s)$ and $\lambda_0 \notin D_0$. If $s' \in g_n^{-1}(D_0)$, then $g_n(s') \neq \lambda_0$ so that $|g(s')| < n$ and $g_n(s') = g(s')$. Thus $s' \in g^{-1}(D_0) \subseteq g^{-1}(D)$ and $g_n^{-1}(D_0) \subseteq g^{-1}(D)$. It follows that $0 < m(g^{-1}(D))$ so that $g_0(s) \in \mathrm{sp}(g)$. Hence g_0 has range in $\mathrm{sp}(g)$.

(iii)   With h in B_u, define $\psi(h)$ to be $M_{h \circ g_0}$. Then $\psi(1) = I$ and $\psi(\imath) = M_{g_0} = M_g$ from Exercise 5.7.47(i), since g and g_0 agree almost everywhere. Moreover, ψ is a homomorphism of B_u into $\mathcal{N}(\mathcal{A})$ from Exercise 5.7.47(ii) and (iii). Suppose $\{h_n\}$ is an increasing sequence in B_u tending pointwise to h. Then $\{M_{h_n \circ g_0}\}$ is an increasing sequence in $\mathcal{S}(\mathcal{A})$ and $M_{h \circ g_0}$ is an upper bound of this sequence from Theorem 5.6.4 and Exercise 5.7.47. If M_k is another upper bound for $\{M_{h_n \circ g_0}\}$ in $\mathcal{S}(\mathcal{A})$, then $M_k \,\hat{-}\, M_{h_n \circ g_0} = M_{k - h_n \circ g_0}$ has nonnegative real spectrum from Proposition 5.6.21. From (i) and (ii), $k - h_n \circ g_0$ is nonnegative almost everywhere. Now $\{h_n \circ g_0\}$ tends pointwise to $h \circ g_0$. Thus $k - h \circ g_0$ is nonnegative almost everywhere. It follows that the essential range of $k - h \circ g_0$ contains no λ less than 0. Hence $M_{k - h \circ g_0} = M_k \,\hat{-}\, M_{h \circ g_0}$ is positive and $M_{h \circ g_0}$ is the least upper bound of $\{M_{h_n \circ g}\}$. Thus ψ is a σ-normal homomorphism of B_u into $\mathcal{A}$. With our initial observations, $M_{h \circ g_0} = \psi(h) = h(M_g)$ for each h in B_u so that $f(M_g) = M_{f \circ g_0}$, from Theorem 5.6.27.   ∎

5.7.49.   Let $\mathcal{H}$ be $L_2(\mathbb{R})$ relative to Lebesgue measure and A be the (unbounded) multiplication operator corresponding to the identity transform $\imath$ (the function $t \to t$) on $\mathbb{R}$ with domain $\mathcal{D}$ consisting of those f in $L_2(\mathbb{R})$ such that $\imath \cdot f \in L_2(\mathbb{R})$.

(i)   Note that A is self-adjoint and that the spectral resolution for A is $\{E_\lambda\}$, where E_λ is the multiplication operator corresponding to the characteristic function of $(-\infty, \lambda]$.

(ii)   Let $\mathcal{D}_0$ be the set of continuously differentiable functions on

$\mathbb{R}$ that vanish outside a finite interval. Note that $\mathcal{D}_0$ is a dense linear submanifold of $\mathcal{H}$. Let D_0 be the operator with domain $\mathcal{D}_0$ that assigns if' $(= i\, df/dt)$ to f. With T the unitary operator defined in Theorem 3.2.31, show that $T^{-1}ATf = D_0 f$ for each f in $\mathcal{D}_0$.

(iii) Conclude that $T^{-1}AT$ $(= D)$ with domain $T^{-1}(\mathcal{D})$ is a self-adjoint extension of D_0.

(iv) Show that $\exp(itD)$ is the unitary operator U_t, where $(U_t f)(p) = f(p - t)$. How does this relate to Stone's theorem?

Solution. (i) With the domain $\mathcal{D}$ described, A is self-adjoint by Theorem 5.6.4. From the discussion just preceding Theorem 5.6.4, $\{E_\lambda\}$ as described is the spectral resolution of A.

(ii) With f in $\mathcal{D}_0$, both f and f' vanish outside a finite interval. By assumption, both are continuous. Theorem 3.2.31 applies; when $p \neq 0$,

$$(Tf)(p) = (2\pi)^{-\frac{1}{2}} \int_{\mathbb{R}} e^{isp} f(s)\, ds = (2\pi)^{-\frac{1}{2}} \lim_{a \to \infty} \int_{-a}^{a} e^{isp} f(s)\, ds$$

$$= (2\pi)^{-\frac{1}{2}} \lim_{a \to \infty} \Big[\frac{e^{iap} f(a) - e^{-iap} f(-a)}{ip} - \frac{1}{ip} \int_{-a}^{a} e^{isp} f'(s)\, ds \Big]$$

$$= -\frac{(2\pi)^{-\frac{1}{2}}}{ip} \int_{\mathbb{R}} e^{isp} f'(s)\, ds = -\frac{1}{ip}(Tf')(p).$$

Thus $ATf = iTf'$ and $T^{-1}ATf = D_0 f$.

(iii) Since A is self-adjoint and T is unitary, $T^{-1}AT$ is self-adjoint with domain $T^{-1}(\mathcal{D})$ and from (ii), $T^{-1}AT$ $(= D)$ is an extension of D_0.

(iv) Since T is unitary and $T^{-1}AT = D$, we have that $\exp itD = T^{-1}(\exp itA)T$. From Exercise 5.7.48, $\exp itA$ is the multiplication operator corresponding to $\exp it\imath$. Suppose $f \in \mathcal{D}_0$. Then from Theorem 3.2.31,

$$(2\pi)^{\frac{1}{2}}((\exp itA)Tf)(p) = (2\pi)^{\frac{1}{2}} e^{itp}(Tf)(p) = e^{itp} \int_{\mathbb{R}} e^{isp} f(s)\, ds$$

$$= \int_{\mathbb{R}} e^{i(t+s)p} f(s)\, ds = \int_{\mathbb{R}} e^{isp} f(s - t)\, ds$$

$$= \int_{\mathbb{R}} e^{isp}(U_t f)(s)\, ds = (2\pi)^{\frac{1}{2}}(TU_t f)(p).$$

Hence $(\exp itA)T = TU_t$ and

$$\exp itD = T^{-1}(\exp itA)T = U_t$$

for each real t.

The one-parameter group $t \to U_t$ has an "infinitesimal generator" according to Stone's theorem, and we have identified it as iD (loosely speaking, as "differentiation").   ∎

5.7.50.   Let f be a jointly continuous function of two complex variables defined on $S_1 \times S_2$, where S_1 and S_2 are subsets of $\mathbb{C}$. Let A be a normal operator such that $\operatorname{sp} A \subseteq S_2$ acting on a Hilbert space $\mathcal{H}$. For each z_1 in S_1, the mapping $z \to f(z_1, z)$ is a continuous (hence, Borel) function defined on $\operatorname{sp} A$ so that $f(z_1, A)$ is a normal operator on $\mathcal{H}$. Define an operator-valued function g on S_1 by $g(z) = f(z, A)$.

(i)   Suppose A is bounded and S_0 is a compact subset of S_1. Show that the restriction of g to S_0 is norm continuous (that is, the mapping $z \to g(z)$ is continuous from S_0 to $\mathcal{B}(\mathcal{H})$ with its norm topology).

(ii)   Suppose S_1 is closed and f is bounded (but no longer that A is bounded). Show that g is strong-operator continuous.

(iii)   Let H be a positive operator on $\mathcal{H}$. Show that $\exp(-izH)$ ($=U_z$) is defined for each z in the closed lower half plane $\mathbb{C}_-$ ($= \{z : \operatorname{Im} z \leq 0\}$), that $\|U_z\| \leq 1$, and that the mapping $z \to U_z$ is strong-operator continuous.

Solution. (i)   Since A is bounded, $\operatorname{sp} A$ is compact. By assumption, S_0 is compact so that $S_0 \times \operatorname{sp} A$ ($= C$) is compact. The restriction of f to C is uniformly continuous. Given a positive ε, choose a positive δ such that $|f(u, v) - f(u', v')| < \varepsilon$ when $\|(u, v) - (u', v')\| < \delta$ and (u, v), $(u', v') \in C$. In particular, $|f(u, v) - f(u', v)| < \varepsilon$ whenever $u, u' \in S_0$, $v \in \operatorname{sp} A$, and $|u - u'| < \delta$. Since the function calculus for A is an isometric linear mapping (see Theorem 4.4.5),

$$\|g(u) - g(u')\| = \|f(u, A) - f(u', A)\| \leq \varepsilon$$

for such u and u'.

(ii)   Let $\mathcal{A}$ be the von Neumann algebra generated by A and let $\{E_n\}$ be a sequence of bounding projections for A in $\mathcal{A}$. (Theorem 5.6.15(i) assures us that such a bounding sequence exists.) From Corollary 5.6.31, $f(z, (A\dot{}E_n)|\mathcal{H}_n) = (f(z, A)\dot{}E_n)|\mathcal{H}_n$ for each n, where $\mathcal{H}_n = E_n(\mathcal{H})$. Since f is bounded, $f(z, A)\dot{}E_n = f(z, A)E_n$. Suppose $x \in \mathcal{H}_n$ and $z_1 \in S_1$. Let S_0 be the intersection of S_1 and a closed disk in $\mathbb{C}$ with center z_1. Since S_1 is closed, S_0 is compact. If z in S_1 is near z_1, then $z \in S_0$. Now $(A^{\hat{}}E_n)|\mathcal{H}_n$ is bounded so that $z \to f(z,(A^{\hat{}}E_n)|\mathcal{H}_n)$ is norm continuous on S_0 from (i). In particular, given a positive ε, with z sufficiently near z_1,

$$\|(f(z,A)-f(z_1,A))x\|$$
$$=\|f(z,(A^{\hat{}}E_n)|\mathcal{H}_n) - f(z_1,(A^{\hat{}}E_n)|\mathcal{H}_n))x\| < \varepsilon.$$

Since $\cup_{n=1}^{\infty}\mathcal{H}_n$ is dense in $\mathcal{H}$ and $\{\|g(z)\| : z \in S_1\}$ has the bound of f as an upper bound, g is strong-operator continuous.

(iii)   Let $\mathcal{A}$ be the (abelian) von Neumann algebra generated by H. Then $\mathcal{A} \cong C(X)$ for some extremely disconnected compact Hausdorff space X. Let h be the function in $\mathcal{S}(X)$ representing H. From Propositions 5.6.20 and 5.6.21, the range of h consists of nonnegative real numbers. By definition of $\exp -izH$, the corresponding function in $\mathcal{N}(X)$ is $\exp -izh$ (that is the normal extension of $p \to \exp -izh(p)$). If $z = t + is$ with t and s real and $s \leq 0$, then since $h(p) \geq 0$,

$$|\exp -izh(p)| = |\exp -ith(p)|\,|\exp sh(p)| = |\exp sh(p)| \leq 1.$$

Thus $U_z \in \mathcal{A}$ and $\|U_z\| \leq 1$. Moreover, if we take for the function f, the mapping $(z_1,z_2) \to \exp -iz_1z_2$ on $\mathbb{C}_- \times \mathbb{R}^+$, the conditions of (ii) are fulfilled with $\mathbb{C}_-$ and $\mathbb{R}^+$ in place of S_1 and S_2 and H in place of A, where $\mathbb{R}^+ = \{t \in \mathbb{R} : t \geq 0\}$, whence g becomes $z \to U_z$ and is, accordingly, strong-operator continuous.   ■

5.7.51.   (i)   With the notation and hypotheses of Exercise 5.7.50(ii), make the following additional assumptions about f:

(1)   for each z_2 in S_2, $z \to f(z,z_2)$ is differentiable at each point z_0 of the interior S_1^0 of S_1 with derivative $f_1(z_0,z_2)$,

(2)   given z_0 in S_1^0, a bounded subset S_2' of S_2, and a positive ε, there are a positive δ, a closed disk D with center z_0 in S_1^0, and a positive C, such that for all z_2 in S_2'

$$|[f(z,z_2) - f(z_0,z_2)](z - z_0)^{-1} - f_1(z_0,z_2)| < \varepsilon,$$

provided $0 < |z - z_0| < \delta$ and $z \in S_1^0$ (that is, $z \to f(z,z_2)$ is differentiable on S_1^0 *uniformly* on bounded subsets of S_2) and such that for all z in $D \setminus \{z_0\}$ and z' in S_2

$$|[f(z,z') - f(z_0,z')](z - z_0)^{-1}| \leq C.$$

(3) $z \to f_1(z_0, z)$ is continuous on S_2 for each z_0 in S_1^0.
Show that for each pair of vectors x, y in $\mathcal{H}$, $z \to \langle g(z)x, y\rangle$ is analytic
on S_1^0 with derivative $\langle f_1(z_0, A)x, y\rangle$ at each z_0 in S_1^0. [*Hint.* With z_0
in S_1^0, let $h(z_1, z_2)$ be $[f(z_1, z_2) - f(z_0, z_2)](z_1 - z_0)^{-1}$ when $(z_1, z_2) \in$
$S_1^0 \times S_2$ and $z_1 \neq z_0$, and let $h(z_0, z_2)$ be $f_1(z_0, z_2)$. Use Exercise
5.7.50(ii) to establish strong-operator continuity of $z \to h(z, A)$.]

(ii)   With the notation and assumptions of Exercise 5.7.50(iii),
show that the function $z \to \langle U_z x, y\rangle$ is analytic in the open lower
half-plane $\mathbb{C}_-^0$ ($= \{z : \mathrm{Im}\, z < 0\}$) for each pair of vectors x, y in $\mathcal{H}$.

(iii)   Show that $z \to \langle U_z x, y\rangle$ is entire for each pair of vectors x
and y in $\mathcal{H}$ when H is bounded. Re-prove the results of (ii) by using
a bounding sequence $\{E_n\}$ of projections for H and considering first
the case where $x, y \in E_n(\mathcal{H})$.

Solution. (i)   Let z_0 and h be as in the hint. We note first
that h is continuous at (z_1, z_2) in $S_1^0 \times S_2$ when $z_1 \neq z_0$ from the
continuity of f at (z_1, z_2) and (z_0, z_2) and the definition of h. From
condition (3), given a positive ε there is a positive δ' such that

$$|h(z_0, z) - h(z_0, z_2)| = |f_1(z_0, z) - f_1(z_0, z_2)| < \frac{\varepsilon}{2}$$

if $z \in S_2$ and $|z - z_2| < \delta'$. Let D_2 be a closed disk with (positive)
radius r and center z_2; let S_2' be $D_2 \cap S_2$. From condition (2), there
is a positive δ'' such that for all z' in S_2'

$$|h(z, z') - h(z_0, z')| = |[f(z, z') - f(z_0, z')](z - z_0)^{-1} - f_1(z_0, z')| < \frac{\varepsilon}{2}$$

provided $0 < |z - z_0| < \delta''$ and $z \in S_1^0$. Let δ be $\min\{\delta', \delta'', r\}$
and suppose $(z, z') \in S_1^0 \times S_2$ and $\|(z, z') - (z_0, z_2)\| < \delta$. Then
$|z' - z_2| < r$ and $z' \in S_2$ so that $z' \in S_2'$. If $z = z_0$, then

$$|h(z, z') - h(z_0, z_2)| = |h(z_0, z') - h(z_0, z_2)| < \frac{\varepsilon}{2} < \varepsilon$$

since $z' \in S_2$ and $|z' - z_2| < \delta \leq \delta'$. If $z \neq z_0$, then $0 < |z - z_0| < \delta \leq$
δ'' so that, by choice of δ'' and since $z' \in S_2'$, $|h(z, z') - h(z_0, z')| < \frac{\varepsilon}{2}$.
But $|h(z_0, z') - h(z_0, z_2)| < \frac{\varepsilon}{2}$, so that $|h(z, z') - h(z_0, z_2)| < \varepsilon$. This
last inequality holds, therefore, for all choices of (z, z') subject to
the conditions noted, and h is continuous at (z_0, z_2). Hence h is
continuous on $S_1^0 \times S_2$.

Let D be the closed disk in S_1^0 with center z_0 and C be the associated constant described in condition (2). It follows from condition (2) that for all z' in S_2, $|h(z_0, z')| = |f_1(z_0, z')| \leq C$ and that $|h(z, z')| \leq C$ when $z \in D \setminus \{z_0\}$. Thus h is bounded on $D \times S_2$. From Exercise 5.7.50(ii), $z \to h(z, A)$ is strong-operator continuous on D. Now with z in $D \setminus \{z_0\}$,

$$h(z, A) = [f(z, A) - f(z_0, A)](z - z_0)^{-1} = [g(z) - g(z_0)](z - z_0)^{-1}$$

and $h(z_0, A) = f_1(z_0, A)$. Thus, with z in $D \setminus \{z_0\}$,

$$(\langle g(z)x, y \rangle - \langle g(z_0)x, y \rangle)(z - z_0)^{-1} = \langle h(z, A)x, y \rangle$$
$$\to \langle h(z_0, A)x, y \rangle = \langle f_1(z_0, A)x, y \rangle$$

as $z \to z_0$ so that $z \to \langle g(z)x, y \rangle$ is differentiable at each point z_0 of S_1^0 with derivative $\langle f_1(z_0, A)x, y \rangle$.

(ii) We adopt the notation of the solution to Exercise 5.7.50(iii) and we show that the function f defined there satisfies condition (1), (2), and (3), of (i). Suppose $z_0 = t_0 + i s_0$ and $s_0 < 0$. Then $z \to \exp - izt$ has the derivative $-it \exp - iz_0 t$ ($= -it \exp - it_0 t \exp s_0 t = f_1(z_0, t)$) at z_0, where $t \in \mathbb{R}^+$, so that f satisfies condition (1).

Suppose next that $0 \leq t \leq a$. Then

$$|[f(z, t) - f(z_0, t)](z - z_0)^{-1} - f_1(z_0, t)|$$
$$= |(e^{-izt} - e^{-iz_0 t})(z - z_0)^{-1} + ite^{-iz_0 t}|$$
$$= |e^{-iz_0 t}[e^{-i(z - z_0)t} - 1 + it(z - z_0)](z - z_0)^{-1}|$$
$$= e^{s_0 t} \left| \frac{(-it)^2}{2!} + \frac{(-it)^3(z - z_0)}{3!} + \cdots \right| |z - z_0|$$
$$\leq t^2 e^{s_0 t} |z - z_0| \left(\frac{1}{2!} + \frac{t|z - z_0|}{3!} + \frac{t^2|z - z_0|^2}{4!} + \cdots \right)$$
$$\leq t^2 e^{s_0 t} |z - z_0| e^{t|z - z_0|} \leq a^2 \cdot 1 \cdot e^{a|z - z_0|} |z - z_0|,$$

and this last term tends to 0 as z tends to z_0.

For the remainder of condition (2), we choose D as the closed disk with center z_0 and radius $-s_0/2$. If $z \in D \setminus \{z_0\}$ and $t \in \mathbb{R}^+$, then

$$|[f(z,t) - f(z_0,t)](z - z_0)^{-1}|$$
$$= |e^{-iz_0 t}[e^{-i(z-z_0)t} - 1](z - z_0)^{-1}|$$
$$= e^{s_0 t}\left| -it + \frac{(it)^2(z - z_0)}{2!} + \cdots \right|$$
$$\leq e^{s_0 t} t \left(1 + \frac{t|z - z_0|}{2!} + \frac{t^2|z - z_0|^2}{3!} + \cdots\right)$$
$$\leq t e^{s_0 t} e^{t|z-z_0|}$$
$$\leq t e^{s_0 t} e^{-s_0 t/2}$$
$$= t e^{s_0 t/2} \leq -2/s_0 e,$$

and we may take C to be $-2/s_0 e$.

For condition (3), we note that $z \to \exp -izt$ has $-it \exp -iz_0 t$ as its derivative at z_0 and that $t \to -it \exp -iz_0 t$ is continuous on $\mathbb{R}^+$ for each z_0 in $\mathbb{C}^0_-$. It follows now that $z \to \langle U_z x, y \rangle$ is analytic on $\mathbb{C}^0_-$.

(iii) Suppose first that H is bounded and $\|H\| = a$. Let x be a unit vector in $\mathcal{H}$. Then, letting z_0 be $t_0 + is_0$, we have that

$$\left| \frac{\langle (U_z - U_{z_0})x, x \rangle}{z - z_0} + \langle iH U_{z_0} x, x \rangle \right|$$
$$= \left| \int_{-a}^a \left(\frac{e^{izt} - e^{-iz_0 t}}{z - z_0} + it e^{-iz_0 t} \right) d\langle E_t x, x \rangle \right|$$
$$\leq e^{a|s_0|} a^2 e^{a|z-z_0|} |z - z_0| \to 0$$

as $z \to z_0$ from the first computation of the solution to (ii) (here t is replaced by $|t|$ in that computation). Thus $z \to \langle U_z x, x \rangle$ is entire when H is bounded. By "polarization" (see 2.4(3)), $z \to \langle U_z x, y \rangle$ is entire for each pair of vectors x, y in $\mathcal{H}$ when H is bounded.

Suppose now that H is positive (though not necessarily bounded) and $x, y \in \mathcal{H}_n \ (= E_n(\mathcal{H}))$. Then

$$U_z x = e^{-izH} E_n x = (e^{-izH} \cdot E_n) x = e^{-iz(H \cdot E_n)|\mathcal{H}_n} x = e^{-izHE_n} x$$

from Corollary 5.6.31. Thus $z \to \langle U_z x, y \rangle$ is entire since HE_n is bounded. If x_0 and y_0 are arbitrary unit vectors in $\mathcal{H}$ and a positive ε

is given, choose n and unit vectors x and y in $\mathcal{H}_n$ such that $\|x - x_0\| < \frac{\varepsilon}{2}$ and $\|y - y_0\| < \frac{\varepsilon}{2}$. With z in $\mathbb{C}_-$,

$$|\langle U_z x, y\rangle - \langle U_z x_0, y_0\rangle| \le |\langle U_z(x - x_0), y\rangle| + |\langle U_z x_0, y - y_0\rangle| < \varepsilon$$

since $\|U_z\| \le 1$ by Exercise 5.7.50(iii). It follows that $z \to \langle U_z x_0, y_0\rangle$ is the uniform limit on $\mathbb{C}_-$ of functions $(z \to \langle U_z x, y\rangle)$ holomorphic in $\mathbb{C}_-^0$. Thus $z \to \langle U_z x_0, y_0\rangle$ is analytic on $\mathbb{C}_-^0$ for each pair of vectors x_0, y_0 in $\mathcal{H}$. ∎

5.7.52.   Let $\mathcal{R}$ be a von Neumann algebra acting on a Hilbert space $\mathcal{H}$ and $t \to \exp(-itH)\ (= U_t)$ be a one-parameter unitary group on $\mathcal{H}$, where H is a self-adjoint operator on $\mathcal{H}$ with domain $\mathcal{D}$. Let x_0 be a unit vector in $\mathcal{D}$. Make the following assumptions about H, U_t, $\mathcal{R}$, and x_0:

(1) $U_t A U_{-t} \in \mathcal{R}$ for each A in $\mathcal{R}$ and each t in $\mathbb{R}$;
(2) $H \ge 0$;
(3) $H x_0 = 0$;
(4) x_0 is generating for $\mathcal{R}$.

(i)   Define U_z as in Exercise 5.7.50(iii) when $z \in \mathbb{C}_-$, and show that $U_z x_0 = x_0$ for each z in $\mathbb{C}_-$.

With A and A' self-adjoint operators in $\mathcal{R}$ and $\mathcal{R}'$, respectively, define $f(z)$ to be $\langle U_z A x_0, A' x_0\rangle$ for z in $\mathbb{C}_-$.

(ii)   Show that $f(t)$ is a real number for real t, and that f is continuous on $\mathbb{C}_-$, analytic on $\mathbb{C}_-^0$, and bounded on $\mathbb{C}_-$.

(iii)   Show that $U_t \in \mathcal{R}$ for each real t.

Solution. (i)   Since $U_z = \exp -izH$ and $H x_0 = 0$, it follows from Remark 5.6.32 that $U_z x_0 = (\exp -iz0)x_0 = x_0$ for each z in $\mathbb{C}_-$.

(ii)   Suppose $t \in \mathbb{R}$. From the definition of f, (i), the self-adjointness of A', the fact that $A' \in \mathcal{R}'$ and condition (1), and (i) and the self-adjointness of A, we have

$$\begin{aligned}
f(t) = \langle U_t A x_0, A' x_0\rangle &= \langle U_t A U_{-t} x_0, A' x_0\rangle \\
&= \langle A' U_t A U_{-t} x_0, x_0\rangle \\
&= \langle U_t A U_{-t} A' x_0, x_0\rangle \\
&= \langle A' x_0, U_t A x_0\rangle = \overline{f(t)}.
\end{aligned}$$

Thus $f(t)$ is real. From Exercises 5.7.50(iii) and 5.7.51(ii), f is continuous on $\mathbb{C}_-$ and analytic in $\mathbb{C}_-^0$. With z in $\mathbb{C}_-$, from Exercise 5.7.50(iii),

$$|f(z)| = |\langle U_z A x_0, A' x_0\rangle| \leq \|U_z A x_0\|\|A' x_0\|$$
$$\leq \|U_z\|\|A x_0\|\|A' x_0\|$$
$$\leq \|A x_0\|\|A' x_0\|.$$

Hence f is bounded on $\mathbb{C}_-$.

(iii)   The Schwarz reflection principle applies to f and there is an entire F such that $F(z) = f(z)$ and $F(\bar{z}) = \overline{f(z)} = \overline{F(z)}$ for z in $\mathbb{C}_-$. Hence, for z in $\mathbb{C}_-$,

$$|F(\bar{z})| = |\overline{F(z)}| = |F(z)| = |f(z)| \leq \|A x_0\|\|A' x_0\|$$

and F is a bounded entire function. From Liouville's theorem, F is constant. In particular, for each real t,

$$\langle A x_0, A' x_0\rangle = f(0) = f(t) = \langle U_t A x_0, A' x_0\rangle = \langle A x_0, U_{-t} A' U_t x_0\rangle.$$

Thus $\langle A x_0, (A' - U_{-t} A' U_t) x_0\rangle = 0$, for all self-adjoint A in $\mathcal{R}$, all self-adjoint A' in $\mathcal{R}'$, and all real t. Since $\mathcal{R}$ is generated linearly by its self-adjoint operators and x_0 is a generating vector for $\mathcal{R}$, $(A' - U_{-t} A' U_t) x_0 = 0$ for each self-adjoint A' in $\mathcal{R}'$. Now $T \to U_{-t} T U_t$ is an automorphism of $\mathcal{B}(\mathcal{H})$ that maps $\mathcal{R}$ onto $\mathcal{R}$ and hence, maps $\mathcal{R}'$ (the operators commuting with $\mathcal{R}$) onto $\mathcal{R}'$. Thus $A' - U_{-t} A' U_t \in \mathcal{R}'$. Since x_0 is generating for $\mathcal{R}$, it is separating for $\mathcal{R}'$, and $A' - U_{-t} A' U_t = 0$. Equivalently, $A' U_t = U_t A'$. Thus U_t commutes with each self-adjoint operator in $\mathcal{R}'$ and $U_t \in \mathcal{R}'' = \mathcal{R}$. ∎
[4,63]

5.7.53.   Let H be a self-adjoint operator acting on a Hilbert space $\mathcal{H}$, $\mathcal{A}$ be the von Neumann algebra generated by H, and $\mathcal{A}_0$ be the von Neumann algebra generated by $\{U_t : t \in \mathbb{R}\}$, where $U_t = \exp(-itH)$.

. (i)   Assume H is bounded and show that the C^*-algebra $\mathfrak{A}$ and $\mathfrak{A}_0$ generated by H (and I) and by $\{U_t : t \in \mathbb{R}\}$, respectively, coincide.

(ii)   Show that $\mathcal{A} = \mathcal{A}_0$.

(iii)   With the notation and assumptions of Exercise 5.7.52, show that $H \eta \mathcal{R}$.

Solution. (i) By definition of U_t, $\mathfrak{A}_0 \subseteq \mathfrak{A}$. From Theorem 4.4.5, $\mathfrak{A} \cong C(\operatorname{sp} H)$, H maps onto the identity transform $\imath$ on $\operatorname{sp} H$, and U_t maps onto the function $\exp -it\imath$ on $\operatorname{sp} H$. Finite linear combinations of the functions $\exp -it\imath$ form a subalgebra $\mathcal{C}_0$ of $C(\operatorname{sp} H)$ stable under complex conjugation and containing the constants. If $\exp -it\lambda = \exp -it\lambda'$ for all real t, then $\exp it(\lambda - \lambda') = 1$ for all real t; and $\lambda = \lambda'$. Thus $\mathcal{C}_0$ separates the points of $\operatorname{sp} H$. The Stone-Weierstrass theorem applies and $\mathcal{C}_0$ is norm dense in $C(\operatorname{sp} H)$. Hence H is a norm limit of finite linear combinations of operators in $\{U_t : t \in \mathbb{R}\}$ and $\mathfrak{A} \subseteq \mathfrak{A}_0$. Thus $\mathfrak{A} = \mathfrak{A}_0$.

(ii) Let $\{E_\lambda\}$ be the spectral resolution of H, F_n be $E_n - E_{-n}$, and $\mathcal{H}_n$ be $F_n(\mathcal{H})$. As in the proof of Theorem 5.6.18, we have that $\{F_n, HF_n : n = 1, 2, \ldots\}$ generates $\mathcal{A}$. The von Neumann algebra $\mathcal{A}_n$ generated by $HF_n|\mathcal{H}_n$ is therefore $\mathcal{A}F_n$ acting on $\mathcal{H}_n$. From Corollary 5.6.31,

$$(U_tF_n)|\mathcal{H}_n = \exp -it[(H^{\cdot}F_n)|\mathcal{H}_n] = \exp -it(HF_n|\mathcal{H}_n);$$

and from (i), $HF_n|\mathcal{H}_n$ and $\{(U_tF_n)|\mathcal{H}_n : t \in \mathbb{R}\}$ generate the same C^*-algebra. Thus $\mathcal{A}F_n$ and $\mathcal{A}_0F_n$ acting on $\mathcal{H}_n$ coincide. If E is a projection in $\mathcal{A}$ contained in F_m, then $E \in \mathcal{A}_0F_n$ when $n \geq m$. Now $F_n \in \mathcal{A}_0'$ so that F_n has a central carrier G_n relative to $\mathcal{A}_0'$ that lies in $\mathcal{A}_0$. From Proposition 5.5.5, the mapping, $A_0F_n \to A_0G_n$ is an isomorphism and hence, from Theorem 4.1.8, an isometry of $\mathcal{A}_0F_n$ onto $\mathcal{A}_0G_n$. Suppose $E = A_nF_n$ with A_n in $\mathcal{A}_0$. Then $1 = \|A_nF_n\| = \|A_nG_n\|$ and $A_nG_n \in \mathcal{A}_0$. Moreover, $A_nG_nF_n = A_nF_n = E$. Replacing A_n by A_nG_n, we may assume that $\|A_n\| = 1$. Since $\cup_{n=1}^\infty \mathcal{H}_n$ is dense in $\mathcal{H}$ and each A_n lies in the unit ball of $\mathcal{A}_0$, E is in the strong-operator closure of that unit ball. Hence $E \in \mathcal{A}_0$. It follows that F_n and the spectral resolution of HF_n are in $\mathcal{A}_0$ for each n. Thus $HF_n \in \mathcal{A}_0$ for each n and $\mathcal{A} = \mathcal{A}_0$.

(iii) From Exercise 5.7.52(iii), $U_t \in \mathcal{R}$ for each real t. Hence the von Neumann algebra $\mathcal{A}_0$ generated by $\{U_t : t \in \mathbb{R}\}$ is contained in $\mathcal{R}$. From (ii), $\mathcal{A}_0$ is the von Neumann algebra generated by H. From Theorem 5.6.18, $H\eta\mathcal{A}_0$. Hence $H\eta\mathcal{R}$.   ∎

5.7.54. Let A be a normal operator acting on a Hilbert space $\mathcal{H}$.

(i) Show that $\exp iA = I$ if $\operatorname{sp}(A) \subseteq \{2\pi n : n \in \mathbb{Z}\}$ and only if A is self-adjoint and $\operatorname{sp} A \subseteq \{2\pi n : n \in \mathbb{Z}\}$.

(ii)   Show that $\exp itA = I$ for all t in $\mathbb{R}$ if and only if $A = 0$.

(iii)   Let A and B be self-adjoint operators on $\mathcal{H}$ such that $\exp itB = \exp itA$ for each real t. Show that $A = B$.

(iv)   Let $t \to U_t$ be a one-parameter unitary group acting on $\mathcal{H}$. Show that there is a *unique* self-adjoint operator H on $\mathcal{H}$ such that $U_t = \exp itH$ for all real t.

Solution.   Let $\mathcal{A}_0$ be the von Neumann algebra generated by A, and suppose $\mathcal{A}_0 \cong C(X)$ with X an extremely disconnected compact Hausdorff space. Let f in $\mathcal{N}(X)$ represent A.

(i)   By definition of $\exp iA$, $\exp if$ has a normal extension in $\mathcal{N}(X)$ that represents $\exp iA$ and this normal extension has the value $\exp if(p)$ at each p in X for which f is defined. If $\exp iA = I$, then $\exp if(p)$ is 1 for each such p, $f(p) \in \{2\pi n : n \in \mathbb{Z}\}$ $(= 2\pi\mathbb{Z})$. From Theorem 5.6.19, A is self-adjoint. Conversely, if $\operatorname{sp} A \subseteq 2\pi\mathbb{Z}$, then the range of f is contained in $2\pi\mathbb{Z}$. Thus $\exp if(p) = 1$ for each p at which f is defined, and $\exp if$ is the constant function 1. Hence $\exp iA = I$ in this case.

(ii)   The normal extension of $\exp itf$ (in $C(X)$) that represents $\exp itA$, takes the value $\exp itf(p)$ at each p in X for which f is defined. Since $\exp itA = I$ for all real t, $\exp itf(p) = 1$ for all real t, and $f(p) = 0$ for each p in X at which f is defined. Thus $f = 0$ and $A = 0$. Of course $\exp it0 = I$ for all real t.

(iii)   If $U_t = \exp itA = \exp itB$, then $A\eta\mathcal{A}$ and $B\eta\mathcal{A}$, where $\mathcal{A}$ is the (abelian) von Neumann algebra generated by $\{U_t : t \in \mathbb{R}\}$, from Exercise 5.7.53(ii). If f and g in $\mathcal{S}(X)$, representing A and B, are defined on $X \setminus Z$ and $X \setminus Z'$, respectively, then $\exp itf$ and $\exp -itg$ have normal extensions in $C(X)$ representing U_t and U_{-t}, respectively. Thus $(\exp itf)(\exp -itg)$ has a normal extension in $C(X)$ equal to the constant function 1 (representing $(I =) U_t U_{-t}$). For each p in $X \setminus (Z \cup Z')$ and t in $\mathbb{R}$,

$$1 = e^{itf(p)}e^{-itg(p)} = e^{it[f(p) - g(p)]}.$$

Hence $f(p) = g(p)$ for p in $X \setminus (Z \cup Z')$. From Lemma 5.6.6, $Z = Z'$ and $f = g$. Thus $A = B$.

(iv)   This is immediate from Stone's theorem (5.6.36) and (iii). ■

5.7.55.   With the notation of Exercise 5.7.52, assume conditions (1), (2) and (3), and in place of (4) assume that x_0 is separating for the center C of $\mathcal{R}$. Show that there is a positive self-adjoint operator K on $\mathcal{H}$ such that $K\eta\mathcal{R}$ and $W_t A W_{-t} = U_t A U_{-t}$ for each A in $\mathcal{R}$ and all real t, where $W_t = \exp(-itK)$ $(\in \mathcal{R})$, and such that $Kx_0 = 0$. [*Hint.* Consider the projection E' with range $[\mathcal{R}x_0]$ and the von Neumann algebra $\mathcal{R}E'$ acting on $E'(\mathcal{H})$.]

Solution.   Let E' be the projection with range $[\mathcal{R}x_0]$ $(= \mathcal{H}_0)$. Since x_0 is separating for C, $C_{E'} = I$. Let $\varphi(A)$ be $AE'|\mathcal{H}_0$. From Proposition 5.5.5, φ is an isomorphism of $\mathcal{R}$ onto $\mathcal{R}E'$ $(= \mathcal{R}_0)$ acting on $\mathcal{H}_0$. With A in $\mathcal{R}$,

$$U_t A x_0 = U_t A U_{-t} U_t x_0 = U_t A U_{-t} x_0 \in \{\mathcal{R}x_0\}.$$

Thus $[\mathcal{R}x_0]$ is stable under each U_t. As $\{U_t : t \in \mathcal{R}\}$ is a self-adjoint family, E' commutes with each U_t. Let V_t be $U_t E'|\mathcal{H}_0$. Then $t \to V_t$ is a one-parameter unitary group on $\mathcal{H}_0$. Let $\mathcal{A}$ be the (abelian) von Neumann algebra generated by $\{U_t : t \in \mathbb{R}\}$ and let $\mathcal{A}_1$ be the (abelian) von Neumann algebra generated by $\mathcal{A}$ and E'. From Exercise 5.7.53(ii), $H\eta\mathcal{A}$ and hence $H\eta\mathcal{A}_1$. Let H_0 be $(H\dot{.}E')|\mathcal{H}_0$. Then H_0 is a self-adjoint operator on $\mathcal{H}_0$ and from Corollary 5.6.31,

$$\begin{aligned}
\exp -itH_0 &= \exp -it[(H\dot{.}E')|\mathcal{H}_0] \\
&= [(\exp -itH)\dot{.}E']|\mathcal{H}_0 = (U_t E')|\mathcal{H}_0 = V_t.
\end{aligned}$$

Passing to the function representation of $\mathcal{A}_1$ and using Propositions 5.6.20 and 5.6.21, we see that $H_0 \geq 0$. Let $\alpha_t(A)$ be $U_t A U_{-t}$ for A in $\mathcal{R}$ and $\beta_t(A_0)$ be $V_t A_0 V_{-t}$ for A_0 in $\mathcal{R}_0$. Then

$$\begin{aligned}
\beta_t(\varphi(A)) &= V_t \varphi(A) V_{-t} \\
&= (U_t A U_{-t})E'|\mathcal{H}_0 = (\alpha_t(A)E')|\mathcal{H}_0 = \varphi(\alpha_t(A)).
\end{aligned}$$

Since each element of $\mathcal{R}_0$ has the form $\varphi(A)$ for some A in $\mathcal{R}$, we see that $V_t A_0 V_{-t} \in \mathcal{R}_0$ for each A_0 in $\mathcal{R}_0$. Note too that x_0 is generating for $\mathcal{R}_0$ and that $H_0 x_0 = 0$. With $\mathcal{R}_0$, $\mathcal{H}_0$, V_t, H_0, and x_0, in place of $\mathcal{R}$, $\mathcal{H}$, U_t, H, and x_0, Exercise 5.7.52(iii) assures us that $V_t \in \mathcal{R}_0$ for each real t. Let $\mathcal{A}_0$ be the (abelian) von Neumann subalgebra of $\mathcal{R}_0$ generated by $\{V_t : t \in \mathbb{R}\}$. From Exercise 5.7.53(ii), $H_0\eta\mathcal{A}_0$. Since the mapping $T \to TE'|\mathcal{H}_0$ is strong-operator continuous, $\varphi^{-1}(\mathcal{A}_0)$ is an abelian von Neumann subalgebra $\mathcal{A}_2$ of $\mathcal{R}$. Let W_t be the unitary operator $\varphi^{-1}(V_t)$ in $\mathcal{A}_2$. The restriction of φ to $\mathcal{A}_2$ has a σ-normal extension ψ mapping $\mathcal{S}(\mathcal{A}_2)$ isomorphically onto $\mathcal{S}(\mathcal{A}_0)$ and there is a positive self-adjoint operator H_2 in $\mathcal{S}(\mathcal{A}_2)$ such that $\psi(H_2) = H_0$. From Proposition 5.6.30, $W_t = \exp{-itH_2}$. Moreover, for each A in $\mathcal{R}$,

$$\varphi(W_t A W_{-t}) = V_t \varphi(A) V_{-t} = \beta_t(\varphi(A)) = \varphi(\alpha_t(A)),$$

so that $U_t A U_{-t} = \alpha_t(A) = W_t A W_{-t}$.

Since $\psi(A) = AE'|\mathcal{H}_0$ for A in $\mathcal{A}_2$, $\psi(B) = (B^{\cdot}E')|\mathcal{H}_0$ for B in $\mathcal{S}(\mathcal{A}_2)$. Thus $(H_2^{\cdot}E')|\mathcal{H}_0 = H_0 = (H^{\cdot}E')|\mathcal{H}_0$. It follows that $x_0 \in \mathcal{D}(H_2)$ and $H_2 x_0 = 0$. Choose H_2 as K and note that $K\eta\mathcal{A}_2$ so that $K\eta\mathcal{R}$. ∎ [63]

BIBLIOGRAPHY

General references

[H] P. R. Halmos, "Measure Theory." D. Van Nostrand, Princeton, New Jersey, 1950; reprinted, Springer-Verlag, New York, 1974.

[K] J. L. Kelley, "General Topology." D. Van Nostrand, Princeton, New Jersey, 1955; reprinted, Springer-Verlag, New York, 1975.

[R] W. Rudin, "Real and Complex Analysis." 2nd ed. McGraw-Hill, New York, 1974.

References

[1] J. F. Aarnes and R. V. Kadison, Pure states and approximate identities, *Proc. Amer. Math. Soc.* **21** (1969), 749–752.

[2] C. A. Akemann, The dual space of an operator algebra, *Trans. Amer. Math. Soc.* **126** (1967), 286–302.

[3] C. A. Akemann and G. K. Pedersen, Ideal perturbations of elements in C^*-algebras, *Math. Scand.* **41** (1977), 117–139.

[4] H. Araki, On the algebra of all local observables, *Res. Inst. Math. Sci. Kyoto* **5** (1964), 1–16.

[5] H. Araki, Some properties of modular conjugation operator of von Neumann algebras and a non-commutative Radon–Nikodým theorem with a chain rule, *Pacific J. Math.* **50** (1974), 309–354

[6] H. Araki, Positive cone, Radon–Nikodym theorems, relative Hamiltonian and the Gibbs condition in statistical mechanics. An application of the Tomita–Takesaki theory, *in* "C^*-algebras and Their Applications to Statistical Mechanics and Quantum Field Theory" (*Proc. Internat. School of Physics "Enrico Fermi," Course LX, Varenna*, D. Kastler, ed., 1973), pp. 64–100, North-Holland Publ., Amsterdam, 1976.

[7] H. Araki, Positive cones for von Neumann algebras, *in* "Operator Algebras and Applications" (*Proc. of Symposia in Pure Math., Vol. 38, Part 2*, R. Kadison, ed., 1980), pp. 5–15 Amer. Math. Soc., Providence, 1982.

[8] R. J. Archbold, On the centre of a tensor product of C^*-algebras, *J. of London Math. Soc.* **10** (1975), 257–262.

[9] R. J. Archbold, An averaging process for C^*-algebras related to weighted shifts, *Proc. London Math. Soc.* **35** (1977), 541–554.

[10] W. Arveson, Subalgebras of C^*-algebras, *Acta Math.* **123** (1969), 141–224.

[11] W. Arveson, On groups of automorphisms of operator algebras, *J. Fnal. Anal.* **15** (1974), 214–243.

[12] W. Arveson, Notes on extensions of C^*-algebras, *Duke Math. J.* **44** (1977), 329–355.

[13] G. Birkhoff, "Lattice Theory." American Mathematical Society Colloquium Publications, Revised Edition, Vol. 25, Amer. Math. Soc., New York, 1948.

[14] J. Bunce, A note on two-sided ideal in C^*-algebras, *Proc. Amer. Math. Soc.* **28** (1971), 635.

[15] J. W. Calkin, Two-sided ideals and congruences in the ring of bounded operators in Hilbert space, *Ann. of Math.* **42** (1941), 839–873.

[16] F. Combes, Sur les états factoriels d'une C^*-algèbre, *Compte Rand. Ser. A–B* **265** (1967), 736–739.

[17] F. Combes, Sur les faces d'une C^*-algèbre, *Bull. Sci. Math.* 2e *Serie* **93** (1969), 37–62.

[18] A. Connes, Une classification des facteurs de type III, *Ann. Sci. École Norm. Sup. Paris* **6** (1973), 133–252.

[19] A. Connes, Charactérisation des espaces vectoriels ordonnés sous-jacents aux algèbres de von Neumann, *Ann. Inst. Fourier, Grenoble* **24** 4 (1974), 121–155.

[20] A. Connes, Classification of injective factors, Cases II_1, II_∞, III_λ, $\lambda \neq 1$, *Ann. of Math.* **104** (1976), 73–115.

[21] A. Connes, On the cohomology of operator algebras, *J. Fnal. Anal.* **28** (1978), 248–253.

[22] A. Van Daele, The Tomita–Takesaki thory for von Neumann algebras with a separating and cyclic vector, *in* "C^*-algebras and Their Applications to Statistical Mechanics and Quantum Field Theory" (*Proc. Internat. School of Physics "Enrico Fermi," Course LX, Varenna*, D. Kastler, ed., 1973), pp. 19–28, North-Holland Publ., Amsterdam, 1976.

[23] A. Van Daele, A new approach to the Tomita–Takesaki theory of generalized Hilbert algebras, *J. Fnal. Anal.* **15** (1974), 378–393.

[24] A. Van Daele, "Continuous Crossed Products and Type III von Neumann Algebras." London Math. Soc. Lecture Note Series 31, Cambridge University Press, London 1978.

[25] J. Dauns and K. Hofmann, Representations of rings by sections, *Mem. Amer. Math. Soc.* **83** (1968).

[26] J. Dixmier, Sur certains espaces considérés par M. H. Stone, *Summa Brasil. Math.* **2** (1951), 151–182.

[27] J. Dixmier, "Les Algèbres d'Opérateurs dans l'Espace Hilbertien." Gauthier-Villars, Paris, 1957; 2nd ed., 1969.

[28] J. Dixmier, Traces sur les C^*-algèbres, *Ann. Inst. Fourier* **13** (1963), 219–262.

[29] J. Dixmier, Existence de traces non normales, *C. R. Acad. Sci. Paris* **262** (1966), 1107–1108.

[30] H. A. Dye, The unitary structure in finite rings of operators, *Duke Math. J.* **20** (1953), 55–69.

[31] E. Effros, Order ideals in a C^*-algebra and its dual, *Duke Math. J.* **30** (1963), 391–411.

[32] G. A. Elliott and D. Olesen, A simple proof of the Dauns–Hofmann theorem, *Math Scand.* **34** (1974), 231–234.

[33] G. Emch, "Algebraic Methods in Statistical Mechanics and Quantum Field Theory." Wiley-Interscience, New York, 1972.

[34] K. Friedrichs, Spektraltheorie halbbeschränkten Operatoren und Anwendung auf die Spektralzerlegung von Differentialoperatoren, *Math. Ann.* **109** (1934), 465–487.

[35] B. Fuglede and R. V. Kadison, On a conjecture of Murray and von Neumann, *Proc. Nat. Acad. Sci. U.S.A.* **37** (1951), 420–425.

[36] J. Glimm, A Stone–Weierstrass theorem for C^*-algebras, *Ann. of Math.* **72** (1960), 216–244.

[37] E. Griffin, Some contributions to the theory of rings of operators, *Trans. Amer. Math. Soc.* **75** (1953), 471–504.

[38] E. Griffin, Some contributions to the theory of rings of operators. II, *Trans. Amer. Math. Soc* **79** (1955), 389–400.

[39] A. Grothendieck, Un résultat sur le dual d'une C^*-algèbre, *J. Math. Pures Appl.* **36** (1957), 97–108.

[40] U. Haagerup, The standard form of von Neumann algebras, Preprint, *Mathematical Institute, University of Copenhagen* 1973.

[41] U. Haagerup, The standard form of von Neumann algebras, *Math. Scand.* **37** (1975), 271–283.

[42] U. Haagerup, All nuclear C^*-algebras are amenable, *Invent. Math.* **74** (1983), 305–319.

[43] P. R. Halmos, The range of a vector measure, *Bull. Amer. Math. Soc.* **54** (1948), 416–421.

[44] H. Halpern, Finite sums of irreducible functionals on C^*-algebras, *Proc. Amer. Math. Soc.* **18** (1967), 352–358.

[45] F. Hansen, Inner one-parameter groups acting on a factor, *Math. Scand.* **41** (1977), 113–116.

[46] L. A. Harris, Banach algebras with involution and Möbius transformations, *J. Fnal. Anal.* **11** (1972), 1–16.

[47] R. Haydon and S. Wassermann, A commutation result for tensor products of C^*-algebras, *Bull. London Math. Soc.* **5** (1973), 283–287.

[48] N. M. Hugenholtz and R. V. Kadison, Automorphisms and quasi-free states of the CAR algebra, *Comm. Math. Phys.* **43** (1975), 181–197.

[49] N. Jacobson and C. E. Rickart, Homomorphisms of Jordan rings, *Trans. Amer. Math. Soc.* **69** (1950), 479–502.

[50] B. E. Johnson, R. V. Kadison and J. R. Ringrose, Cohomology of operator algebras, III. Reduction to normal cohomology, *Bull. Soc. Math. France* **100** (1972), 73–96.

[51] B. E. Johnson and J. R. Ringrose, Derivations of operator algebras and discrete group algebras, *Bull. London Math. Soc.* **1** (1969), 70–74.

[52] R. V. Kadison, A representation theory for commutative topological algebra, *Mem. Amer. Math. Soc.* **7** (1951).

[53] R. V. Kadison, Isometries of operator algebras, *Ann. of Math.* **54** (1951), 325–338.

[54] R. V. Kadison, A generalized Schwarz inequality and algebraic invariants for operator algebras, *Ann. of Math.* **56** (1952), 494–503.

[55] R. Kadison, Isomorphisms of factors of infinite type, *Canad. J. Math.* **7** (1955), 322–327.

[56] R. V. Kadison, Operator algebras with a faithful weakly-closed representation, *Ann. of Math.* **64** (1956), 175–181.

[57] R. V. Kadison, Unitary invariants for representations of operator algebras, *Ann. of Math.* **66** (1957), 304–379.

[58] R. V. Kadison, The trace in finite operator algebras, *Proc. Amer. Math. Soc.* **12** (1961), 973–977.

[59] R. V. Kadison, Transformations of states in operator theory and dynamics, *Topology* **3** (1965), 177–198.

[60] R. V. Kadison, Derivations of operator algebras, *Ann. of Math.* **83** (1966), 280–293.

[61] R. V. Kadison, Lectures on operator algebras, *in* "Cargese Lectures in Theoretical Physics" F. Lurçat, ed., Gordon and Breach, London, 1967.

[62] R. V. Kadison, Strong continuity of operator functions, *Pacific J. Math.* **26** (1968), 121–129.

[63] R. V. Kadison, Analytic methods in the theory of operator algebras, *in* "Proc. of the Washington Consortium Lectures—Lectures in Modern Analysis and Applications II" (*Lecture Notes in Mathematics*, C. T. Taam, ed., Vol. 140), 8–29, Springer-Verlag, Heidelberg, 1970.

[64] R. V. Kadison, A note on derivations of operator algebras, *Bull. London Math. Soc.* **7** (1975), 41–44.

[65] R. V. Kadison, Limits of states, *Comm. Math. Phys.* **85** (1982), 143–154.

[66] R. V. Kadison, Diagonalizing matrices, *Amer. J. Math.* **106** (1984), 1451–1468.

[67] R. V. Kadison and D. Kastler, Pertubations of von Neumann algebras I. Stability of type, *Amer. J. Math.* **93** (1972), 38–54.

[68] R. V. Kadison and G. K. Pedersen, Means and convex combinations of unitary operators, *Math. Scand.* **57** (1985), 249–266.

[69] R. V. Kadison and J. R. Ringrose, Derivations and automorphisms of operator algebras, *Comm. Math. Phys.* **4** (1967), 32–63.

[70] R. R. Kallman, A generalization of free action, *Duke Math. J.* **36** (1969), 781–789.

[71] I. Kaplansky, The structure of certain operator algebras, *Trans. Amer. Math. Soc.* **70** (1951), 219–255.

[72] I. Kaplansky, A theorem on rings of operators, *Pacific J. Math.* **1** (1951), 227–232.

[73] I. Kaplansky, Modules over operator algebras, *Amer. J. Math.* **75** (1953), 839–858.

[74] E. C. Lance, On nuclear C^*-algebras, *J. Fnal. Anal.* **12** (1973), 157–176.

[75] A. Liapounoff, Sur les fonctions-vecteurs complétement additives, *Bull. Acad. Sci. U.R.S.S., Sér Math.* **4** (1940), 465–478.

[76] F. J. Murray and J. von Neumann, On rings of operators, *Ann. of Math.* **37** (1936), 116–229.

[77] F. J. Murray and J. von Neumann, On rings of operators, II, *Trans. Amer. Math. Soc.* **41** (1937), 208–248.

[78] F. J. Murray and J. von Neumann, On rings of operators. IV, *Ann. of Math.* **44** (1943), 716–808.

[79] J. von Neumann, Zur Algebra der Funktionaloperationen und Theorie der normalen Operatoren, *Math. Ann.* **102** (1930), 370–427.

[80] J. von Neumann, Über Funktionen von Funktionaloperatoren, *Ann. of Math.* **32** (1931), 191–226.

[81] J. von Neumann, On infinite direct products, *Compositio Math.* **6** (1939), 1–77.

[82] J. von Neumann, On rings of operators. III, *Ann. of Math.* **41** (1940), 94–161.

[83] T. Ogasawara, Finite dimensionality of certain Bannach algebras, *J. Sci. Hiroshima Univ. Ser. A* **17** (1954), 359–364.

[84] T. Ogasawara, A theorem on operator algebras, *J. Sci. Hiroshima Univ. Ser. A* **18** (1955), 307–309.

[85] C. Pearcy and J. R. Ringrose, Trace-preserving isomorphisms in finite operator algebras, *Amer. J. Math.* **90** (1968), 444–455.

[86] G. K. Pedersen, A decomposition theorem for C^*-algebras, *Math. Scand.* **22** (1968), 266–268.

[87] G. K. Pedersen, Some operator monotone functions, *Proc. Amer. Math. Soc.* **36** (1972), 309–310.

[88] J. R. Ringrose, Automatic continuity of derivations of operator algebras, *J. London Math. Soc.* **5** (1972), 432–438.

[89] J. R. Ringrose, Derivations of quotients of von Neumann algebras, *Proc. London Math. Soc.* **36** (1978), 1–26.

[90] J. R. Ringrose, On the Dixmier approximation theorem, *Proc. London Math. Soc.* **49** (1984), 37–57.

[91] B. Russo and H. A. Dye, A note on unitary operators in C^*-algebras, *Duke Math. J.* **33** (1966), 413–416.

[92] S. Sakai, A characterization of W^*-algebras, *Pacific J. Math.* **6** (1956), 763–773.

[93] S. Sakai, On linear functionals of W^*-algebras, *Proc. Japan Acad.* **34** (1958), 571–574.

[94] S. Sakai, On a conjecture of Kaplansky, *Tôhoku Math. J.* **12** (1960), 31–33.

[95] S. Sakai, Derivations of W^*-algebras, *Ann. of Math.* **83** (1966), 287–293.

[96] S. Sakai, Derivations of simple C^*-algebras, *J. Fnal. Anal.* **2** (1968), 202–206.

[97] S. Sakai, Derivations of simple C^*-algebras, II, *Bull. Soc. Math. France* **99** (1971), 259–263.

[98] S. Sakai, Automorphisms and tensor products of operator algebras, *Amer. J. Math.* **97** (1975), 889–896.

[99] J. T. Schwartz, Two finite, non-hyperfinite, non-isomorphic factors, *Comm. Pure Appl. Math.* **16** (1963), 19–26.

[100] I. E. Segal, Irreducible representations of operator algebras, *Bull. Amer. Math. Soc.* **53** (1947), 73–88.

[101] I. E. Segal, Two-sided ideals in operator algebras, *Ann. of Math.* **50** (1949), 856–865.

[102] S. Sherman, The second adjoint of a C^*-algebra, *Proc. Int. Congress of Mathematicians, Cambridge, 1950* Vol. 1, p. 470.

[103] S. Sherman, Order in operator algebras, *Amer. J. Math.* **73** (1951), 227–232.

[104] W. F. Stinespring, Positive functions on C^*-algebras, *Proc. Amer. Math. Soc.* **6** (1955), 211–216.

[105] M. H. Stone, Boundedness properties in function lattices, *Canad. J. Math.* **1** (1949), 176–186.

[106] E. Størmer, Positive linear maps of operator algebras, *Acta Math.* **110** (1963), 233–278.

[107] E. Størmer, Two-sided ideals in C^*-algebras, *Bull. Amer. Math. Soc.* **73** (1967), 254–257.

[108] E. Størmer, Positive linear maps on C^*-algebras, *in* "Foundations of Quantum Mechanics and Ordered Linear Spaces"(Advanced Study Institute Held in Marburg, A. Hartkä mper and H. Neumann, eds., 1973) (*Lecture Notes in Physics*, Vol. 29), 85–106, Springer-Verlag, Berlin, 1974.

[109] Z. Takeda, Conjugate spaces of operator algebras, *Proc. Japan Acad.* **30** (1954), 90–95.

[110] M. Takesaki, On the singularity of a positive linear functional on operator algebra, *Proc. Japan Acad.* **35** (1959), 365–366.

[111] M. Takesaki, On the cross-norm of the direct product of C^*-algebras, *Tôhoku Math. J.* **16** (1964), 111–122.

[112] M. Takesaki, "Tomita's Theory of Modular Hilbert Algebras and Its Applications." Lecture Notes in Mathematics, Vol. 128, Springer-Verlag, Heidelberg, 1970.

[113] M. Takesaki, "Theory of Operator Algebras I." Springer-Verlag, New York, 1979.

[114] J. Tomiyama, On the projection of norm one in W^*-algebras, *Proc. Japan Acad.* **33** (1957), 608–612.

[115] J. Tomiyama, Tensor products and projections of norm one in von Neumann algebras, *Seminar notes of the Math. Institute, University of Copenhagen* (1970).

[116] S. Wassermann, Extension of normal functionals on W^*-tensor products, *Math. Proc. Camb. Philos. Soc.* **78** (1975), 301–307.

[117] S. Wassermann, The slice map problem for C^*-algebras, *Proc. London Math. Soc.* **32** (1976), 537–559.

[118] K. Yosida, On vector lattice with a unit, *Proc. Imp. Acad. Tokyo* **17** (1941), 121–124.

INDEX

A

Algebra
  of bounded operators, 188–192, 205–206
  countably decomposable von Neumann, 247
  finite-dimensional C*, 147–149
  $l_1(\mathbb{Z})$, 115–117, 120–121
  $L_1(\mathbb{R})$, 128–129
  $L_1(\mathbb{T}_1, m)$, 121-128, 129-131
  multiplication, 231, 245–246, 248–250
  quotient, 196–197
  semi-simple, 106
  simple, 235–236

Anti-homomorphism, 113

Approximate identity (in C*-algebras), 166-168, 196-197

Approximation theorem, Stone-Weierstrass, 138

Archimedian partially ordered vector space, 182

B

Banach lattice, 184-186

Banach module, 201

Banach-Orlicz theorem, 33

Banach space
  non-separable, 30–31
  reflexive, *see* Reflexive (Banach space)
  separable, 30–31, 35–36

Banach space, examples of
  c, 14, 15, 17, 20, 21, 188–189
  c_0, 14, 15, 16, 17, 20, 21, 188–189
  l_∞, 14, 19, 20, 21, 188–189
  l_1, 16, 17, 19, 20, 22
  $l_1(\mathbb{Z})$ (as a Banach algebra), 115–117, 120–121
  l_p, 24–26, 115–117
  L_∞, 30–32
  L_1, 31-32
  $L_1(\mathbb{R})$ (as a Banach algebra), 128–129
  $L_1(\mathbb{T}_1, m)$ (as a Banach algebra), 121–128, 129–131
  L_p, 26–30

β-compactification, 89–90, 188–189, 222-224

Borel function calculus
    for bounded multiplication operators, 232–233
    for unbounded multiplication operators, 249–250

Boundedly complete lattice, *see* Lattice

C

C^*-algebra, simple, 235–236

Central carrier, 240

Character
    of $\mathbb{T}_1$, 123–126
    of $\mathbb{Z}$, 117–120

Closed subspace, 4

Compact linear operator, 55, 56, 58–66, 72, 95, 98,
        99, 100–104, 190, 192, 204

Compact self-adjoint operator, 59–66

Complexification
    of a real Banach algebra, 85–86
    of a real Hilbert space, 41
    of a real linear space, 4–5
    of a real normed space, 4–5

Convolution, 115–117, 121–123

Countably decomposable von Neumann algebra, 247

Countably decomposable projection, 246

D

Definite state, 150–151, 165–166

Derivation, 200–203

Direct sum
    of Banach algebras, 86–87
    of operators, 51–52

Dual group
    of $\mathbb{T}_1$, 123
    of $\mathbb{Z}$, 117

E

Eigenvalue, 229

Essential range (of a measurable function), 249–250

Exponential unitary, 140–145

Extension
  of pure states, 178–180
  of states, 178–180
Extreme point, 49-50, 180, 217–218
Extremely disconnected space, 90, 218–221, 223–224, 225, 227, 228–229

F

Faithful state, 150
Function calculus, *see* Borel function calculus
Function representation (of a Banach lattice), 184–186

G

Generalized nilpotent, 95–99, 103–104
Generating vector, 241–242, 243–246

H

Hahn–Jordan decomposition, 158
Hilbert–Schmidt operator, 71–73, 103–104
Hölder's inequality, 24, 27

I

Ideal, 86–88, 166–171, 172, 196–200
Invariant mean, 91

L

Lattice, 186
  Banach, 184–186
  boundedly complete, 218–220, 223–224, 225, 227
  sublattice, 137
l_2-independent, 243–245
Linear functional, 2
  positive (on a partially ordered vector space), 177–182
Linear operator, *see* Operator
Linear transformation, *see* Operator
Locally finite group, 91

M

Meager set, 224–227

Multiplication algebra, 231, 245–246, 248–250

Multiplication operator
    bounded, 50–51, 231, 232–233
    unbounded, 248–250

Multiplicative linear functional
    on $l_\infty(\mathbb{Z})$, 87–88
    on $l_1(\mathbb{Z})$, 118–120
    on $L_1(\mathbb{T}_1, m)$, 124–127

Multiplicity (of an eigenvalue), 64, 65, 101

N

Non-separable
    Banach space, 30–31
    Hilbert space, 45

Norm topology, 5–6

Normal operator, 108–109

Normal state, 227

Null space (of an unbounded operator), 77–78

O

Operator
    compact, *see* Compact linear operator
    compact self-adjoint, 59–66
    Hilbert–Schmidt, 71–74, 103–104
    multiplication, *see* Multiplication operator
    normal, 108–109
    tensor product, 74–75
    unbounded, 75–78

Operator-monotonic increasing, 173–176

Order unit, 177–184

P

Partially ordered vector space, 177–186
    archimedian, 182

Plancherel's theorem, 120, 127

Point spectrum, 229

Polar decomposition, 172–173

approximate, 171-172

Positive linear functional (on a partially ordered
          vector space), 177–184

Positive square root, 64, 66

Principle of uniform boundedness, 38

Projection, countably decomposable, 246

Pure state
    of $\mathcal{B}(\mathcal{H})$, 203–205
    extension of, 178–180

Q

Quotient C*-algebra, 196–197

R

Radical (of a commutative Banach algebra), 106

Range projection, 77, 78

Reflexive (Banach space), 8–10, 20–21, 24–25, 26,
          28–29,36

Regular open set, 224–226

Representation (of $\mathcal{B}(\mathcal{H})$), 190–192

S

Semi-norm, 1

Semi-simple, 106

Separable Banach space, 30–31, 35–36

Separating vector, 243–245

σ-ideal, 224

Simple C*-algebra, 235–236

Square root (in a Banach algebra), 132–134, 137

State
    definite, 150–151, 165–166
    extension of, 178–180
    faithful, 150
    pure, *see* Pure state
    vector, 152, 187–188, 204

Stone's theorem, 251, 260

Stone–Weierstrass theorem, 138

Strong-operator continuity (of functions), 237–240

Strong–operator topology, 207–209, 212, 214, 215, 216, 247

Sublattice, 137

Sublinear functional, 1

Subspace, closed, 4

Support functional, 1

T

Tensor products of operators, 74–75

Topology
    norm, 5–6
    strong-operator, *see* Strong-operator topology
    weak, 5–6
    weak-operator, 207–217

Totally disconnected, 220, 223

Trace, normalized, 152

Transformation, linear, *see* Operator

Translation invariant subspace
    in $L_1(\mathbb{R})$, 128–129
    in $L_1(\mathbb{T}_1, m)$, 129–131

U

Uniform boundedness principle, 38

Uniformly convex Banach space, 10, 41

Unitary, exponential, 140–145

Unitary group, 142–143

V

Vector state, 152, 187–188, 204

W

Weak convergence, 36–38, 43

Weak * convegence, 6, 31–32, 37–38

Weak * continuous
    linear functional, 12
    linear operator, 12

Weak topology, 5-6

Weak-operator topology, 207–217

Wiener's Tauberian theorems, 128